Cell Surface Engineering
Fabrication of Functional Nanoshells

RSC Smart Materials

Series Editors:
Hans-Jörg Schneider, *Saarland University, Germany*
Mohsen Shahinpoor, *University of Maine, USA*

Titles in this Series:

How to obtain future titles on publication:
A standing order plan is available for this series. A standing order will bring delivery of each new volume immediately on publication.

For further information please contact:
Book Sales Department, Royal Society of Chemistry, Thomas Graham House, Science Park, Milton Road, Cambridge, CB4 0WF, UK
Telephone: +44 (0)1223 420066, Fax: +44 (0)1223 420247
Email: booksales@rsc.org
Visit our website at www.rsc.org/books

Cell Surface Engineering
Fabrication of Functional Nanoshells

Edited by

Rawil F Fakhrullin
Kazan Federal University, Republic of Tatarstan, Russian Federation
Email: kazanbio@gmail.com

Insung S Choi
KAIST, Daejeon, Republic of Korea
Email: ischoi@kaist.ac.kr

and

Yuri Lvov
Louisiana Tech University, LA, USA
Email: ylvov@coes.latech.edu

THE QUEEN'S AWARDS
FOR ENTERPRISE:
INTERNATIONAL TRADE
2013

RSC Smart Materials No. 9

Print ISBN: 978-1-84973-902-3
PDF eISBN: 978-1-78262-847-7
ISSN: 2046-0066

A catalogue record for this book is available from the British Library

Published by The Royal Society of Chemistry,
Thomas Graham House, Science Park, Milton Road,
Cambridge CB4 0WF, UK

Registered Charity Number 207890

For further information see our web site at www.rsc.org

Printed and bound in Great Britain by CPI Group (UK) Ltd, Croydon, CR0 4YY

Rawil F. Fakhrullin would like to dedicate this book to his family, who have enthusiastically supported his work.

Insung S. Choi dedicates this book to Jiyoung, Jeanne, and Sun.

Foreword

The book "Cell Surface Engineering: Fabrication of Functional Nanoshells," which is edited by Rawil F. Fakhrullin, Insung S. Choi and Yuri M. Lvov, is a very interesting book. It covers a variety of cutting-edge topics. These include functional multilayered polyelectrolyte assemblies on biological cells, direct deposition of functional nanomaterials onto biological cells, bioinspired encapsulation of living cells with inorganic nanoshells, and cell surface engineering using layer-by-layer nanofilms. The book also discussses approaches for characterizing living cells encapsulated with nanomaterials and examines the cytocompatibility and toxicity of cell surface coatings. Other interesting topics covered in the book include microelectronic devices based on nanomaterial-carrier cells, magnetic decoration and labelling of prokaryotic and eukaryotic cells, artificial multicellular assemblies from cells interfaced with polymers and nanomaterials, and formation of artificial spores mimicking bacterial endospores. The book concludes by examining the future of cell surface engineering. Overall, this book provides an excellent summary and analysis of the area of cell surface engineering and should be of broad interest to cell biologists, engineers, chemists and materials scientists.

Robert S. Langer
David H. Koch Institute Professor
Department of Chemical Engineering
Massachusetts Institute of Technology
77 Massachusetts Ave
Cambridge, MA 02139, USA

RSC Smart Materials No. 9
Cell Surface Engineering: Fabrication of Functional Nanoshells
Edited by Rawil F Fakhrullin, Insung S Choi and Yuri Lvov
© The Royal Society of Chemistry 2014
Published by the Royal Society of Chemistry, www.rsc.org

Contents

RSC Smart Materials No. 9
Cell Surface Engineering: Fabrication of Functional Nanoshells
Edited by Rawil F Fakhrullin, Insung S Choi and Yuri Lvov
© The Royal Society of Chemistry 2014
Published by the Royal Society of Chemistry, www.rsc.org

CHAPTER 1

Introduction

RAWIL F. FAKHRULLIN,*[a] YURI M. LVOV[b] AND
INSUNG S. CHOI[c]

[a] Bionanotechnology Group, Institute of fundamental medicine and
biology, Kazan (Idel buye/Volga region) Federal University, Kreml uramı
18, Kazan, Republic of Tatarstan, Russian Federation; [b] Institute for
Micromanufacturing, Louisiana Tech University, 911 Hergot Ave., Ruston,
LA 71272, USA; [c] Center for Cell-Encapsulation Research, Department of
Chemistry, KAIST, Daejeon 305-701, Korea
*Email: kazanbio@gmail.com

Bionanotechnology strives to fabricate functional biohybrid structures by
synergistically combining biomacromolecules, cells, or multicellular assemblies with a wide range of nanomaterials, to which interesting functionalities are deliberately designed and introduced at the nanometer scale.
The equipment of micrometer-sized cells with tiny devices, expanding the
uses of the cells in biotechnology and biomedicine, also requires the
nanoengineering of individual cells. Cells, having the typical sizes of dozens
of micrometers, are relatively large if compared with nanometer-sized particles and films; therefore, the nanomodification of living cells opens new
pathways to perform the selective control of cell properties. One can dream
of designing a comfortable "smart dress" for cells, which protects the cells
from hostile external environments and also provides useful toolkits that the
cells may learn to use for their benefit. In addition to the application aspect,
the "smart dress" coatings can be applied to the detailed investigation of
fundamental biological properties, such as enzyme activity, membrane
permeability, or viability preservation in cells, among many others.

RSC Smart Materials No. 9
Cell Surface Engineering: Fabrication of Functional Nanoshells
Edited by Rawil F Fakhrullin, Insung S Choi and Yuri Lvov
© The Royal Society of Chemistry 2014
Published by the Royal Society of Chemistry, www.rsc.org

The recent progress in bionanotechnology has been based extensively on inspiration from nature and adopted the mechanisms and structures found in living creatures for elaboration of man-made materials and devices. The cells are the biological units, which are used by nature to build all known living organisms, from micro-organisms to human beings.[1] The complexity of cellular machinery, consisting of precisely orchestrated ensembles of enzymes, nucleic acids, and other biological macromolecules, is yet to be understood. However, each and every cell has a *cellular membrane*, acting as the thin barrier that protects the cell from the outer environments, ensures the internal integrity, and facilitates the controllable transport of the chemical species and particles. The cellular membranes are as diverse as the cells themselves, displaying a number of surface molecules, which can be used as a fingerprint profile of a certain biological species. Generally, they are built of membrane proteins, lipids and carbohydrates, complementing one another to support the shape, integrity and functionality of the cells. In addition to lipid bilayer membranes, some cells (*i.e.* most of bacteria, fungi, and plants) develop a thick protective coating, termed the *cell wall*, which reinforces the membranes and makes the cell-wall-protected cell rigid and resistant to external impacts. Other species rely solely on subtle cellular membranes, while sometimes employing exoskeleton-like structures like solid shells or building external skin-like structures. The enormous number of biological events is associated with the surfaces of membranes and cell walls, which are responsible for ion pumps, feeding, excretion, neuro-transmission, antibody/antigen recognition, cell division, *etc.* Virtually all of the cellular processes are controlled or regulated directly or indirectly by cellular surfaces that interface with the outer surroundings. Here, we regard all the cellular coatings as external *cell surfaces*, which exhibit certain surface chemistries. Like biological systems, any changes in cell-surface chemistry will inevitably lead to changes in cell functioning, and provide the cells with otherwise unavailable functionalities. Recent progresses allow for modification/functionalization of biological cells, viewed as colloid microparticles (albeit their very complex nature) in the first approximation, with the wide range of chemical techniques in a controllable fashion.

In this book, we define *cell surface engineering* as a chemical methodology aimed to deliberately modify, control, functionalize or alter the nanoscale surface chemistry of biological cells by the directed deposition of macro-molecules or nanomaterials as artificial shells on the nanometer scale.[2] The ultimate goal of the cell surface engineering is to fabricate an artificial nanocoating on the cellular surfaces without modifying the genome of the cells by using purely chemical architecture approaches. The cell nanocoating generates various types of artificial cell-in-shell structures, where single isolated cells or cell aggregates are coated with either flexible or solid shells. These shells are typically fabricated to endow the cells with additional functionalities of interest. Alternatively, the cells can be used as mere sac-rificial templates for fabrication of cell-mimicking hollow microcapsules or as parts of microelectronic devices. Historically, the layer-by-layer (LbL)[3] deposition technique was used first to form nanolayers over sacrificial cells, but in the recent decade, a number of other approaches have emerged, offering new avenues in cell surface engineering.

We start this book with the introduction of three most popular approaches in cell surface engineering: 1) multilayered polyelectrolyte assemblies; 2) direct deposition of nanomaterials, and 3) bioinspired encapsulation with inorganic shells. These approaches have been successfully applied to fabricate nanocoatings on numerous representative model organisms. Next, the book reviews the experimental techniques to characterize and image the nanomaterial-encapsulated cells and to assess the viability and biological functionality of the surface-engineered cells. Then, we focus on applications of the nanocoated cells. The topics include the microelectronic devices fabricated with nanocoated cells including magnetically labeled prokaryotic and eukaryotic cells in analytical applications, and the LbL-coated cells in biomedical applications, such as tissue engineering and biosensors. Finally, the book covers the fabrication of artificial multicellular assemblies, where cells interfaced with polymers and nanomaterials are used as building blocks, and the formation of artificial spores mimicking bacterial endospores. These areas are truly multidisciplinary, utilizing the practical skills in nano- and colloidal sciences to chemically mimic and control biological processes, such as of evolution of multicellularity and cryptobiosis.

References

1. *Membrane Structural Biology With Biochemical and Biophysical Foundations.* Mary Luckey (ed.), Cambridge University Press, Cambridge, UK, 2008, ISBN: 9780521856553.
2. E. Saxon and C. R. Bertozzi, *Science*, 2000, **287**, 2007–2010.
3. R. F. Fakhrullin and Y. M. Lvov, *ACS Nano*, 2012, **6**(6), 4557–4564.

CHAPTER 2

Functional Multilayered Polyelectrolyte Assemblies on Biological Cells

BEN WANG[a,b]

[a] Cancer Institute (Key Laboratory of Cancer Prevention and Intervention, Ministry of Education), The Second Affiliated Hospital, Zhejiang University School of Medicine, Hangzhou, Zhejiang 310009, China; [b] Institute for Translational Medicine, School of Medicine, Zhejiang University, Hangzhou, Zhejiang 310029, China
Email: benwang@zju.edu.cn

2.1 Layer-by-Layer (LbL) Polyelectrolyte Assembly

2.1.1 Self-Assembly and Emerging of LbL Polyelectrolyte Assembly

Nature exploits self-organization of multiple materials to produce biopolymer fibers, cell membranes, the flagellar motor, viruses, hard tissue and other multiple-scale organic–inorganic hybrid structures in many ways. Self-assembly is one of the forces behind the bottom-up construction of well-ordered structures at the nanometer scale. It typically occurs through reversible interactions that slowly arrange building blocks into the most thermodynamically favored structure, which is the fundamental process, and generates structural organization across scales.[1] This process relies on molecular recognition between building blocks through noncovalent interactions, such as van der Waals and electrostatic forces, hydrogen bonding

RSC Smart Materials No. 9
Cell Surface Engineering: Fabrication of Functional Nanoshells
Edited by Rawil F Fakhrullin, Insung S Choi and Yuri Lvov
© The Royal Society of Chemistry 2014
Published by the Royal Society of Chemistry, www.rsc.org

and π–π stacking, which provide the thermodynamic driving force to form, and determine the structure of highly ordered nanostructures.

In the second half of the 20th century, more and more attention was given to the design of thin solid films at the molecular level because of their potential applications in biology and medicine. The electrostatic attraction between oppositely charged molecules seemed to be a good candidate as a driving force for multilayer buildup, because it has the least steric demand of all chemical bonds. Based on the early pioneering work of Iler in 1966,[2] Decher and coworkers developed an approach to coat charged surfaces through LbL adsorption of chain-like molecules equipped with ionic groups at the end, polyelectrolytes, or other charged materials, such as nanoparticles, from aqueous solution in the 1990s.[3–6]

2.1.2 Physics and Molecular Properties of LbL Polyelectrolyte Assembly

Strong electrostatic attraction occurs between a charged surface and an oppositely charged molecule in solution.[7] This phenomenon has long been known to be a factor in the adsorption of small organics and polyelectrolytes, but it has rarely been studied with respect to the molecular details of layer formation. In principle, the adsorption of molecules carrying more than one equal charge allows for charge reversal on the surface, which has two important consequences. One is that repulsion of equally charged molecules occurs, and thus self-regulation of the adsorption and restriction to a single layer. The other one is the ability of an oppositely charged molecule to be adsorbed in a second step on top of the first one. Cyclic repetition of both adsorption steps leads to the formation of multilayer structures.

A number of external parameters, which can be varied during the deposition process, are known to influence the resulting layer structure. These are the salt content of the deposition solutions,[8–11] the concentration[12] and charge density of the polyions (either varied by charge dilution in copolymers or as a result of the pH in weak polyion solutions),[13] the polyion rigidity, the molecular weight and also the surface charge density.

2.1.3 Types of LbL Assembly

Multiple electrostatic bonds causing a strong attraction are generally discussed as being responsible for the formation and stability of polyelectrolyte membranes. However, in order to explain the phenomenological behavior of layer formation, not only the Coulomb attraction, but additional contributions to the free energy of the complex have to be considered. Therefore, different kinds of LbL assembly can be classified based on this principle.

2.1.3.1 *Conventional LbL Assembly*

A solid substrate with a positively charged planar surface is immersed in a solution containing an anionic polyelectrolyte and a monolayer of the polyanion is adsorbed. This process of multilayer formation is based on the attraction of opposite charges, and thus requires a minimum of two oppositely charged molecules. Consequently, one is able to incorporate more than two molecules into the multilayer, simply by immersing the substrate in as many solutions of polyeletrolytes as desired, as long as the charge is reversed from layer to layer. Even aperiodic multilayer assemblies can easily be prepared.[4]

2.1.3.2 Hydrogen-Bond-Mediated LbL Assembly

The concept of electrostatically driven assembly of multilayer structures allows for the incorporation of a wealth of different materials. However, polymers as multifunctional materials also offer the choice of building up layered structures through other types of interaction. One of the most commonly studied, nonelectrostatic interactions used in LbL assembly to date is hydrogen bonding. By exploiting this interaction, uncharged materials can be successfully incorporated into multilayer films. The pioneering studies in LbL multilayer assembly based on hydrogen bonding were reported independently by Stockton and Rubner,[14] and Zhang's group[15] in 1997. Stockton and Rubner investigated the use of polyaniline in alternation with a variety of water-soluble macromolecules, such as poly(vinyl-pyrrolidone) (PVPON), poly(vinyl alcohol) (PVA), polyacrylamide, and PEO, in which the oxygen atoms on the polymer backbone can be hydrogen-bonding acceptors, or poly(N-isopropylacrylamide), in which both an acceptor (carbonyl) and donor (amide) are present. Besides, film deposition is possible with polymer pairs containing side groups with carbazole and dinitrophenyl units that can form charge-transfer complexes.[16,17]

2.1.3.3 Covalent Bonding Based LbL Assembly

Covalent bonds can be used to assemble LbL films having high stability due to the covalent bonds formed, and therefore these do not disassemble with changes in pH or ionic strength. Bergbreiter and coworkers performed the first example of sequential covalent assembly of polymers, using a copolymer of maleic anhydride reacted in alternation with a polyamidoamine dendrimer.[18] Blanchard and coworkers also investigated approaches to prepare multilayer films using a sequential covalent strategy.[19,20]

Inspired by dopamine self-polymerizing at alkaline pH to form adherent polymer coatings on a large variety of substrates,[21] synthetic polymers with catechol and amine functionalities may be useful as "universal" LbL primers. The synthetic catecholamine polymers adsorb to virtually all surfaces and can serve as a platform for LbL assembly in a surface-independent fashion.[22] Besides, click chemistry refers to a set of covalent reactions with high reaction yields that can be performed under extremely mild conditions. This technique provides a simple and general method for the assembly of polyelectrolyte films of controlled thickness and that the click moiety provides stable crosslinks within the films.[23]

2.1.3.4 Colloid-Involved LbL Assembly

The assembly of oppositely charged nanoparticles without polymers was first reported in 1966.[2] The variety of inorganic shapes and compositions of the nanocomponents available for the LbL assembly process has led to an exceptional growth in the fabrication of LbL composites. Various assembly approaches have been employed to assemble polymers or nanoparticles in an ordered manner and to investigate the scope of potential applications. Polymers and inorganic nanocrystals have been studied in detail to create unique architectures inspired from Nature by manipulating the specific interactions.[24,25] The LbL approach for assembling polymers with inorganic nanoparticles provides the opportunity to combine the electronic, optical, and magnetic properties of inorganic nanostructures with unique physical responses of macromolecules.[26,27] For example, Rubner and coworkers used LbL assembly for oppositely charged nanoparticles without polymers, which exhibited antireflection, antifogging, and self-cleaning properties.[28] LbL assembly of biomolecules with inorganic nanocomponents also leads to a new direction in the field of biomedical research, and the development of new technologies for diagnostic and therapeutic applications.[29–32]

2.2 LbL Polyelectrolyte Assembly on Cells

2.2.1 Cell-Templated LbL Assembly for Polyelectrolyte Shell

Biological cells possess a wide variety of shapes and sizes, thus, using them as templates would allow the production of capsules with a wide range of morphologies. In the pioneering work on this topic, *Escherichia coli* (*E. coli*) and red blood cells (RBCs) are used as templates to produce hollow polyelectrolyte multilayered microcapsules consisting of poly(allylamine hydrochloride) (PAH) and poly(styrene sulfonate) (PSS) (Figure 2.1).

This technique employs the step-wise self-assembly of polyelectrolyte multilayers on the biological templates with subsequent dissolution of the biological core, yields polyelectrolyte microcapsules of controlled size and shape that essentially replicate the morphology of original cells. The nature of the polyelectrolyte species, as well as solution properties and digestion procedures of the biological templates seem to influence the final properties of the microcapsule. Alternating adsorption of PAH and PSS onto charged latex particles always results in a reversal of the surface charge, independent of the layer number.[33,34] However, alternating adsorption of the same polyelectrolytes onto the RBC surface produces a reversal of the surface potential only from the third layer onwards, with increasing potential differences up to the seventh layer. This difference is attributed to the more complex structure of a biological surface, where the negative surface charge

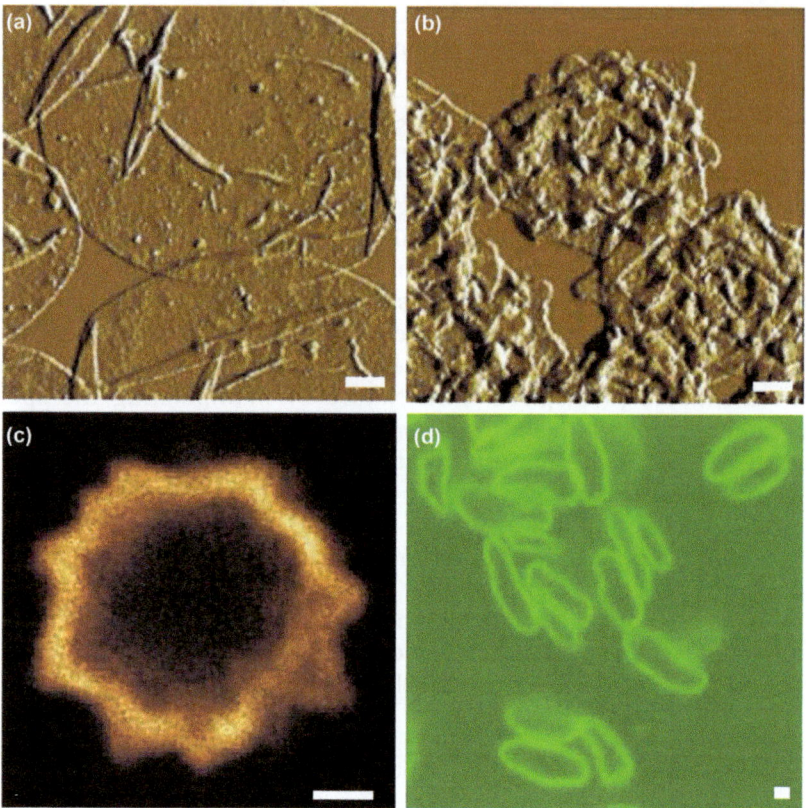

Figure 2.1 Cell-templated LbL assembly for polyelectrolyte shell.[40] (a) AFM images of polyelectrolyte shells composed of nine layers of PSS/PAH that are templated on discocytes, and (b) echinocytes, respectively. (c) The confocal microscopy image of a polyelectrolyte shell consisting of 11 layers templated on an echinocyte. The outer layer is FITC-labeled PAH. (d) Confocal images of polyelectrolyte capsules templated on discocytes with assembling five layers of PSS/PAH. The shells are exposed to FITC-labeled albumin solution (1 mg mL^{-1}). Scale bar is 1 μm.

is largely provided by sialic acid residues attached to glycoproteins. Hence, for biological particles, the surface charge is distributed nonhomogeneously in a layer of several nanometers thickness.[35–37] Therefore, it is possible that the first layers of adsorbing polyelectrolytes do not fully compensate for this spatially distributed surface charge of the cell surface. The permeability of the polyelectrolyte microcapsules can be controlled by means of the outer milieu conditions and the polyelectrolyte species selected.[38,39] Appropriate variation of these experimental parameters allows adjustment of the threshold for molecular permeability from a few Daltons up to more than tens of kDa or a pore size range of 0.1–15 nm. An additional advantage is the exceptional stability of the fabricated shells in various solvents, thereby offering the potential for stable micro- or even nanoemulsions, and the production of a new class of colloids with supermonodispersity and with a variety of shapes. Two interesting prospects for further development are the construction of polyelectrolyte microcapsules on biological cell templates other than RBC or *E. coli*, which could result in an extremely wide variety of geometric and physicochemical properties, and the functional properties, for example controllable permeability, of the shells templated on biological cells are also preserved.[40]

As indicated above, the digestion of the cell template generates osmotically active solutes that initially cause swelling of the polyelectrolyte shell. However, this initial swelling is followed by a return of the shell's shape and size to essentially the original size and shape of the biological templates. These polyelectrolyte shells exhibit elastic mechanical properties that cause a return to the undeformed shape once diffusion has eliminated the osmotic stress. They also exhibit a bending modulus comparable in magnitude to native RBC.[41,42] In contrast, microcapsules fabricated with spherical melamine templates behave like plastic particles that undergo nonrecoverable deformation. This development is a significant step forward in the fabrication of well-defined, micrometer-scale supramolecular structures for various applications, such as DNA and proteins encapsulation within hollow multilayered microcapsules built on biological cells.[43] However, strong interactions with the adsorbing polyelectrolyte molecules can lead to membrane rupture with subsequent breakdown of the template structure during the coating procedure. Therefore, in the previous study, human RBCs and *E. coli* bacteria were used, which were chemically stabilized by glutardialdehyde treatment prior to the LbL construction.

2.2.2 Cell-Directed LbL Assembly for Cellular Shell

2.2.2.1 *Microorganisms*

2.2.2.1.1 Fungi. *Saccharomyces cerevisiae* (*S. cerevisiae*) is an interesting class of eukaryotic cells whose genome covers 12 million base pairs and about 6000 genes, and shares a common life cycle and cell structure with higher eukaryotes.[44] Moreover, it is the first eukaryotic organism whose genome was completely sequenced.[45] Mutants are useful for studying eukaryotic DNA replication, transcription, RNA processing, protein sorting, and regulation of cell division. The first example of single living cell coating was conducted in a yeast cell model by Diaspro and coworkers in 2002 (Figure 2.2a).[46] A single living yeast cell is encapsulated by the alternate adsorption of oppositely charged polyelectrolytes. They provide evidences of the shell and cell integrity after the coating procedure by exploiting fluorescence techniques. The morphological and mechanical properties of encapsulated yeast cells are investigated by AFM, and the results indicate an increase in polyelectrolyte coating stability when decreasing the solution ionic strength. An evaluation of the viability of encapsulated cells is obtained by confocal laser scanning microscopy (CLSM) measurements.

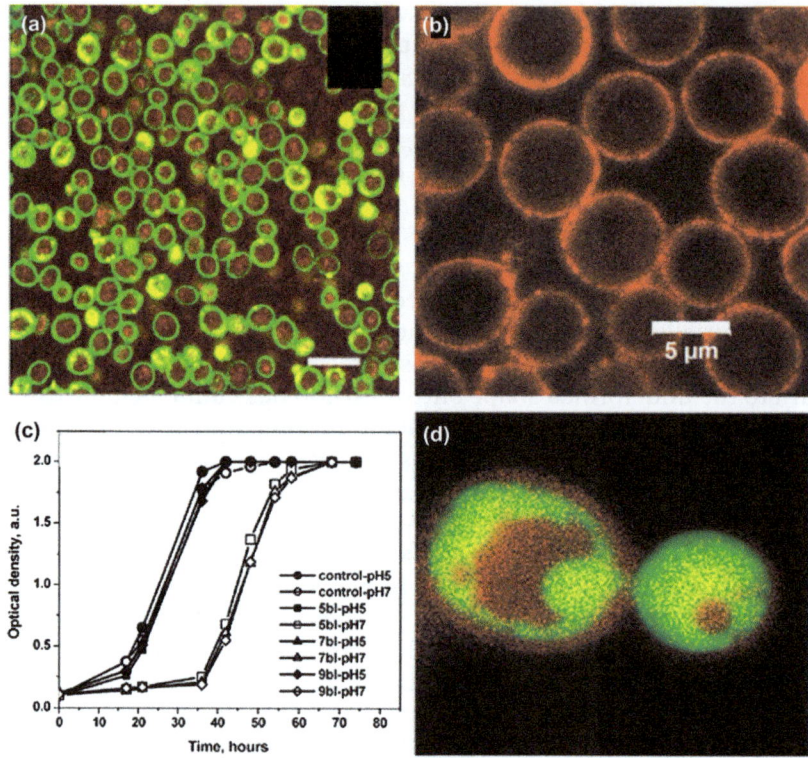

Figure 2.2 Single yeast cell encapsulation in nano-organized polyelectrolyte shells. (a) Confocal images of freshly coated cells with overlapping of reflection (red) and fluorescence (green);[46] Scale bar: 10 μm. (b) (PMAA-co-NH$_2$)$_9$-coated yeast cells at pH 5.0. (c) Growth kinetics of (PMAA-co-NH$_2$)-coated yeast cells grown in cell media adjusted to pH 5.0 and 7.0 with different shell thicknesses. (d) Expression of yEGFP 17 h for the (PMAA-co-NH$_2$)$_7$ coated yeast cells after induction with 2% galactose in media at pH 5.0.[52]

CLSM analysis indicates that cells preserve their subcellular structure and duplication capability after encapsulation.[47] It is illustrated that hybrid polyelectrolyte cells could provide a cheap model system in a wide range of biophysical and biotechnological applications, due to the tunable properties of the polyelectrolyte shell. Since then, individual living fungi cells have been encapsulated on microstructured surfaces by first coating cells with polyelectrolytes using the LbL methods and subsequently immobilized on patterned surfaces. It is found that coating did not kill the cells and coated GFP-expressing cells still function after immobilization, which are checked by fluorescence microscopy.[48] However, capsules prepared under these deposition conditions are still permeable to lysosomal enzymes, leading to degradation of the yeast inside the intact capsules.[49]

In contrast to ionic polyelectrolytes pairs, truly nonionic hydrogen-bonded LbL coatings for cell surface engineering will have much better bio-compatibility. These ultrathin, highly permeable polymer membranes are constructed on living cells without the cationic component typically employed to increase the stability of LbL coatings. Without the cytotoxic cationic PEI prelayer, the viability of encapsulated cells drastically increases to 94%, in contrast to 20% viability in electrostatically bonded LbL shells. Moreover, the long-term growth of encapsulated cells is not affected, thus facilitating efficient function of protected cells in hostile environment.[50] Engineering surfaces of living cells with natural or synthetic compounds can mediate intercellular communication and provide a protective barrier from hostile environments.

The main purposes of the LbL coating of yeast cells are biological protection against hostile environments. However, modulation of cellular metabolism, such as control cell division, has not yet been realized, mainly because of the relative fragility of electrostatically formed LbL shells, although the salient advantages of LbL processes include cytocompatibility of materials and processes, facile thickness tunability, and "functionalizability". Tsukruk and coworkers report a simple LbL method of encapsulating individual yeast cells in nonionic, hydrogen-bonded polymer shells composed of tannic acid and poly(N-vinylpyrrolidone) (PVPON), and of controlling cell division by tuning the shell thickness.[51] They also form a tough pH-responsive LbL shell on individual yeast cells by crosslinking the amine-bearing poly(methacrylic acid) (PMMA-co-NH$_2$) in the LbL layers of PVPON and PMMA-co-NH$_2$ (Figures 2.2b, c and d).[52] The yeast cell viability is maintained, and cell growth is delayed by raising the pH value above the isoelectric point of the polymer shell, which causes pH-triggered swelling of the shell. These results indicate that LbL shells have great potential for the generation of artificial spores; further modifications of LbL processes would allow incorporation of robustness into the shell without losing the advantages of LbL self-assembly, such as cytocompatibility, tunability, and "functionalizability".

LbL coating of single cells can also be as starting point of chemical functionalization for 2D or 3D cellular assembly. Electrostatic LbL self-assembly of polyelectrolytes multilayers could act as a shell material itself, but also as a catalytic template for subsequent formation of inorganic shells. Tang and coworkers pioneered this field and *in situ* precipitation of calcium phosphates on the LbL treated *S. cerevisiae* is induced by control the interfacial energy on the cell surface.[53,54] The viability of the cells is maintained after the encapsulation (Figure 2.3a). The enclosed cells become inert (stationary phase) and their lifetime can be extended. Furthermore, the mineral shell protects the cell under harsh conditions (Figure 2.3b). The encapsulated *S. cerevisiae* can even survive the attack of the lytic enzyme, zymolyase. The shell can also be used as a scaffold for chemical and biological functionalization. For example, *S. cerevisiae* becomes magnetic by the incorporation of Fe$_3$O$_4$ nanoparticles in the mineral layer. Their work demonstrates

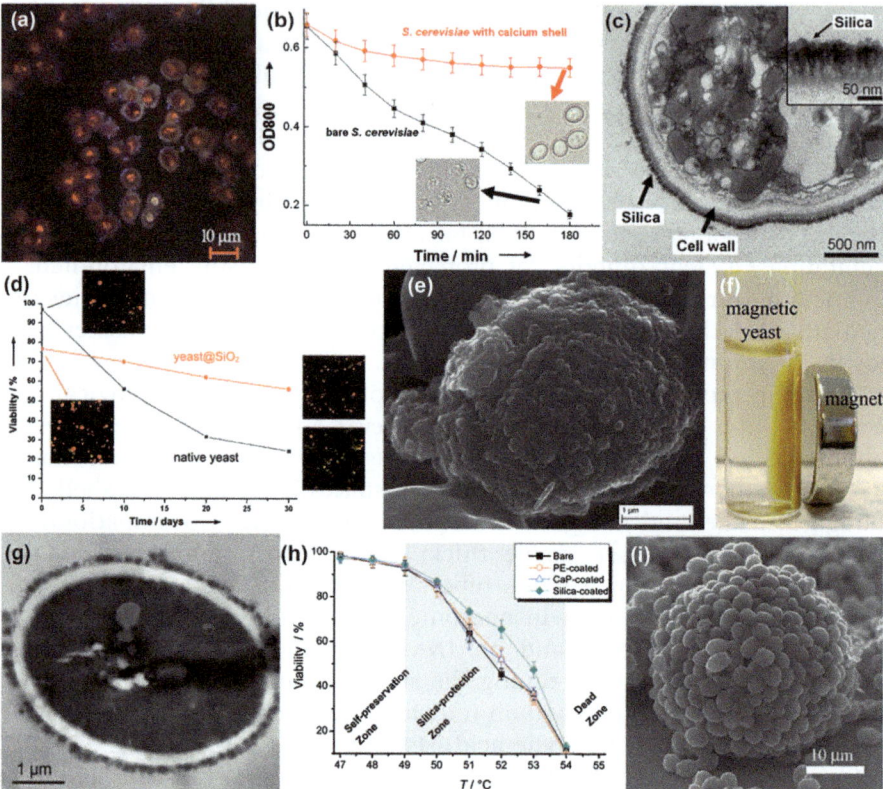

Figure 2.3   Different pathways for cellular shell engineering based on LbL assembly. (a) The fluorescent image of yeast cell with calcium mineral
shell that show mineral shell (yellow)-cell wall (green)-vacuole structure
(red), respectively. Red spots imply that the cells are alive.[53] (b) The
artificial shell also serves as an enhanced safeguard to protect the living
cells against foreign aggression, such as zymolyase that can digest the
cell wall and the cells cannot maintain homeostasis and burst without
the protection of the artificial shells.[53] (c) TEM micrographs of microtome-sliced yeast encapsulated in silica by biomimetic silicification on
the catalytic basis of LbL medication.[57] (d) Silica encapsulation also
give better viability of yeast in the long-term storage test.[57] (e) One
yeast cell coated with PAH-stabilized Fe$_3$O$_4$ nanoparticles and gets
the magnetic properties accordingly (f).[62] (g) TEM micrograph of
yeast cell with LbL silica layers. The organelles and walls of silica-
coated cells remained almost unchanged after heat exposure.[63]
(h) Viability of yeast cells with and without silica coating following
thermal treatment at different temperatures shows that extracellular
silica nanocoat confers thermotolerance on individual cells.[63] (i) PAH-
coated yeast cells could form an unusual structure of celloidosomes,
which is similar to the primitive forms of multicellular species,
Volvox.[64]

that the artificial shell has a great potential in the storage, protection, delivery/sorting, and modification of living cells.[55] Furthermore, insights from systems biology combined with an understanding of the molecular mechanisms of functional shells will facilitate the tailoring of "supercells" through biomimetic encapsulation.[56]

Subsequently, tough biomineralized shells of silica (Figure 2.3c),[57] and calcium carbonate,[58] were typically formed by introduction of catalytic organic templates onto cells by Choi group, Fakhrullin and coworkers separately. They also show that the biomimetic silicification involving biocompatible processes could increase the long-term viability of individual yeast cells (Figure 2.3d). Furthermore, chemical functional groups could be introduced directly to the biomimetically formed silica by adding silanol derivatives that contain functional groups in the course of polycondensation of silicate derivatives, and the biotin-functionalized yeast@SiO_2 is further exploited for immobilization onto defined surfaces, which is an important step for the fabrication of cell-based sensors.[59]

Polyelectrolytes facilitate adhesion of nanoparticles on cells, thus providing the stability and responsibility of the sandwich-like polyelectrolyte/nanoparticle coating and suppressing nanoparticle internalization through the cell walls into the cytoplasm. Layer-by-layer assembly with metal nanoparticles, such as Au and Ag, was fabricated on the surface of fungi cells.[60] The LbL nanoparticle coating is similar to deposition of linear polycations and polyanions, but some of the polyelectrolyte layers are replaced by a layer of properly charged nanoparticles. It is often necessary to finish shells with linear polyelectrolyte as outer layers in order to fix the multilayer architecture and preventing nanoparticles loss. Magnetic nanorods and carbon nanotubes have also been deposited on cells (Figures 2.3e and f).[61,62] Nanotubes load with drugs could potentially provide sustained drug release and long-lasting cell treatment. Yeast cells are also coated with a uniformly thin and continuous layer of biocompatible silica through a layer-by-layer chemical modification approach. The silica nanoshell forms an extracellular shield that endows these cells with enhanced defense against high temperature without significantly compromising cell viability (Figures 2.3g and h).[63]

Fabrication of three-dimensional multicellular clusters mimicking the structure of primitive multicellular organisms is yet another avenue for LbL-coated cells. Functionalization of simple microbial model cells with polyelectrolytes helps to restore the likely environmental conditions facilitating the transition from unicellularity to multicellularity. With LbL-coated yeast cells, spherical, needle-shaped, and cubic shaped multicellular living assemblies, called cellosomes can be produced, and that represent a man-made artificial model of multicellular organisms.[64,65] These cellosomes are viable for several weeks and resembled natural colonial micro-organisms, such as Volvox species (Figure 2.3i). It is expected that similar approaches might be employed to fabricate more complex cellular structures consisting of different types of cells, and even human cells.

2.2.2.1.2 Bacteria. Lvov and coworkers report that bacterial spores are encapsulated in organized nanofilms using layer-by-layer assembly in order to assess the biomaterial as a suitable core and determined the physiological effects of the coating. The shells are constructed on *Bacillus subtilis* spores using biocompatible polymers polyglutamic acid, polylysine, albumin, lysozyme, gelatin A, protamine sulfate, and chondroitin sulfate. The assembly process was monitored by measuring the electrical surface potential, ζ-potential, of the particles at each stage of assembly. CLSM and scanning electron microscopy (SEM) confirmed the formation of uniform coatings on the spores. The ultrathin coating surface charge and thickness could be selectively tuned by using appropriate polymers and the number of bilayers assembled. The coated spores are viable, but the kinetics and extent of germination are changed compared with control spores in all instances.[66] These experiments give the insight to design various bioinspired systems inspired by Nature. For example, the spores can be made dormant for one certain period using the LbL encapsulation approach and can return to being active when needed.

Another method for encapsulation of living micro-organisms is by using the preparation of hollow polymer microspheres based on the pre-precipitation of porous calcium carbonate cores with an average size of 5 μm. The microspheres filled with individual living *E. coli* cells are prepared by LbL deposition of different polyelectrolytes and proteins on the porous calcium carbonate cores leading to the formation of matrix-like complexes of the compounds followed by calcium carbonate core dissolution using EDTA. Both the influence of the encapsulation process as well as of the used polyelectrolytes on the survival rate of the cells are determined by CLSM and microtiter plate fluorescence tests. Around 40% of the cells are alive after the encapsulation process. Cultivation tests indicate that the lag phase of cells treated with polyelectrolytes increases and the encapsulated E. coli cells are able to produce green fluorescent protein inside the microcapsules, which proves the viabilities are kept well after LbL deposition.[67]

The encapsulation of probiotic *Lactobacillus acidophilus* (*L. acidophilus*) through LbL self-assembly of polyelectrolytes chitosan (CHI) and carboxymethyl cellulose (CMC) has been investigated by Raichur and coworkers to enhance its survival in adverse conditions encountered in the GI tract. The survival of encapsulated cells in simulated gastric and intestinal fluids is significant when compared to nonencapsulated cells. On sequential exposure to simulated gastric and intestinal fluids for 120 min, almost complete death of free cells is observed. However, for cells coated with three nanolayers of PEs (CHI/CMC/CHI), about 33 log % of the cells (6 log cfu/500 mg) survived under the same conditions. The enhanced survival rate of encapsulated *L. acidophilus* can be attributed to the impermeability of polyelectrolyte nanolayers to large enzyme molecules like pepsin and pancreatin that cause proteolysis and to the stability of the polyelectrolyte nanolayers in gastric and intestinal pH. The PE coating also serves to reduce viability losses during freezing and freeze-drying. About 73 and 92 log % of uncoated and coated cells survived after freeze-drying, and the losses occurring between freezing and freeze-drying were found to be lower for the coated cells.[68]

2.2.2.1.3 Viruses. In the course of evolution, Nature designed viruses, small particles that defy categorization as actual living entities or just small assemblies of biomolecules. Viruses actually are nanocomposites consisting of only a few polymeric species and sometimes equipped with a lipid membrane. They carry genetic information for their replication but need a host cell to accomplish reproduction. The virus envelope, composed of less than a handful of macromolecular species, nevertheless, bears all the functions needed to recognize and enter a host cell. This is exploited in various biotechnological applications, such as protein production, vaccine design, development of genetic libraries, gene therapy and so on. If it were possible to engineer colloidal particles with surfaces bearing the functionality of viruses, this would be a novel means for the delivery of a variety of materials into cells and tissues. Donath and coworkers described how such virus-modified particles have been fabricated by the LbL approach originally introduced for macroscopic surfaces, and how their functionality has been proved.[69] Through the consecutive adsorption of oppositely charged polyions on colloidal particles, multilayers can be fabricated.

Their thickness and composition can be tuned on the nanometer scale. When this synthetic method is applied to colloids with soluble cores, capsules can be fabricated. Various functions can be added by either employing functional species for adsorption or by subsequent modification. A possible strategy would be to fabricate composite particles or capsules equipped on their surface with all the necessary virus functions for passage of cell membrane. Taking an example, this approach could be used to deliver a cocktail of material packed into a small colloidal entity into cells, which is not easy to do with the existing delivery devices.

At the same time molecular biology, in particular, virology, provides a variety of tools for genetic engineering. Techniques for the manipulation of viruses are well established. Various proteins and peptides can be expressed at the virus surface.[56,69] These modified viruses thus provide great flexibility in the choice of biologically engineered surfaces. For example, Tang and coworkers used a combination of genetic engineering and biomineralization techniques to produce a thermostable vaccine.[56] The self-directed biomineralized vaccine can be used efficiently after storage at ordinary temperatures, which significantly increases the efficacy of immunization systems and lowers the cost of vaccine delivery and storage. Combining the assembly of virus-like LbL colloids and capsules with genetic engineering would thus open up novel pathways for fabricating functional bionanocomposites of complex but fully controlled interfacial composition. They may be useful as combinatorial entities in a variety of biomedical and biotechnological applications such as diagnostics, vaccination, and delivery.[70]

On the other hand, viral nanoparticles have received great attention in current nanotechnology. They can be used as potential templates or building

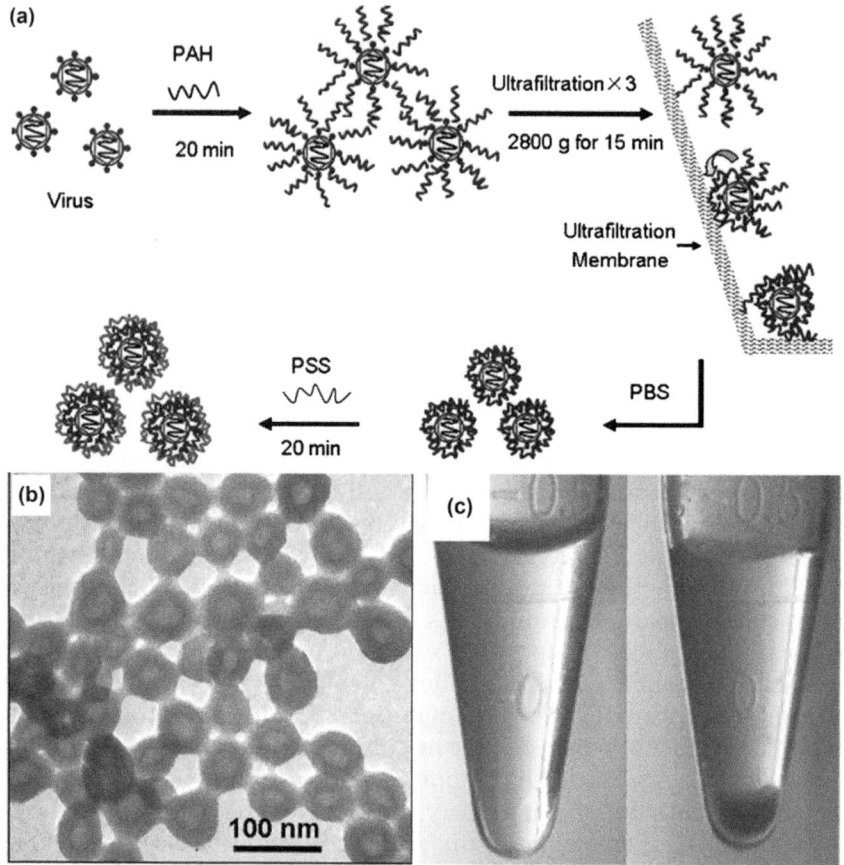

Figure 2.4   (a) Scheme for the process of preparation of virus/PAH-PSS hybrid complex. (b) TEM micrograph of virus/PAH-PSS without stain. (c) Virus could not be separated and concentrated by normal centrifugation due to the LbL treatment.[74]

blocks and incorporated with alternative materials to design multi-functional biocompositions.[71–73] Tang and coworkers developed a facile strategy for single-virus coating by polyelectrolytes assembly with high efficiency and biosecurity (Figure 2.4).[74] The modified viruses have different physical and biological properties from the native ones, such as the ability for direct transmission electron microscope observation, convenient separation and concentration, rapid delivery, and suppression of infectivity.

2.2.2.2   Mammal Cells

Formation of robust thin artificial shells is more difficult for mammalian cells. The surfaces of mammalian cells are not protected or reinforced by a polysaccharide layer, which makes them more fragile towards the osmotic shocks upon exposure to polyelectrolyte solutions and more difficult to treat using chemical approaches. However, development of such shells would have a great impact on many technologically important areas. For instance, a durable artificial shell could find applications in cell therapy, where the shell's physical stability is an important issue for implantation of encapsulated cells into the body.[75,76] Because of the weakness of mammalian cells in the presence of harsh conditions, such as abnormal pH, ion concentrations, toxicants, and temperature, careful selection of materials and processes is required to encapsulate individual mammalian cells. As the viability of different cells may depend in a completely different way from the presence of the same polymer, a careful selection of polyelectrolytes is recommended to choose those less cytotoxic.[77] Chaikof *et al.* reduced the cytotoxicity of coating materials by adding biocompatible poly(ethylene glycol) parts to poly L-lysine (PLL*g*-PEG) for coating pancreatic islets (Figure 2.6a).[78] In fact, mostly biogenic and biocompatible polyelectrolytes (*i.e.* polypeptides, polysaccharides) are currently used for functionalization of animal cells, whereas synthetic polymers are used to encapsulate microbial cells. Although the deposition of polyelectrolytes onto animal cells is a more complex procedure, it is still possible to engineer the surface of such cells with polyelectrolyte shells without fixing them, thus preserving cellular viability.[79] Special attention has been given to surface engineering of isolated human cells, since it is believed that multilayered films of various architectures may be used to attenuate the functions of the cellular membranes and, as a consequence, considerably contribute to the field of biosensors,[80] tissue engineering[77] and cell-based therapy (Figure 2.5).[81]

Recent reports demonstrate some examples of polyelectrolyte-mediated surface modification of various cells. Erythrocytes are encapsulated with alginate (AL) and chitosan-grafted-phosphorylcholine (CH-PC) further surrounded by two bilayers of alginate and poly L-lysine-grafted-polyethylene glycol (PLL-PEG).[82] In another paper, the encapsulation of mouse mesenchymal stem cells is shown *via* the sequential introduction of isolated cells into polyelectrolyte (poly L-lysine and hyaluronic acid) solutions.[83] Matsusaki *et al.* develop a highly biocompatible pairing for polyelectrolyte membranes, with gelatin and fibronectin employing the coating of substrate-assembled cells,[84] inspired by strong interactions between cells and extracellular matrices in organs and tissues (Figure 2.5b); they successfully constructed densely packed multicellular tissue-mimicking clusters in a tunable way by accumulating the fibronectin–gelatin-coated single fibroblast cells.[85] After the formation of the first fibroblast monolayer on glass substrates, the cells were coated with seven consecutive nanolayers of polyelectrolytes, followed by seeding of the second layer of cells. The procedure described was repeated until four layers of fibroblasts are assembled on substrates. This experimental approach is similar to the first reported LbL assembly of PAH/PSS on glass substrates, differing only by the preliminary deposition of cells onto planar glass surfaces.[86] Yet another original

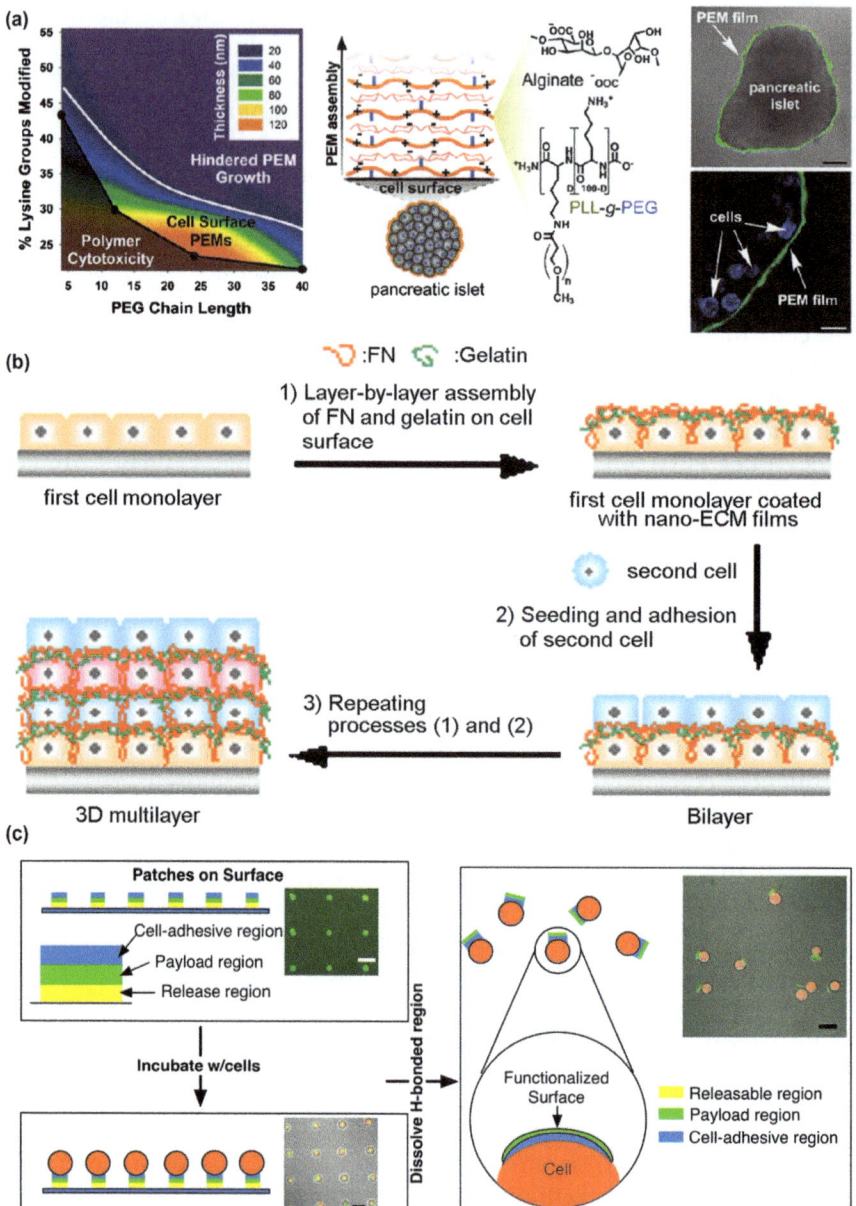

Figure 2.5 Layer-by-layer assembly on mammal cells. (a) Cell surface engineering by polyelectrolyte multilayer films with modular biological functionalities and tunable physicochemical properties which have been engineered to abrogate cytotoxicity.[78] (b) Fabrication of 3D tissue-like cellular multilayers with fibronectin (FN) and gelatin for nanometer-sized extracellular matrix films.[84] (c) Functional polyelectrolyte multilayer (PEM) patches can be attached to a fraction of the surface area of individual living lymphocytes by using emerging photolithographic patterning techniques.[81]

technique is reported, where the living lymphocytes are functionalized *via* LbL assembly combined with a photolithographic-patterning technique (Figure 2.5c).[81] The peculiar feature of this work is the selective functionalization of the cellular membrane, *i.e.* just a certain part of the cell is coated with polymers, while the bulk of the cell remains intact.

The development of biosensors also requires the surface modification of human cells, for instance, human breast cancer cells are encapsulated in multilayers built from various polycations (PEI, PAH, PLL or PDDA) and polyanions (PSS or PAA).[87] The LbL deposition of polymers was applied not only to single cells, but also to multicellular assemblies such as pancreatic islets. Polymer-encapsulated islets are believed to find applications in cell-based therapy of type I diabetes as therapeutic transplants with increased immune response resistance with polymer shells acting as a barrier between the transplant and the immune system. Isolated islets consisting of numerous cells assemble in spherical aggregates of around 200 mm are coated with PAH/PSS, PDDA/PSS and PLL-PEG/AL.[78] The comparatively large diameters of pancreatic islets outline a modified coating procedure. Islets are retained in cell-culture inserts with small pores (12 mm), which are filled with polyelectrolytes, incubated and then drained and washed with appropriate buffers. This allows numerous centrifugation steps to be avoided, which may lead to unnecessary aggregations.

2.2.2.3 Multicellular Animals

Although the modification and engineering of the cell surfaces with polymers/nanoparticles became a versatile tool to control the physiological parameters and the spatial distribution of the modified cells, this technique has been limited exclusively to unicellular organisms so far. Since the multicellular organisms are regarded as considerably more complex than unicellular species, both in terms of physiology and behavior, the encapsulation of multicellular organisms with polyelectrolytes and nanoparticles may provide the scientific community with a powerful instrument to study ecology, behavior, and motility of microscopic invertebrates *via* tailoring functional moieties using LbL assembly. Fakhrullin and coworkers report the surface modification of microscopic live multicellular nematodes *Caenorhabtidis elegans* with polyelectrolyte multilayers (pure and doped with 20-nm gold nanoparticles) and the direct magnetic functionalization of nematodes with biocompatible magnetic nanoparticles. Magnetically functionalized nematodes can be effectively separated and moved using an external magnetic field. At the same time, the surface-functionalized nematodes preserve their viability and reproduction (Figure 2.6a).[88]

Inspired by strategies in the living systems generated by evolution, Tang and coworkers induced an extra UVB-adsorbed coat on the chorion (eggshell surrounding embryo) of zebrafish, during the blastula period. Short and long-term UV exposure experiments show that the artificial

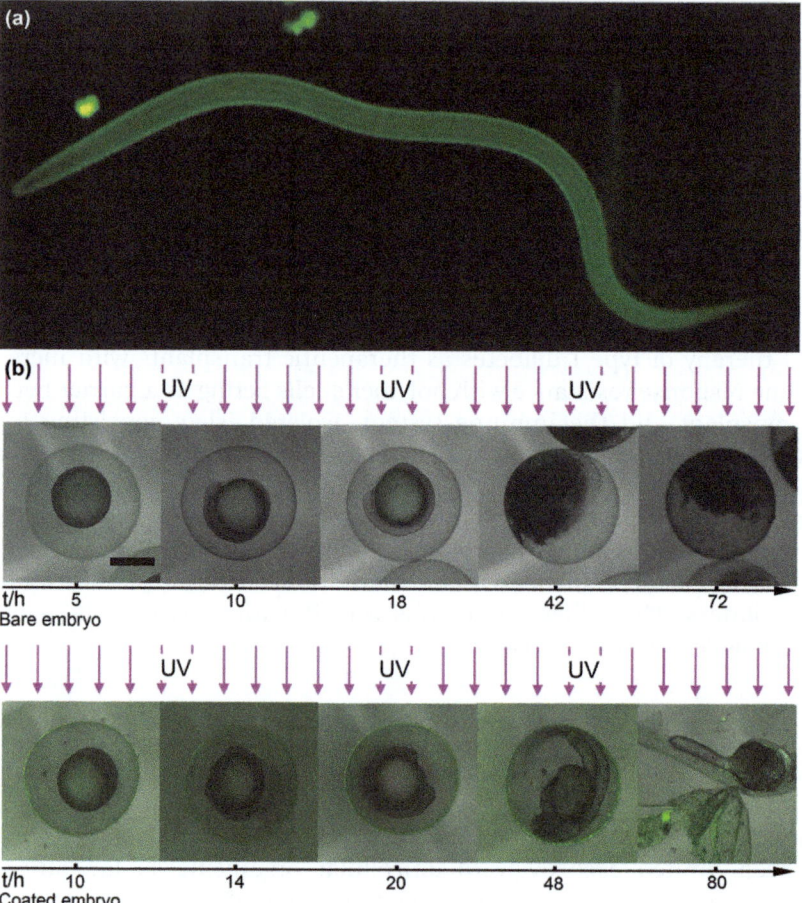

Figure 2.6 Interfacing multicellular organisms with polyelectrolyte or derivatived shells. (a) The surface modification of microscopic live multicellular nematodes *Caenorhabtidis elegans* with polyelectrolyte multilayers, while preserving their viability and reproduction.[88] (b) The coated embryo of Zebrafish with mineral shell can develop and hatch successfully within the man-made shield under enhanced UV exposure but the bare one cannot accomplish the embryonic development.[89]

mineral-shell reduces the UV radiation effectively and the enclosed embryos become more robust. In contrast, the uncoated embryos cannot survive under the enhanced UVB condition (Figure 2.6b).[89] It is suggested that an engineered shell of functional materials onto biological units can be developed as a strategy to shield lives to counteract negative changes of global environment, or to provide extra protection for the living units in biological research.

2.3 The Toxicity of Multilayered Polyelectrolyte Assembly

Since the layer-wise polyelectrolyte deposition offers the opportunity to modify surfaces for biomedical applications, interactions and toxicity between polyelectrolytes and living cells become critical. From the past results we assume that positive charges damage the cell layer more than negative ones. For example, Goodman *et al.* find that gold nanoparticles functionalized with positive side chains (quaternary ammonium group) have a moderate cytotoxic effect on Cos-1 cells, red blood cells, or bacteria (*Escherichia coli*), while negatively charged side chains (carboxylbate group) are non-toxic.[90] This is in good agreement with the findings for the positive and negative charged outermost layer on gold nanoparticles and demonstrates that quaternary ammonium groups show a slightly lower cytotoxicity. The decrease in response of the cells from a polyelectrolyte solution to polyions assembled in layers with counterions implies that only free charges damage the cell walls and induce apoptosis.[77]

The nature of cell death induced by the polycations was investigated by the nuclear morphology after DAPI staining and the inhibition of the toxic effects by the caspase inhibitor, zVAD-fmk. These assays yield the following comparable results and rank of the polymers with regard to cytotoxicity, poly(ethylenimine) = poly(l-lysine) > poly(diallyl-dimethyl-ammonium chloride) > diethylaminoethyl-dextran > poly(vinyl pyridinium bromide) > starburst dendrimer > cationized albumin > native albumin. The magnitude of the cytotoxic effects of all polymers are found to be time and concentration dependent. The molecular weight as well as the cationic charge density of the polycations are confirmed as key parameters for the interaction with the cell membranes and consequently, the cell damage. This is not a detectable indication for apoptosis by evaluating the nature of cell death induced by poly (ethylenimine), suggesting that the polymer induced a necrotic cell reaction.[91]

Interactions of polycationic polymers with supported 1,2-dimyristoyl-sn-glycero-3-phosphocholine (DMPC) lipid bilayers and live cell membranes (KB and Rat2) are also investigated by using AFM, cytosolic enzyme assays, CLSM, and a fluorescence-activated cell sorter (FACS). Polycationic polymers PLL, PEI, and diethylaminoethyl-dextran (DEAE-DEX) and sphere-like poly(amidoamine) (PAMAM) dendrimers are employed because of their importance for gene and drug delivery. AFM studies indicate that all the polycationic polymers cause the formation and expansion of pre-existing defects in supported DMPC bilayers in the concentration range of 1–3 µg/mL. By way of contrast, hydroxyl-containing neutral linear PEG and poly(vinyl alcohol) (PVA) do not induce hole formation or expand the size of pre-existing defects in the same concentration range. All polymers tested are not toxic to KB or Rat2 cells up to a 12 µg/mL concentration. In the concentration range of 6–12 µg/mL, however, significant amounts of the cytosolic enzymes lactate dehydrogenase (LDH) and luciferase (LUC) are released. PEI, which possesses the greatest density of charged groups on its chain, shows the most dramatic increase in membrane permeability. In addition, treatment with polycationic polymers allows the small dye molecules propidium iodide (PI) and fluorescein (FITC) to diffuse in and out of the cells. CLSM images also show internalization of PLL labeled with FITC. In contrast, controls of membrane permeability using the neutral linear polymers PEG and PVA show dramatically less LDH and LUC leakage and no enhanced dye diffusion. Taken together, these data are consistent with the hypothesis that polycationic polymers induce the formation of transient, nanoscale holes in living cells and that these holes allow a greatly enhanced exchange of molecules across the cell membrane.[92]

2.4   Conclusions and Perspective

In this chapter, we have described recent progress in LbL nanocoating to modify biological cells in order to provide them with new structural and functional features. Cellular shells with defined multicomponent compositions may contain polyelectrolytes, proteins, DNA, and nanoparticles in locations defined with nanometer precision, thus making assemblies with predetermined functionality. This vision is inspired in part by Nature design, such as bacterial spore formation, and artificial composite or hybrid shells enable robust cells in harsh conditions and allow cells to flourish in favorable environments. The LbL shells also can protect cells against mechanical, chemical or biological threats, additional selectivity in cellular membrane permeability, and new magnetic, electrical, and optical properties for microorganisms. The emergence of nanosize robotic devices that may be fabricated with biocells through architectural shell formation and "supercells" with new instrumentation and more productive living will be approached in the future.

So far, a series of methods for endowing the single cells with polyelectrolyte coatings, hard shells and other nanoparticles assemblies on their surfaces have been proposed, allowing such cell-in-shell complex to perform new functions, distinguished from the original ones. The current efforts are mostly focused on preserving the viability of the surface-functionalized cells, taking advantage from both the cellular and polyelectrolyte/nanoparticle shell functionality. Major cell types have been employed as the biological parts of the hybrid assembly of living–nonliving complexes, including microbial, plant and human cells and even multicellular animals. However, there is still plenty of room for the further work, which promises yet more fascinating combination of cells and materials. The currently established protocols for cellular surface functionalization are flexible and are easily modified to tailor virtually various types of materials to wide brand of cells. We suppose that the future work in this area will be focused on (i) attenuating the properties of the cells by functional materials, to exploit functional cells and find their medical and even clinical usages; (ii) developing protocol for single-cell quarantine and furthermore for cell–cell communication study; and (iii) developing feasible ways for *ex vivo* 3D cell culture or to produce 3D cell agglomerates. These directions are very promising from the practical point of view, for instance, new types of nanoparticles, including natural inorganic vessels like clay nanotubes or multifunctional composites and bioinspired nanomaterials, can be assembled on cells either as multi-functional containers carrying a payload or as versatile biosorbents. Surface-modified cells may find applications in biofuel cells, where the nanocoatings may enhance the naturally occurring enzymes. Another important direction is the use of polymers in mimicking of extracellular barriers, which may help in studying drugs uptake in the cellular level. The LbL method in conjunction with living cells, especially for human cells, may be applied to deposit any combination of polymers, pure and doped with enzymes, DNA, nanoparticles, *etc.*, which may potentially help to study the transmembrane traffic of nutrients, hormones and neuromediators. This would allow the effective use of functionalized cells in cell-based therapies, where the external coating would provide the therapeutic cells with protection, a load of nutrients, enhanced adhesion and the ability to be guided directly into the region of interest within the body. Stem cells as the unlimited source of therapeutic cells may get more opportunities inspired by these areas. Tissue engineering may also benefit from using the cell–materials complex since the tailored functionalities will provide the cells with the necessary micro-environment that is important for the colonization and the subsequent differentiation of cells in 3D microenvironment. Importantly, special attention should be given to investigation of the toxicity of cellular shells. The cooperative effects of the compositions of LbL films on human cells are to be taken into account. The possible toxic effects may be manifested not only in the functionalized cells, but also in the adjacent regions within the target tissue/organ.

Overall, the unlimited combinations of living cells engrafted with polyelectrolytes or nanoparticles offer an emerging opportunity for the yet undiscovered findings. In summary, we are in the beginning of a novel and emerging field of integrating nanomaterials, biology and medicine demonstrated by the interfacing of living cells with polymers/nanomaterials whose untapped potential for new applications will inspire many researchers and will lead to explosive growth in this area.

References

1. G. M. Whitesides, *Science*, 2007, **295**, 2418–2421.
2. R. Iler, *J. Colloid Interf. Sci.*, 1966, **21**, 569–594.
3. G. Decher and J. Hong, *Berichte der Bunsengesellschaft für physikalische Chemie*, 1991, **95**, 1430–1434.
4. G. Decher and J. D. Hong, *Makromolekulare Chemie. Macromolecular Symposia*, 1991.

5. G. Decher, J. D. Hong and J. Schmitt, *Thin Solid Films*, 1992, **210**, 831–835.
6. G. Decher, *Science*, 1997, **277**, 1232–1237.
7. M. Stuart, G. Fleer, J. Lyklema, W. Norde and J. Scheutjens, *Adv. Colloid Interfac.*, 1991, **34**, 477–535.
8. M. Ullner and C. E. Woodward, *Macromolecules*, 2002, **35**, 1437–1445.
9. Y. Hayashi, M. Ullner and P. Linse, *J Chem. Phys.*, 2002, **116**, 6836.
10. A. Dedinaite and M. Ernstsson, *J. Phys. Chem. B*, 2003, **107**, 8181–8188.
11. M. A. Dahlgren, A. Waltermo, E. Blomberg, P. M. Claesson, L. Sjoestroem, T. Aakesson and B. Joensson, *J. Phys. Chem.*, 1993, **97**, 11769–11775.
12. E. Poptoshev, M. Rutland and P. Claesson, *Langmuir*, 1999, **15**, 7789–7794.
13. O. J. Rojas, M. Ernstsson, R. D. Neuman and P. M. Claesson, *Langmuir*, 2002, **18**, 1604–1612.
14. W. Stockton and M. Rubner, *Macromolecules*, 1997, **30**, 2717–2725.
15. L. Wang, Z. Wang, X. Zhang, J. Shen, L. Chi and H. Fuchs, *Macromol. Rapid Commun.*, 1997, **18**, 509–514.
16. Y. Shimazaki, M. Mitsuishi, S. Ito and M. Yamamoto, *Langmuir*, 1997, **13**, 1385–1387.
17. Y. Shimazaki, M. Mitsuishi, S. Ito and M. Yamamoto, *Langmuir*, 1998, **14**, 2768–2773.
18. Y. Liu, M. L. Bruening, D. E. Bergbreiter and R. M. Crooks, *Angew. Chem. Int. Ed. Eng.*, 1997, **36**, 2114–2116.
19. P. Kohli and G. Blanchard, *Langmuir*, 2000, **16**, 4655–4661.
20. P. Kohli and G. Blanchard, *Langmuir*, 2000, **16**, 8518–8524.
21. H. Lee, S. M. Dellatore, W. M. Miller and P. B. Messersmith, *Science*, 2007, **318**, 426–430.
22. H. Lee, Y. Lee, A. R. Statz, J. Rho, T. G. Park and P. B. Messersmith, *Adv. Mater.*, 2008, **20**, 1619–1623.
23. G. K. Such, J. F. Quinn, A. Quinn, E. Tjipto and F. Caruso, *J. Am. Chem. Soc.*, 2006, **128**, 9318–9319.
24. Y. Ofir, B. Samanta and V. M. Rotello, *Chem. Soc. Rev.*, 2008, **37**, 1814–1825.
25. F. Caruso, R. A. Caruso and H. Möhwald, *Science*, 1998, **282**, 1111–1114.
26. C. B. Murray, C. Kagan and M. Bawendi, *Annu. Rev. Mater. Sci.*, 2000, **30**, 545–610.
27. R. Gangopadhyay and A. De, *Chem. Mater.*, 2000, **12**, 608–622.
28. D. Lee, M. F. Rubner and R. E. Cohen, *Nano Lett.*, 2006, **6**, 2305–2312.
29. Z. Tang, Y. Wang, P. Podsiadlo and N. A. Kotov, *Adv. Mater.*, 2006, **18**, 3203–3224.
30. K. C. Wood, J. Q. Boedicker, D. M. Lynn and P. T. Hammond, *Langmuir*, 2005, **21**, 1603–1609.
31. D. S. Koktysh, X. Liang, B. G. Yun, I. Pastoriza-Santos, R. L. Matts, M. Giersig, C. Serra-Rodríguez, L. M. Liz-Marzán and N. A. Kotov, *Adv. Funct. Mater.*, 2002, **12**, 255–265.

32. B. Thierry, F. M. Winnik, Y. Merhi, J. Silver and M. Tabrizian, *Biomacromolecules*, 2003, **4**, 1564–1571.
33. S. W. Keller, S. A. Johnson, E. S. Brigham, E. H. Yonemoto and T. E. Mallouk, *J. Am. Chem. Soc.*, 1995, **117**, 12879–12880.
34. E. Donath, D. Walther, V. Shilov, E. Knippel, A. Budde, K. Lowack, C. Helm and H. Möhwald, *Langmuir*, 1997, **13**, 5294–5305.
35. K. A. Sharp and D. E. Brooks, *Biophys. J.*, 1985, **47**, 563–566.
36. E. Donath and A. Voigt, *J. Colloid Interf. Sci.*, 1986, **109**, 122–139.
37. E. Donath, A. Budde, E. Knippel and H. Bäumler, *Langmuir*, 1996, **12**, 4832 4839.
38. H. Bäumler, G. Artmann, A. Voigt, R. Mitlöhner, B. Neu and H. Kiesewetter, *J. Microencapsul.*, 2000, **17**, 651–655.
39. O. I. Vinogradova, *J. Phys. Condens. Mat.*, 2004, **16**, R1105.
40. E. Donath, S. Moya, B. Neu, G. B. Sukhorukov, R. Georgieva, A. Voigt, H. Bäumler, H. Kiesewetter and H. Möhwald, *Chem.-Eur. J.*, 2002, **8**, 5481–5485.
41. E. A. Evans, *Meth. Enzymol.*, 1989, **173**, 3–35.
42. B. Neu, A. Voigt, R. Mitlöhner, S. Leporatti, C. Gao, E. Donath, H. Kiesewetter, H. Möhwald, H. Meiselman and H. Bäumler, *J. Microencapsul.*, 2001, **18**, 385–395.
43. O. Kreft, R. Georgieva, H. Bäumler, M. Steup, B. Müller-Röber, G. B. Sukhorukov and H. Möhwald, *Macromol. Rapid Commun.*, 2006, **27**, 435–440.
44. D. Botstein and G. R. Fink, *Science*, 1988, **240**, 1439–1443.
45. A. Goffeau, B. G. Barrell, H. Bussey, R. W. Davis, B. Dujon, H. Feldmann, F. Galibert, J. D. Hoheisel, C. Jacq, M. Johnston, E. J. Louis, H. W. Mewes, Y. Murakami, P. Philippsen, H. Tettelin and S. G. Oliver, *Science*, 1996, **274**, 546–567.
46. A. Diaspro, D. Silvano, S. Krol, O. Cavalleri and A. Gliozzi, *Langmuir*, 2002, **18**, 5047–5050.
47. T. Svaldo-Lanero, S. Krol, R. Magrassi, A. Diaspro, R. Rolandi, A. Gliozzi and O. Cavalleri, *Ultramicroscopy*, 2007, **107**, 913–921.
48. S. Krol, M. Nolte, A. Diaspro, D. Mazza, R. Magrassi, A. Gliozzi and A. Fery, *Langmuir*, 2004, **21**, 705–709.
49. S. Krol, O. Cavalleri, P. Ramoino, A. Gliozzi and A. Diaspro, *J. Microsc.*, 2003, **212**, 239–243.
50. J. L. Carter, I. Drachuk, S. Harbaugh, N. Kelley-Loughnane, M. Stone and V. V. Tsukruk, *Macromol. Biosci.*, 2011, **11**, 1244–1253.
51. V. Kozlovskaya, S. Harbaugh, I. Drachuk, O. Shchepelina, N. Kelley-Loughnane, M. Stone and V. V. Tsukruk, *Soft Matter*, 2011, **7**, 2364–2372.
52. I. Drachuk, O. Shchepelina, M. Lisunova, S. Harbaugh, N. Kelley-Loughnane, M. Stone and V. V. Tsukruk, *ACS Nano*, 2012, **6**, 4266–4278.
53. B. Wang, P. Liu, W. Jiang, H. Pan, X. Xu and R. Tang, *Angew. Chem. Int. Ed.*, 2008, **47**, 3560–3564.
54. B. Wang, P. Liu, Z. Liu, H. Pan, X. Xu and R. Tang, *Biotechnol. Bioeng.*, 2013.

55. C. W. Yung, J. Fiering, A. J. Mueller and D. E. Ingber, *Lab Chip*, 2009, **9**, 1171–1177.
56. G. Wang, R.-Y. Cao, R. Chen, L. Mo, J.-F. Han, X. Wang, X. Xu, T. Jiang, Y.-Q. Deng and K. Lyu, *Proc. Natl. Acad. Sci. U. S. A.*, 2013, **110**, 7619–7624.
57. S. H. Yang, K. B. Lee, B. Kong, J. H. Kim, H. S. Kim and I. S. Choi, *Angew. Chem. Int. Ed.*, 2009, **48**, 9160–9163.
58. R. F. Fakhrullin and R. T. Minullina, *Langmuir*, 2009, **25**, 6617–6621.
59. S. H. Yang, E. H. Ko, Y. H. Jung and I. S. Choi, *Angew. Chem. Int. Ed.*, 2011, **50**, 6115–6118.
60. R. F. Fakhrullin, A. I. Zamaleeva, M. V. Morozov, D. I. Tazetdinova, F. K. Alimova, A. K. Hilmutdinov, R. I. Zhdanov, M. Kahraman and M. Culha, *Langmuir*, 2009, **25**, 4628–4634.
61. A. I. Zamaleeva, I. R. Sharipova, A. V. Porfireva, G. A. Evtugyn and R. F. Fakhrullin, *Langmuir*, 2010, **26**, 2671–2679.
62. R. F. Fakhrullin, J. Garcia-Alonso and V. N. Paunov, *Soft Matter*, 2010, **6**, 391–397.
63. G. Wang, L. Wang, P. Liu, Y. Yan, X. Xu and R. Tang, *ChemBioChem*, 2010, **11**, 2368–2373.
64. R. F. Fakhrullin, M.-L. Brandy, O. J. Cayre, O. D. Velev and V. N. Paunov, *Physical Chemistry Chemical Physics*, 2010, **12**, 11912–11922.
65. M.-L. Brandy, O. J. Cayre, R. F. Fakhrullin, O. D. Velev and V. N. Paunov, *Soft Matter*, 2010, **6**, 3494–3498.
66. S. S. Balkundi, N. G. Veerabadran, D. M. Eby, G. R. Johnson and Y. M. Lvov, *Langmuir*, 2009, **25**, 14011–14016.
67. J. Flemke, M. Maywald and V. Sieber, *Biomacromolecules*, 2012, **14**, 207–214.
68. A. J. Priya, S. P. Vijayalakshmi and A. M. Raichur, *J. Agricul. Food Chem.*, 2011, **59**, 11838–11845.
69. M. J. Schnell, L. Buonocore, E. Kretzschmar, E. Johnson and J. K. Rose, *Proc. Natl. Acad. Sci. U. S. A.*, 1996, **93**, 11359–11365.
70. M. Fischlechner, O. Zschörnig, J. Hofmann and E. Donath, *Angew. Chem.*, 2005, **117**, 2952–2955.
71. M. Fischlechner and E. Donath, *Angew. Chem. Int. Ed.*, 2007, **46**, 3184–3193.
72. M. Uchida, M. T. Klem, M. Allen, P. Suci, M. Flenniken, E. Gillitzer, Z. Varpness, L. O. Liepold, M. Young and T. Douglas, *Adv. Mater.*, 2007, **19**, 1025–1042.
73. C. Mao, D. J. Solis, B. D. Reiss, S. T. Kottmann, R. Y. Sweeney, A. Hayhurst, G. Georgiou, B. Iverson and A. M. Belcher, *Science*, 2004, **303**, 213–217.
74. X. Wang, Y. Deng, H. Shi, Z. Mei, H. Zhao, W. Xiong, P. Liu, Y. Zhao, C. Qin and R. Tang, *Small*, 2010, **6**, 351–354.
75. Y. Teramura and H. Iwata, *Soft Matter*, 2010, **6**, 1081–1091.
76. J. T. Wilson and E. L. Chaikof, *Adv. Drug Deliver. Rev.*, 2008, **60**, 124–145.

77. M. Chanana, A. Gliozzi, A. Diaspro, I. Chodnevskaja, S. Huewel, V. Moskalenko, K. Ulrichs, H. J. Galla and S. Krol, *Nano Lett.*, 2005, **5**, 2605–2612.
78. J. T. Wilson, W. Cui, V. Kozlovskaya, E. Kharlampieva, D. Pan, Z. Qu, V. R. Krishnamurthy, J. Mets, V. Kumar and J. Wen, *J. Am. Chem. Soc.*, 2011, **133**, 7054–7064.
79. H. Ai, M. Fang, S. A. Jones and Y. M. Lvov, *Biomacromolecules*, 2002, **3**, 560–564.
80. M. Germain, P. Balaguer, J.-C. Nicolas, F. Lopez, J.-P. Esteve, G. B. Sukhorukov, M. Winterhalter, II. Richard-Foy and D. Fournier, *Biosens. Bioelectron.*, 2006, **21**, 1566–1573.
81. A. J. Swiston, C. Cheng, S. H. Um, D. J. Irvine, R. E. Cohen and M. F. Rubner, *Nano Lett.*, 2008, **8**, 4446–4453.
82. S. Mansouri, Y. Merhi, F. o. M. Winnik and M. Tabrizian, *Biomacromolecules*, 2011, **12**, 585–592.
83. N. G. Veerabadran, P. L. Goli, S. S. Stewart-Clark, Y. M. Lvov and D. K. Mills, *Macromol. Biosci.*, 2007, **7**, 877–882.
84. M. Matsusaki, K. Kadowaki, Y. Nakahara and M. Akashi, *Angew. Chem.*, 2007, **119**, 4773–4776.
85. A. Nishiguchi, H. Yoshida, M. Matsusaki and M. Akashi, *Adv. Mater.*, 2011, **23**, 3506–3510.
86. K. Kadowaki, M. Matsusaki and M. Akashi, *Langmuir*, 2010, **26**, 5670–5678.
87. M. Germain, P. Balaguer, J. C. Nicolas, F. Lopez, J. P. Esteve, G. B. Sukhorukov, M. Winterhalter, H. Richard-Foy and D. Fournier, *Biosens. Bioelectron.*, 2006, **21**, 1566–1573.
88. R. T. Minullina, Y. N. Osin, D. G. Ishmuchametova and R. F. Fakhrullin, *Langmuir*, 2011, **27**, 7708–7713.
89. B. Wang, P. Liu, Y. Tang, H. Pan, X. Xu and R. Tang, *PloS ONE*, 2010, **5**, e9963.
90. C. M. Goodman, C. D. McCusker, T. Yilmaz and V. M. Rotello, *Bioconjugate Chem.*, 2004, **15**, 897–900.
91. D. Fischer, Y. Li, B. Ahlemeyer, J. Krieglstein and T. Kissel, *Biomaterials*, 2003, **24**, 1121–1131.
92. S. Hong, P. R. Leroueil, E. K. Janus, J. L. Peters, M.-M. Kober, M. T. Islam, B. G. Orr, J. R. Baker and M. M. Banaszak Holl, *Bioconjugate Chem.*, 2006, **17**, 728–734.

CHAPTER 3

Direct Deposition of Nanomaterials onto Cells

ALSU I. ZAMALEEVA,*[a,b] RENATA T. MINULLINA,[a]
JOSHUA R. TULLY,[c] MARIA R. DZAMUKOVA,[a]
SVETLANA A. KONNOVA[a] AND EKATERINA A. NAUMENKO[a]

[a] Biomaterials and nanomaterials group, Department of Microbiology, Kazan (Idel buye/Volga region) Federal University, Kreml uramı 18, Kazan, Republic of Tatarstan, 420008, Russian Federation; [b] Dendritic cell and antigen presentation group, Institut Curie (INSERM U932) 26 Rue d'Ulm, 75005 Paris, France; [c] Nanoassembly group, Institute for Micromanufacturing, Louisiana Tech University 911 Hergot Ave., 72272 Ruston, LA, USA
*Email: dr.alsu@gmail.com

3.1 Introduction

Interest in the interactions of nanoparticles with biological cells continues to grow, progressively covering more fields of science such as biomedicine, optoelectronics, ecology, and many others. There are two primary methods that are utilized to combine nanoparticles with biological cells. The first one is *delivering nanoparticles inside of cells* in order to create new therapeutic agents for targeted drug and gene delivery, efficient systems for high-resolution imaging, or to study intracellular processes.

Another approach is *modification of the cell surface with nanomaterials* to obtain "hybrid systems" for bioelectronic applications or to provide living cells with a new phenotype depending on the unique physical and chemical

RSC Smart Materials No. 9
Cell Surface Engineering: Fabrication of Functional Nanoshells
Edited by Rawil F Fakhrullin, Insung S Choi and Yuri Lvov
© The Royal Society of Chemistry 2014
Published by the Royal Society of Chemistry, www.rsc.org

properties of the chosen particles. The direct deposition of nanoparticles onto the cells seems to be the simplest route but it has some limitations. These limitations are caused by random endocytosis pathways and direct translocation across the plasma membrane that cells usually use for non-specific absorption of exogenous molecules or nanoparticles. Therefore, it can be quite challenging to selectively modify only the cell surface and avoid the penetration of nanoparticles inside the cell. In this chapter we will discuss the approaches that are currently used for direct deposition of nano-materials on the surface of biological cells.

The term *"direct deposition"* will be defined as a one-step deposition of nanomaterials onto the surface of the cell. However, this one-step process does not include any work necessary to prepare the nanoparticles for de-position, nor does it include any modifications to the surface of the cell that would be required for successful deposition. Numerous types of nano-materials are available and/or one could synthesize them depending on scientific needs, whereas the biological template for deposition will be re-stricted to two types of cell surface. For prokaryotes and some eukaryotes the template will be the cell wall and for mammalian cells it will be the cell membrane. Obviously, the structure and biochemical composition of the cell surface plays an important role for nanomaterial deposition and their spatial arrangement. Therefore, we will give a brief introduction to the biochemical composition of surfaces of organisms widely used for deposition of nano-materials onto their surface such as bacteria, fungi, and mammalian cells. In general, all biological organisms have a cell membrane that serves as a fundamental component of cell protection from its external environment and is normally composed of a phospholipid bilayer with embedded chol-esterol, glycosylated surface proteins, and transmembrane proteins. In mammalian cells it is thin (5–10 nm) and fragile, therefore the deposition process should be carried out in a precise manner to avoid the internal-ization of nanomaterials inside of the cell. Chemically, the plasma mem-brane is composed of a wide variety of proteins, lipids in the form of cholesterol, phospholipid and sphingolipid, and carbohydrates in the form of glycoproteins. The phospholipids phosphatidylserine and phosphatidyli-nositol have negatively charged head groups that provide the negative charge of cellular membranes, besides, the carboxyl groups of glycoproteins tend to dissociate protons at physiological pH and also contribute to the surface charge. However, differences in the type and amount of lipids, carbohydrates and proteins exist between various types of cells, also the amount of these chemicals may vary between each monolayer of the plasma membrane.

Microorganisms like fungi, bacteria and plants have an additional layer of protection. These cells have another membrane composed mostly of com-plex polysaccharides that makes this layer a thick, tough and rigid structure in comparison to the cell membrane. The cell-wall composition and struc-ture varies not only between different domains of living organisms, such as with prokaryotic and eukaryotic cells, but also between different families of bacterial cells (see Figure 3.1). Gram-positive bacteria have a relatively thick

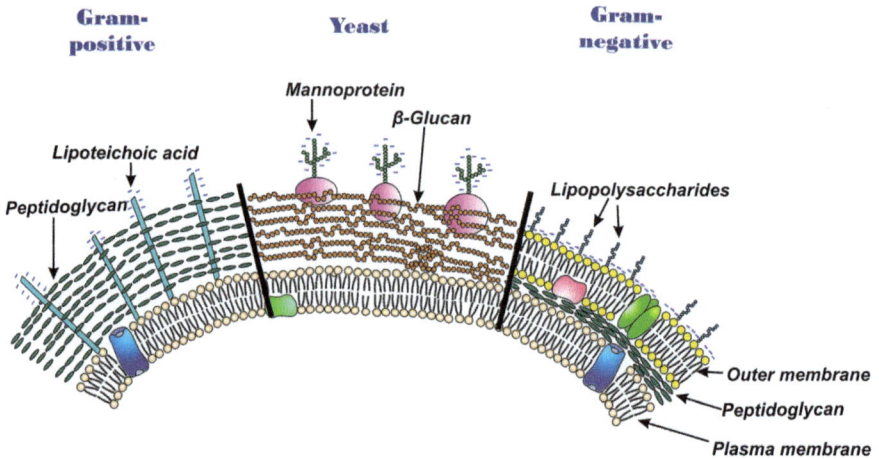

Figure 3.1 Composition of Gram-positive, Gram-negative and yeast cell wall.

(20–80 nm) homogeneous layer of peptidoglycan and large amounts of lipoteichoic acids that pierce the cell wall from the membrane. In contrast Gram-negative bacteria contain a thin layer of peptidoglycan this is buried in between an outer and inner plasma membrane. The yeast cells that are often used in research studies as a model eukaryote have a chemically simpler cell wall than bacterial peptidoglycan. The *Saccharomyces cerevisiae* yeast cell wall is made of beta-glucan polysaccharides, mannoproteins and a small amount of chitin.

Overall, the differences in biochemical composition and organization of surface structures determine the diversity of physicochemical characteristics of cell surfaces such as its hydrophobicity, adhesivity, electromagnetic and surface-charge properties. All of these parameters will strongly influence the deposition, distribution and spatial arrangement of nanomaterials on the cell surface.

3.2 Cell–Nanoparticle Hybrids

Unicellular microorganisms with a wide variety of morphologies and sizes provide convenient templates for organized assembly of nanomaterials for development of microdevices that can be used in bioelectronics. Different types of cells, including bacteria, fungi, algae cells and viruses, have been used as templates for the deposition of nanomaterials because these microorganisms typically have a relatively thick cell walls composed mostly of polysaccharides and proteins. This particular combination of molecules provides the mechanical strength and flexibility required for nanoparticle layer assembly. Using cells as templates allows one to control the spatial orientation of nanoparticles and even drive the formation or synthesis of nanoparticles on the cell surface. Furthermore, it possible to combine the unique processes of living cells with specific nanoparticles to create novel nanomaterials.

One of the common techniques for direct deposition of nanoparticles is based on electrostatic interactions with the cell surface. The composition of the cell wall provides a homogeneous negatively charged surface where cationic nanoparticles can be easily attached that will lead to the formation of structured layers on the cell surface. Using this approach, a monolayer of gold nanoparticles was deposited onto live *Bacillus cereus* bacterial cells to fabricate conductive bridges. In this experiment, microbial cells that have a high affinity for lysine were immobilized onto lysine-coated substrates and washed with a sodium hydroxide solution to remove the extracellular matrix and enhance the negative surface charge. Then, the bacteria-modified chip was incubated in a lysine-stabilized gold nanoparticle solution that resulted in a deposition of percolating clusters of gold nanoparticles exclusively on the cells and not on the similarly charged substrate. Although the deposition is based on electrostatic interactions, several factors greatly influence the deposition process. It was shown that only the negative charges of physical flat surface are not able to form percolating layers of nanoparticles and their spatial arrangement plays a key role in this process. In the case of the Gram-positive bacterium, the negatively charged teichoic acids within its wall have high mobility and brush-like structure that can attract and wrap positively charged gold nanoparticles which results in a three-dimensional organization of nanoparticles on the cell surface. Further, by regulating the incubation time it was possible to control the density of percolating layers. For example, increasing the incubation time of the bacteria-immobilized chip in the solution of nanoparticles from 30 min to 8 h created denser percolating layers. Interestingly, the cells viability and life cycle also play an important role in the deposition process of nanoparticles. Only bacteria in the logarithm phase of growth could drive the formation percolating clusters of gold nanoparticles, whereas bacteria that had been cultured for 48 h were still able to form bridges but without percolating layers. This technique of nanoparticles deposition does not cause damaging effects and nanoparticle-modified bacteria preserved their viability. Thus, an active hybrid bioelectronic device was fabricated where the electrical properties of gold nanoparticles was controlled by actuating the peptydoglican layer of the bacterium.[1,2]

A similar approach was used to modify the surface of *Bacillus cereus* with gold nanoparticles of different shapes such as nanorods and nanospheres[3] (Figure 3.2a and b). To achieve deposition the gold nanoparticles were prepared by a seed-mediated method and stabilized with cetyltrimethylammonium bromide (CTAB). Cetyltrimethylammonium bromide is a cationic surfactant used to synthesize gold nanoparticles of different shapes. The advantages of CTAB-coated gold nanoparticles are their stable positive charge, which does not dependent on pH, and their ability to be directly deposited onto the cells within a short period of time of about 15 min. Another commonly used positively charged molecule, known as poly(allylamine) hydrochloride, was utilized to deposit magnetic nanoparticles onto the surface of different cell types including fungi *S. cerevisiae*,[4] microalgae *C. pyrenoidosa*[5] and mammalian *HeLa* cells.[6]

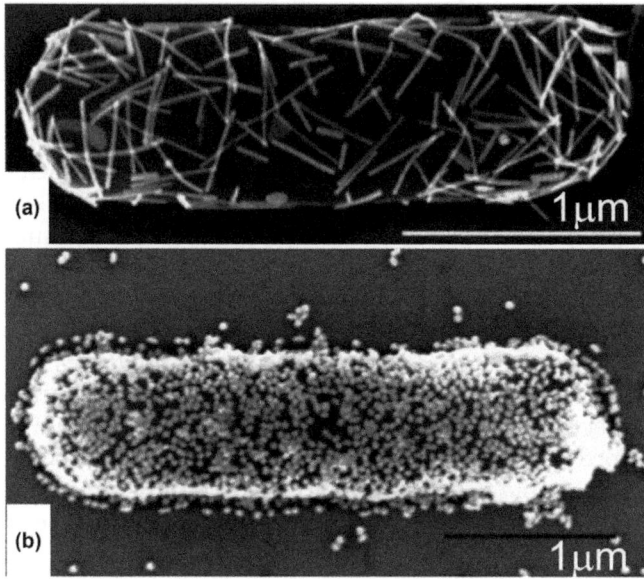

Figure 3.2   Percolating monolayer of gold nanorods (25 nm in diameter and
400 nm long) (a) and nanospheres (45 nm diameter) (b) is formed on
the surface of gram-positive bacteria *Bacillus cereus*.
Reprinted with permission from ref. 3. Copyright 2005 The American
Chemical Society.

Natural biopolymers also may serve as mediators for nanoparticles deposition onto cells and they have some advantages over synthetic components, like biocompatibility. For example, DNA-mediated deposition of
gold nanoparticles has been demonstrated using the diatoms *Navicula* and
Synedra genera as templates. The cells were pretreated with an acidic Piranha
solution to remove the organic components of cell walls and were further
activated by amino-silans. Amine-functionalized diatoms were conjugated
with fluorophore-labeled thiolated oligonucleotides using a crosslinker.
13-nm gold nanoparticles, modified with complementary oligonucleotides,
were then deposited onto the diatoms. Driven by hybridization between
complementary sequences, the nanoparticles assembled onto the diatoms
were shown to be highly specific and reversible at the same time because
increasing the temperature beyond the melting point of double-stranded
DNA led to dehybridization and the subsequent detachment of nanoparticles. The nanoscale coating on the diatoms allowed for detailed characterization of their surface, which could be used for the classification of
these unicellular algae.[7] Ti (5 nm) and Ag (30 nm) nanoparticles were deposited onto diatoms cell walls *via* evaporation, which yielded microshell
replicas displaying elaborate diatom surface features.[8]
    Modification of the cell surface with synthetic inorganic nanoparticles
can provide the cells with additional functionality which does not exist in

Nature. For example, Safarik *et al.* imparted new functions of paramagnetism to yeast cells *Kluyveromyces fragilis*[9] and *Saccharomyces cerevisae*[10] by the direct deposition of magnetic iron oxide nanoparticles on the cell wall. In the initial step the cells were washed with acetic acid or saline solution to remove the extracellular compounds from the cell wall and reduce the nonspecific binding. Then, the cell suspension was incubated with magnetic nanoparticles that led to magnetically responsive yeast hybrids. It was observed that the nanoparticle stabilization agents influenced the efficiency and incubation time for magnetization of the cells.[11] In the case of magnetic nanoparticles that were stabilized by sodium citrate, the obtained magnetic coating on the cell surface was thin and poor and required a longer incubation time than other agents. Using tetramethylammonium hydroxide or perchloric acid-stabilized nanoparticles made the process of cell magnetization faster and more efficient. Magnetic nanoparticle–cell hybrids were applied to the development of biosorbents, biocatalysts, and new cell isolation techniques. The magnetized cells were shown to effectively absorb heavy metal ions from solution, including toxic mercury,[10] and were easily separated using an external magnetic field. The catalytic functions of magnetic nanoparticles interfaced cells have been demonstrated by hydrogen peroxide decomposition and sucrose conversion by intracellular enzymes. This fact indicates that the cell magnetization is nontoxic and modified cells preserve their metabolic activity.[11]

From the studies described above we can see that by using a suitable stabilizing mediator it is possible to deposit nanoparticles of great variety of shapes and compositions. Therefore, different types of 1D and 2D nanostructures could also be used to modify the cell surface. Among these alternative nanomaterials multilayered polyelectolyte-based nanofilms were extensively used to modify cell surfaces to obtain hollow micro- and nanocapsules for living cell encapsulation. In general, they were deposited onto the cells *via* layer-by-layer assembly of oppositely charged polyelectrolytes and were also utilized as mediators for nanoparticles deposition.[12] Since the deposition of nanofilms often cannot be completed in a single step this technique is beyond the scope of this chapter and was discussed in more details in Chapter 2.

Carbon nanomaterials are of particular interest as materials for cell functionalization due to their unique electronic, optical, thermal, mechanical, and chemical properties. Biocompatible carbon nanotubes were interfaced with the cytoplasmic membrane of Chinese hamster ovary cells *via* carbohydrate receptors. The carbon nanotubes were first coated with mucinlike glycoproteins to mimic the surface chemistry of a cell followed by specific interactions *via* the carbohydrate-binding protein lectin, which is capable of crosslinking cells and glycoproteins. The most important observation was that these coated tubes were found to be noncytotoxic while uncoated tubes were cytotoxic.[13]

The group of Maheshwari demonstrated the interaction of graphene sheets with yeast cells for a "dual-use technology" application.[14] In one case an "electronic cell" was obtained by electromechanical coupling between the graphene sheets and the yeast surfaces that allowed them to study cellular processes such as cell growth, division, and response of cells to physiological stress (*i.e.* osmotic shock). They were able to detect changes in the graphene-modified cell's volume that lead to straining of the sheets and could be detected through the electrical signal produced by graphene. Using this simple biosensor, the dynamic response of the yeast cells to stress caused by exposure to alcohol was observed. At the same time the thermal stability and high modulus of the graphene shell enhanced the mechanical stability of the cell wall and protected it from harsh environmental conditions. One thing to note is that the coating is impermeable but does not interfere with transport processes or metabolic activity of the living cell due to partial coverage of the cell.[15] For yeast cell-wall modification graphene oxide sheets first were coupled with Ca^{2+} ions and gold nanoparticle branched-chain assemblies and were then immobilized onto living yeast. The graphene oxide of the coating was then reduced using glucose and ammonium hydroxide to form graphene sheets.

This approach was explored even further by the electrochemical synthesis of ZnO nanorods directly on the graphene functionalized living cells. Using graphene as an electrical conductor, it was possible to control the surface potential of the cell, which resulted in anisotropic growth of radially projecting ZnO rods. This new photosensor, based on the semiconducting properties of ZnO nanorods, showed improved efficiency compared to detectors built on planar solid surfaces due to the 3D architecture and the radial growth of the nanorods.[16] Thus, a cell can serve not only as a structural template but their biological features can drive nanoparticle assembly. In the case of the previously mentioned experiment, the surface potential of the cell was responsible for the unique structure of ZnO nanorods. Another example of this approach is fabrication of high-performance mesoporous $LiFePO_4$-yeast cell hybrid systems by a sol-gel method using living cells both as a biotemplate and a carbon source. In the first step of incubation, the yeast cells are combined with a ferrous gluconate solution that contains positively charged iron cations that are attracted to the COO^- and OPO_3^{2-} groups of various biomacromolecules of the cell wall, which induces the formation of an iron cation layer on the cell surface. The sequential addition of $(NH_4)_2HPO_4$, LiOH and ascorbic acid (a source of carbon) followed by the sol-gel mineralization resulted in densely packed $LiFePO_4$ nanoparticles assembled with mesoporous biocarbon into highly ordered networks. The obtained cell-based nanostructured system revealed high electronic conductivity which makes it a promising tool for rechargeable batteries.[17]

In summary, the first part of the chapter focuses on the nonspecific electrostatic and hydrophobic interactions between presynthesized nanomaterials and extracelullar structures of the cell wall that play a major role in forming a homogenous nanostructured coating of the cell. The obtained cell–nanomaterial systems often do not preserve cell viability after assembly because these hybrids aim at developing nanodevices for electronics, new advanced materials for ecology, and other fields where the active metabolic activity of the cell is not required. Nevertheless, the viability of the cell is important for initiation of nanomaterials deposition, spatial arrangement, and to maintain the 3D structure of cell–nanomaterial hybrids.

3.3   *In Situ* Deposition of Nanoparticles onto Cells

An alternative way to deposit nanoparticles onto cell surfaces is to synthesize them *in situ*. In this method the nanoparticles are synthesized or assembled into ordered structures directly on the cell surface. Precursors of nano-particles as well as presynthesized nanoparticles could be used for this type of deposition. This has been shown using viruses as biological templates. Their unique properties such as the highly ordered structure of the protein capsid and their small sizes makes viruses the perfect template for the de-velopment of nanoscale smart materials. The approach using nanoparticle precursors was used by Tseng *et al.* to design a tobacco mosaic virus based prototype of a new digital memory device. For this, platinum ions were mixed with a tobacco mosaic virus (TMV) protein capsid suspension to ac-tivate the viral surface *via* electrostatic interactions. Further addition of a reducing agent (dimethylamine borane) to this mixture resulted in Pt nanowire formation. The obtained nanoparticles were uniformly distributed on the virus surface and had an average size of 10 nm.[18] In contrast, Rh, Pd and Ru nanoparticles were presynthesized first by reduction of metal salts by sodium borohydrate and were deposited afterwards on the viral surface of M13 bacteriophages. Nanowire-like structures were obtained within an hour of incubation at room temperature.[19]

"Cell friendly" *in situ* synthesis of gold nanoparticles has been demon-strated by modifying the cell surface of different *E.coli* strains by growing them in a medium containing Au^{3+} precursors.[20] It is interesting to note that the formation of gold nanoparticles occurred without the addition of reducing agents. Possibly, the macromolecules of the cell wall (lipopoly-saccharides, glycoproteins) may serve as binding sites for Au^{3+} ions, while the aldehyde groups from the reducing sugars can act as electron donors to reduce Au^{3+} to Au^0. In this gold-nanoparticle–bacterium hybrid system, the light-absorption properties of gold nanoparticles in the near-infrared region were retained and have been successfully utilized for photothermal therapy of cancer. In this therapy, malignant lung cancer cells were bound to gold-nanoparticle-coated bacteria through receptor–antibody interaction and ex-posed to a laser. Gold nanoparticles effectively absorbed the near-infrared light and converted it to heat, which spread to nearby cancer cells and burned their cell walls causing death. Approaches of *in situ* synthesis of nanoparticles on the cell surface are summarized in Figure 3.3.

Interestingly, some natural phenomena can aid *in situ* synthesis of nanoparticles to engineer sophisticated and highly ordered cell–nanomaterial structures. For instance, the mycelium formation of fila-mentous fungi from asexual spores in the presence of nanoparticles or their

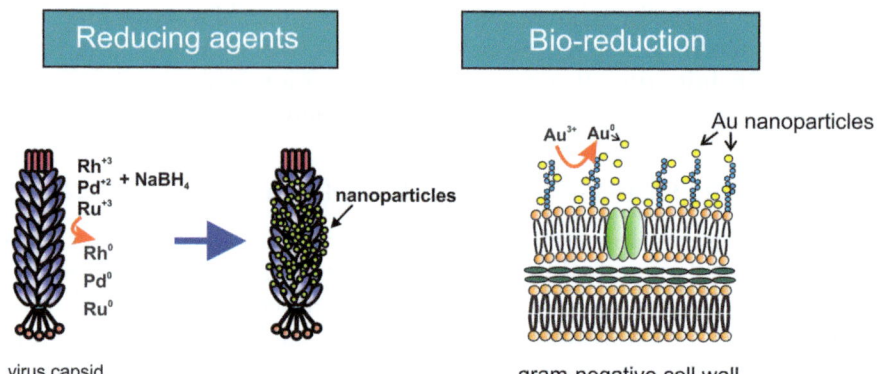

Figure 3.3   Approaches of *in situ* synthesis of nanoparticles. Left – example of metal-ion reduction by the addition of chemical reducing agents lead to deposition of metal nanoparticles on virus surface. Right – reduction process on the cell surface of Gram-negative bacteria.

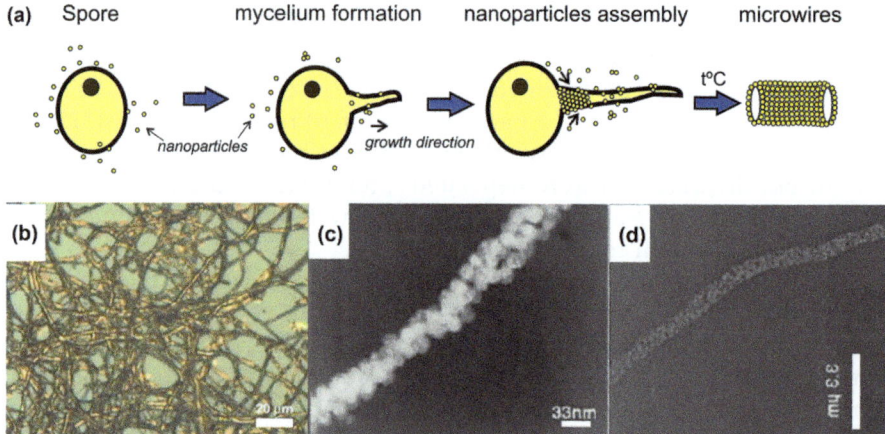

Figure 3.4   (a) Self-assembly process of nanoparticles onto fungi mycelium; (b) Optical microscopy image (reflected light mode) of a typical gold-coated mycelium obtained by growing the fungus in a solution of colloidal gold nanoparticles. (c) Image of a thin section of an *Aspergillus niger* hypha loaded with Au nanoparticles embedded in epoxy resin. (d) TEM image of the surface of an individual *Streptomyces venezuelaehypha* (*ca.* 0.8 mm in diameter) loaded with Au nanoparticles Wiley-VCH. Reprinted with permission from ref. 22 (b) and ref. 21 (c and d). Copyright 2007 and 2003, Wiley.

precursors allows one to obtain wire-like hybrid structures (Figure 3.4a). A fungal spore, in this case, serves as a living template for metal nanoparticle deposition. When a spore is placed into the growth medium it starts to grow and form elongating protrusions (hyphae), whereas if the medium contains nanoparticles or their precursors the hyphae were reported to accumulate them. Based completely on natural fungi growth, this type of nanoparticle deposition results in the formation of relatively long wire-like structures with a uniform and dense coating on the surface of the hyphae (Figure 3.4b–d). Metal nanoparticles, including gold nanoparticles, with different surface modifications such as stabilization with alkylthiol-capped oligonucleotides,[21] glutamate[22] and citrate,[23] silver, palladium, and platinum nanoparticles have been utilized for fabrication of highly ordered fungi-templated hybrid microwires. In addition, a variety of fungal genera, such as *Aspergillus*,[21,22] *Trichoderma*, *Fusarium*, and *Penicillium*,[23] have also been successfully employed. This indicates the versatility of this technique and it could potentially be applied for other species of fungi and nanoparticles. Nanoparticle precursors could also be used for *in situ* synthesis of nanoparticles. This was demonstrated by bioreduction of Ag^+ ions on the fungus *Verticillum*.[24] In this study exposure of the *Verticillum* spores to aqueous $AgNO_3$ solution caused formation of silver nanoparticles with an average diameter of 25 nm. The nanoparticles were shown to assemble below the fungal cell wall, which led the authors to suggest that the process of Ag^+ ions reduction is driven by intracellular-surface-related reactions.

As can be seen, several mechanisms can participate in *in situ* synthesis of nanoparticles. Electrostatic and hydrophobic interactions between nanoparticles and the cell wall are important for efficient immobilization of nanoparticle. Also important are the nutrients of the medium and cellular metabolites that can act as reducing agents and drive the assembly of the nanoparticles from their precursors or the formation of a bulk metal coating from the nanoparticles.

3.4 Ligand-Specific Nanomaterials Deposition onto Cells

The direct deposition of nanomaterials onto cells based on biocompatibility and high specificity between the interfacing nanomaterials and cells provides new possibilities to study the biochemical structure of cell surfaces, selective extracellular drug delivery, and other related research. The specificity in this case could be gained from the interaction of the surface ligands of nanomaterials with extracellular receptors, channels, and other proteins embedded into the phospholipid bilayer of the cytoplasmic membrane. In comparison to a cell wall, that is tough and quite rigid, the cytoplasmic membrane is soft, fragile, and semipermeable. The membrane controls the transport of substances in and out of the cell; therefore the deposition of nanomaterials onto the cell membrane should be carried out in a controllable manner for the specific sites of binding to avoid the internalization of nanomaterials. The electrostatic and hydrophobic interactions could also significantly influence the cell-surface/nanomaterial interface. On artificial lipid membranes, consisting of 1,2 distearoyl-sn-glycero-3-phosphocholine bilayers, it was shown that the interaction between lipid membrane and gold nanoparticles depends on the nanoparticles surface charge. Positively charged gold nanoparticles modified by [N,N,N-trimethyl (11-mercaptoun-decyl) ammonium] showed the tendency to pass through the fluid phase of the lipid membrane and became trapped between the hydrophobic bilayers while mercaptoundecanoic-acid-coated gold particles, which had a negative charge, did not penetrate the lipid membrane.[25]

Another research group proposed a new technique to study the ad-sorption/internalization process of gold nanoparticles (AuNP) onto the cells depending on their surface charges. Neutral, positive, and negatively charged gold nanoparticles (modified with poly(vinyl alcohol), poly(allylamine hydrochloride) polymers, and citric acid, respectively) were incubated with breast cancer cells (SK-BR-3) followed by treatment with an etching solution containing I_2 and KI at low molar concentrations so that it could selectively dissolve AuNPs on the cell surface without killing the cells. The results indicate that the uptake and adsorption onto the surface for positively charged gold nanospheres was much higher than for neutral or negatively charged ones.[26] These studies show that the process of nano-material deposition can be regulated through the surface chemistry of both components in the deposition process. Factors like the z-potential of the cell membrane, the chosen nanomaterials, and the ionic strength/pH of the medium, seem to play an important role in selective modification of the cell surface with nanomaterials.

The cell plasma membrane contains several extracellular structures in-cluding receptors, channels, and glycocalyx that can be used as targets for selective nanoparticles deposition. Cell surface receptors regulate com-munication between the cell and its environment through extracellular signaling ligands which bind to the receptor and trigger intracellular changes. Thus, a receptor may serve as an anchor for ligand-functionalized nanoparticle attachment. Click and surface chemistry methods allow one to efficiently conjugate biomacromolecules to the surface of nanoparticles for further deposition onto the cell surface receptors. For example, magnetic nanoparticles were covalently coupled to the metal-binding proteins lacto-ferrin and ceruloplasmin for targeting cell surface receptors (Figure 3.5a). On human dermal fibroblasts, it was shown that protein functionalized particles were attached to the cell membrane due to ligand–receptor inter-actions (Figure 3.5c and d) while bare magnetic nanoparticles were in-ternalized into the cytoplasm possibly *via* endocytosis (Figure 3.5a). The overexpression of the receptors for these ligands on the mammalian cells led to a homogeneous and quite dense distribution of magnetic nanoparticles on the cell surface.[27]

In another experiment, urokinase plasminogen activator receptor (uPAR), the specific surface biomarker of several types of cancer, was targeted by modified magnetic nanoparticles. For this purpose, the stabilizing oleic acid coating of the magnetic nanoparticles was first replaced by silane amine (3-aminopropyl-trimethoxysilane) and reacted with NHS-PEG-N$_3$ (*N*-Hydroxy-succinimide-polyethylene glycol-azide) thereby providing azide groups

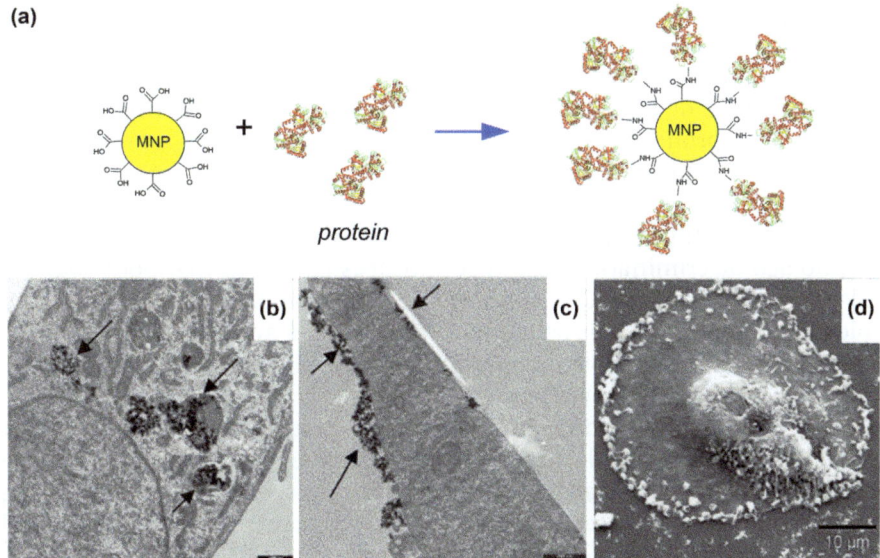

Figure 3.5 Ligand–receptor interaction between magnetic nanoparticles (MNP) coupled with proteins (lactoferrin or ceruloplasmin). (a) Scheme of protein coupled MNP preparation. (b) TEM image of human fibroblast cells after incubation with uncoated MNP. Plain uncoated showing nanoparticle internalization (see arrows). (c) lactoferrin coated nanoparticles showing nanoparticles at the cell surface (see arrows); (d) SEM image of human fibroblasts incubated with lactoferrin coated nanoparticles.
Reprinted with permission from ref. 27. (b–d) Copyright 2004, Elsevier.

tailored for click chemistry. The Cu^{1+}-catalyzed Huisgen cycloaddition reaction between azide-functionalized MNP and alkyne-modified uPAR-binding protein resulted in the formation of stable covalent bonds between them. The MNP-decorated uPAR-binding protein maintained its activity and function and was able to specifically target receptors in uPAR-transfected HEK 293 cells.[28]

The biology of a living cell offers a number of mechanisms that can be utilized for nanoparticles deposition to the cell's surface. Calcium is a ubiquitous messenger that plays an important role in regulation of cellular processes. Using ion channels yeast cells can continuously take up excess Ca^{2+} ions and store them in vacuoles, enabling them to grow in a medium with widely ranging Ca^{2+} concentration.[29] This phenomenon was employed for the deposition of citrate-capped 10 nm gold nanoparticles on living yeast cells.[30] First, gold nanoparticles were bound to Ca^{2+} ions through electrostatic interactions between carboxylic groups of the citrate molecule and divalent cations, which led to their self-assembly into aggregates having a branched necklace-like morphology. Next, yeast cells were incubated in the Ca^{2+}–AuNP solution resulting in a dense deposition of gold nanoparticles on the cell surface. It is probable that specific domains of receptor proteins with a high affinity to Ca^{2+}-ions serve as binding sites for Ca^{2+}–Au nanoparticles. Interestingly, it was found the deposition of Ca^{2+}–AuNP depended on the generation of the yeast cells. The younger cells showed a denser deposition of nanoparticles on their surfaces whereas older cells were less prone to Ca^{2+}–Au immobilization resulting in a poor coating by Ca^{2+}–AuNP. This effect can be explained by the higher metabolic activity of younger cells. Mg^{2+} ions were also utilized in this immobilization strategy and showed a similar uniform deposition of Mg^{2+}–Au on yeast cells. However, no age discrimination was observed. It is proposed that both younger and older cells possessed a similar Mg^{2+} ions uptake rate. Thus, using biologically important ions as mediators represents another attractive route for selective deposition of nanoparticles onto cells due to its biocompatibility, low toxicity, and specificity.

The development of nanotechnology enables scientist to create new smart materials with controllable properties particularly with high affinity to the cell surface. Janus nanoparticles, called nanocorals, were engineered using nanosphere lithography techniques for specific attachment to the cells and sensing *via* surface-enhanced Raman spectroscopy (SERS). This polystyrene-based nanocoral is divided into two distinct parts with different functionalities. One half of the nanocoral serves for antibody mediated cell targeting while the other half, coated with a gold layer, enhances the SERS signal. As a result, anti-HER-2 antibody-modified nanocorals were able to specifically target breast cancer cells that overexpress human epidermal growth factor receptor 2 (HER-2).[31]

New strategies to control the deposition and orientation of nanomaterials onto the cell are relevant for the development of two fields. The functionalization of the cell surface to create cell–nanomaterial hybrids and efficient drug delivery. Though these two fields have different goals, they are connected by the cell surface because the interaction and orientation of nanomaterials on the cell surface can drastically alter their internalization into the cytoplasm. Cellular backpacks[32] and polymeric microtubes[33] are new smart materials that are able to specifically attach to the cell membrane and further promote delivery of their active agents into mammalian cells. Designed by photolithographic techniques, layer-by-layer assembly, and spray deposition the cellular backpacks represent multilayered disk-shaped structures doped with magnetic nanoparticles to provide mechanical rigidity and loaded with fluorescein-labeled bovine albumin as a prototype of an active agent or a drug. The final layers of hyaluronic acid and chitosan mediate the backpack's deposition onto a macrophage's cell membrane due to specific interactions of hyaluronic acid with CD44 cell receptors. As a result, the obtained cellular backpacks showed strong attachment to macrophages surfaces without internalization and were able to release a model protein in a controllable manner *in vitro*, thus providing a promising new tool for efficient drug delivery.[32] Extending this approach, nanometer thickness films were folded into microtubes using a sacrificial track-etched membrane as a template. Three types of chemically nonuniform microtubes

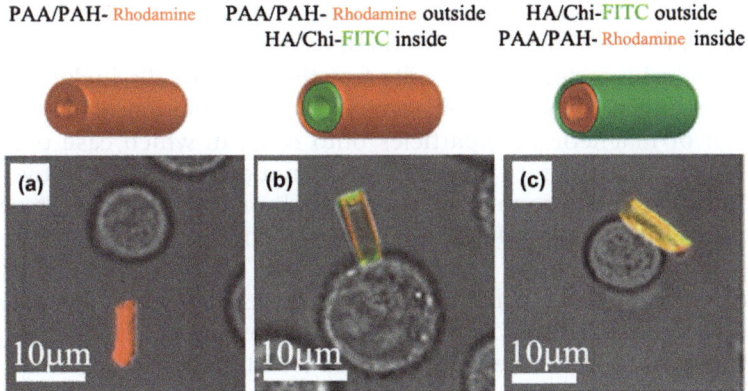

Figure 3.6 Confocal images of B-cells tube interactions: (a) (PAA/PAH-Rhoda-mine)$_{40}$; (b) (PAA/PAH-Rhodamine)$_{40}$ outside and (HA/FITC-Chi)$_{40}$ inside; (c) (HA/FITC-Chi)$_4$ outside and (PAA/PAH-Rhodamine)$_{40}$ inside. Chi – chitosan, HA - hyaluronic acid.
Reprinted with permission from ref. 33. (a–c) Copyright 2013, Wiley.

were fabricated with different adherent properties. Microtubes coated with a final bilayer of poly(acrylic acid)/poly(allylamine hydrochloride) and Rhodamine dye did not attach to cell surfaces (Figure 3.6a), whereas modification of polymer tubes with hyaluronic acid and chitosan (HA/FITC-Chi) resulted in their binding to the cell surface. Moreover the position of the adhesive domain in the tube makes it possible to control the orientation of the microtubes on the cell surface. Polymer tubes with HA/FITC-Chi layers in the interior attached preferentially ends-on (Figure 3.6b) while HA/FITC-Chi layers on the outside leaded to side-on (Figure 3.6c) deposition of microtubes.[33]

Combining molecular mechanisms given by Nature with nanotechnology allows scientists to study biological processes of living cells, to develop early diagnostic systems and fabricate new smart delivery materials.

3.5 From the Nanomaterials Deposition onto Cells to Emerging Technologies

The deposition of nanomaterials onto the cells in a biocompatible way changes the cell surface characteristics and provides cells with new non-native features and functionalities without affecting their viability that allows use of these cells in some biological applications as described earlier. Furthermore, the modified living cells can be integrated into other cell-based approaches leading to so-called emerging technologies, and results in a significant improvement and development of the cell-based applications. As an example of emerging technology, the utilization of magnetically modified cells as tools for tissue engineering and new cell-delivery approaches will be described below.

Magnetic nanostructures have been used in a wide range of diverse applications, including magnetically modified cell based areas.[34] The most common techniques for magnetically functionalizing cells include ligand-specific labeling of cells for their separation and isolation, or nonspecific adsorption of magnetic nanoparticles onto cells, in which case the penetration of nanoparticles into the cytoplasm is not excluded. Alternatively, the deposition of magnetic nanoparticles selectively on the cell surface results in a uniform coating of the cell and prevents penetration of nanoparticles inside the cells that could be potentially cytotoxic or interfere with intracellular processes. Following this strategy, a direct technique for magnetic functionalization of living cell surfaces was developed, using the yeast *S. cerevisiae* as a model.[4] The novelty of this technique was quite simple and elegant – commonly used iron oxide nanoparticles were coated by the polyelectrolyte (poly)allylamine hydrochloride (PAH), which stabilized the nanoparticles and provided them with a positive charge. These superparamagnetic iron oxide nanoparticles (SPIONs) had a very strong affinity to negatively charged surfaces and therefore their incubation with the cell suspension produced very fast magnetization of the cell surfaces. The technique was successfully applied to other cell-wall-protected cells, such as bacteria,[35] unicellular algae cells,[5] and it was shown to be equally effective and biocompatible.

Later, the cell surface functionalization by SPIONs was extended to isolated human cells.[6] Taking into account the fragile nature of human cellular membranes, the magnetization was performed in phosphate-buffered saline, and the osmotic concentration of the suspension of magnetic nanoparticles was adjusted using NaCl to avoid damaging the cell by osmotic pressure during the incubation. Using isolated HeLa cells, it was shown that the magnetic functionalization results in a uniform layer of nanoparticles around the individual cells that do not pierce the cellular membrane (Figure 3.7a). The viability studies of SPION-functionalized cells confirmed

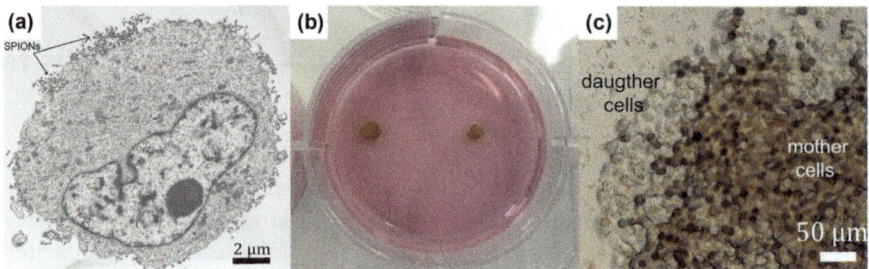

Figure 3.7   (a) TEM image of SPIONs-functionalized HeLa cells; (b) photograph of SPIONs-functionalized HeLa cells, micro-organised by magnet in a culture well; (c) optical microscopy image of SPIONs-functionalized HeLa cells in culture, showing the process of division.
Reprinted with permission from ref. 6 (b). Copyright 2011, American Chemical Society.

the high biocompatibility of the magnetization process. The modified HeLa cells remained alive and were able to divide and grow in culture and form colonies (Figure 3.7b).

The unique feature of SPION-modified cells is their ability to respond to an external magnetic field that makes it possible to manipulate the magnetically functionalized cells. In particular, the individual cells can be spatially organized into multicellular clusters, delivered to other target organisms or deposited for oriented growth in a controllable manner using conventional magnets. Moreover, the magnetic field can penetrate deep down into the tissues and this opens up new perspectives for human tissue engineering and cell-based therapies. Thus, magnetically labeled adenocarcinomic human alveolar epithelial cells (A549) and human skin fibroblasts (HSF) were assembled into a two-layered porous tissue mimicking lung tissue.[36] The surfaces of human cells were separately functionalized by the cationic PAH-stabilized MNP, as described above for HeLa cells. Then, both modified A549 and HSF cells were sequentially deposited onto the scaffold-free solid support using the 3-mm cylindrical magnets for precise positioning of the cells (Figure 3.8a). After incubation, the assembled system under conditions (5% CO_2, 37 °C) required for normal growth and proliferation of human cells, the magnet was removed and free-standing two-layered multicellular clusters were obtained (Figure 3.8b). The porous morphology of these clusters resembles the morphology of human lung tissue (Figure 3.8c), which suggests that the surface modification of A549

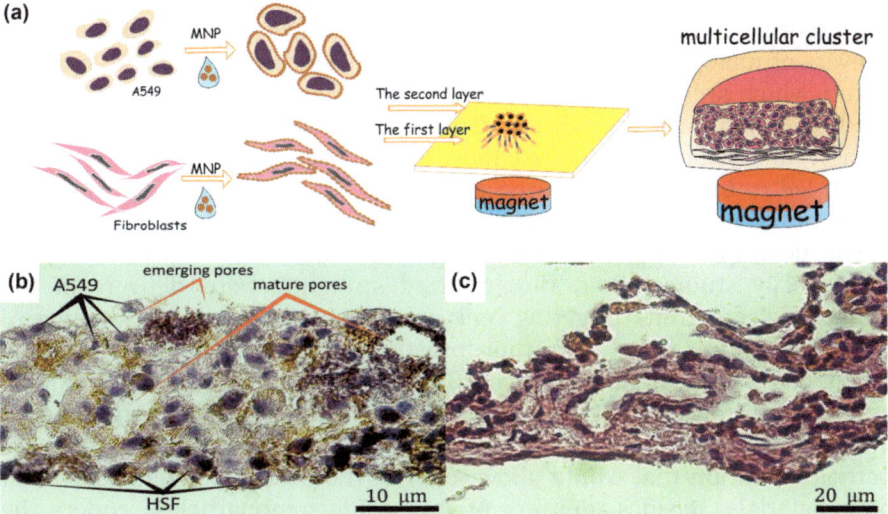

Figure 3.8 (a) Scheme of the lung-mimicking tissue fabrication using magnetically functionalized HSF (the first layer) and A549 (the second layer) cells; (b) optical microscopy images of the hematoxulin-eosin stained slice of obtained multicellular clusters; (c) human lung tissue slice. Adapted from ref. 36.

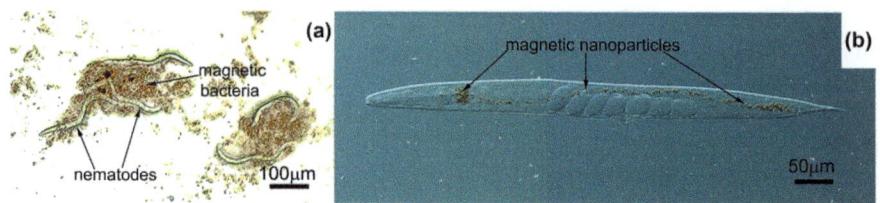

Figure 3.9   (a) *C. elegans* nematodes feed on MNP-coated *E.coli,* (b) optical microscopy images illustrating the distribution of MNP inside the digestive
tract of the worms.

and HSF cells does not inhibit their functionality to reconstitute the original
tissue. Besides, the presence of magnetic nanoparticles makes the multicellular clusters magnetically responsive and could be delivered within the
body and further manipulated by an external magnetic field.

Recently, a new emerging application combining the *in vivo* toxicity assessment of nanomaterials with magnetic functionalization of cells was reported.[37] The multicellular nematode *Caenorhabditis elegans* is widely used
as an *in vivo* model for toxicological studies,[38] particularly for nanomaterials
toxicity assessment.[39–41] However, the nanotoxicological studies are limited
by the methods of nanomaterial delivery into the nematode's body. The
typical procedure is based on the exposure of the microworms to the
nanoparticles in a growth or incubation media. This method has some
disadvantages, such as the aggregation of nanoparticles, spontaneous low-
efficiency uptake of nanoparticles by *C. elegans* and unspecific attachment of
the nanoparticles to the microworm's surface. A novel method of the direct
delivery of nanoparticles into *C. elegans* was suggested by Nature: bacteria
E. coli, which are usually used to feed on the microworms, were functionalized by cationic PAH-stabilized MNPs and added to the growth medium.
Hungry microworms swallowed these magnetized bacteria – nanobaits –
which led to the digestion of the microbial cells and consecutive accumulation of the magnetic nanoparticles in the worm's intestine (Figure 3.9).

The developed approach will bring two major advantages for *in vivo*
nanotoxicity studies: first, the localized straightforward delivery of the
nanoparticles permits targeting only one organ, *i.e.* the digestive tract of
worms in this case, and to further investigate the influence of the affected
organ on the whole organism. Second, the active process of feeding results in
efficient controllable uptake of nanoparticles in comparison to random
uptake of nanoparticles during the incubation of nematodes with nanoparticles solution that would allow study of the dose-dependent toxicity of
nanomaterials. Besides, delivery of iron oxide nanoparticles renders the
microworms magnetically responsive and therefore they could be further
magnetically manipulated for a separation from nonlabeled nematodes or
other applications. Finally, the cell-mediated delivery of nanoparticles
C. elegans is a versatile technique and could be applied to different types
on nanomaterials, such as nanofilms, silver, gold nanoparticles or other

nanostructures, suitable for surface functionalization of micro-organisms bait.

3.6 Conclusions

To conclude, the main principles and methods of the direct deposition of nanomaterials onto cells are described in this chapter. Similar to other nanobiotechnological cell-based approaches, the direct deposition of nanomaterials aims at combining the unique physicochemical properties of the nanomaterials with cell features and functions. However, "the direct deposition" is directed to the functionalization of a cell surface with nano-materials to create new hybrid materials, enhance native cell functionalities or provide new features not found in Nature.

The interaction of nanomaterials with the cell surface is mostly mediated by the cell surface biochemical groups, therefore electrostatic and hydro-phobic forces are essential for the assembly of the nanoparticles on the cell. This makes the process of the direct deposition versatile because using postsynthetic methods of nanoparticles modification it will be possible to bind any type of nanoparticle to any type of cells. So far, major cell types including prokaryotes and eukaryotes have been decorated with different nanomaterials, such as metals, metal oxide nanoparticles, carbon nanos-tructures, and have been used in wide range of diverse applications, ranging from technology and electronics to biomedicine. Currently, the projects in this area are mostly focused on the development of new smart materials for cell surface functionalization, fabrication of 3D hybrid cell–nanoparticle systems and utilization of nanomodified cells in emerging technologies. We believe that the development of these directions will lead to the progress of cell-based technologies, new sources of energy and new discoveries in fun-damental science.

Acknowledgements

This work was funded by the subsidy of the Russian Government to support the Program of Competitive Growth of Kazan Federal University among World's Leading Academic Centers. The authors acknowledge the funding by RFBR 12-03-93939-G8, RFBR 12-04-33290_mol_ved_a and RFBR 14-04-01474_a grants.

References

1. V. Berry, S. Rangaswamy and R. Saraf, *Nano Lett.*, 2004, **4**, 939–942.
2. V. Berry and R. F. Saraf, *Angew. Chem.*, 2005, **117**, 6826–6831.
3. V. Berry, A. Gole, S. Kundu, C. J. Murphy and R. F. Saraf, *J. Am. Chem. Soc.*, 2005, **127**, 17600–17601.
4. R. F. Fakhrullin, J. García-Alonso and V. N. Paunov, *Soft Matter*, 2010, **6**, 391–397.

5. R. F. Fakhrullin, L. V. Shlykova, A. I. Zamaleeva, D. K. Nurgaliev, Y. N. Osin, J. García-Alonso and V. N. Paunov, *Macromolec. Biosci.*, 2010, **10**, 1257–1264.
6. M. R. Dzamukova, A. I. Zamaleeva, D. G. Ishmuchametova, Y. N. Osin, A. P. Kiyasov, D. K. Nurgaliev, O. N. Ilinskaya and R. F. Fakhrullin, *Langmuir*, 2011, **27**, 14386–14393.
7. N. L. Rosi, C. S. Thaxton and C. A. Mirkin, *Angew. Chem.*, 2004, **116**, 5616–5619.
8. E. K. Payne, N. L. Rosi, C. Xue and C. A. Mirkin, *Angew. Chem. International Edition*, 2005, **44**, 5064–5067.
9. I. Safarik, L. F. T. Rego, M. Borovska, E. Mosiniewicz-Szablewska, F. Weyda and M. Safarikova, *Enzyme Microb. Technol.*, 2007, **40**, 1551–1556.
10. H. Yavuz, A. Denizli, H. Güngüneş, M. Safarikova and I. Safarik, *Sep. Purific. Technol.*, 2006, **52**, 253–260.
11. M. Safarikova, Z. Maderova and I. Safarik, *Food Res. Int.*, 2009, **42**, 521–524.
12. R. F. Fakhrullin, A. I. Zamaleeva, M. V. Morozov, D. I. Tazetdinova, F. K. Alimova, A. K. Hilmutdinov, R. I. Zhdanov, M. Kahraman and M. Culha, *Langmuir*, 2009, **25**, 4628–4634.
13. X. Chen, U. C. Tam, J. L. Czlapinski, G. S. Lee, D. Rabuka, A. Zettl and C. R. Bertozzi, *J. Am. Chem. Soc.*, 2006, **128**, 6292–6293.
14. R. Kempaiah, A. Chung and V. Maheshwari, *ACS Nano*, 2011, **5**, 6025–6031.
15. R. Kempaiah, S. Salgado, W. L. Chung and V. Maheshwari, *Chem. Commun.*, 2011, **47**, 11480–11482.
16. J. Tam, S. Salgado, M. Miltenburg and V. Maheshwari, *Chem. Commun.*, 2013, **49**, 8641–8643.
17. X. Zhang, X. Zhang, W. He, C. Sun, J. Ma, J. Yuan and X. Du, *Colloids Surfaces B: Biointerfaces*, 2012.
18. R. J. Tseng, C. Tsai, L. Ma, J. Ouyang, C. S. Ozkan and Y. Yang, *Nature Nanotechnol.*, 2006, **1**, 72–77.
19. K. N. Avery, J. E. Schaak and R. E. Schaak, *Chem. Mater.*, 2009, **21**, 2176–2178.
20. W.-S. Kuo, C.-M. Wu, Z.-S. Yang, S.-Y. Chen, C.-Y. Chen, C.-C. Huang, W.-M. Li, C.-K. Sun and C.-S. Yeh, *Chem. Commun.*, 2008, 4430–4432.
21. Z. Li, S. W. Chung, J. M. Nam, D. S. Ginger and C. A. Mirkin, *Angew. Chem.*, 2003, **115**, 2408–2411.
22. A. Sugunan, P. Melin, J. Schnürer, J. G. Hilborn and J. Dutta, *Advanced Materials*, 2007, **19**, 77–81.
23. N. C. Bigall, M. Reitzig, W. Naumann, P. Simon, K. H. van Pée and A. Eychmüller, *Angew. Chem. Int. Ed.*, 2008, **47**, 7876–7879.
24. P. Mukherjee, A. Ahmad, D. Mandal, S. Senapati, S. R. Sainkar, M. I. Khan, R. Parishcha, P. Ajaykumar, M. Alam and R. Kumar, *Nano Lett.*, 2001, **1**, 515–519.
25. S. Tatur, M. Maccarini, R. D. Barker, A. Nelson and G. Fragneto, *Langmuir*, 2013, **29**, 6606–6614.

26. E. C. Cho, J. Xie, P. A. Wurm and Y. Xia, *Nano Lett.*, 2009, **9**, 1080–1084.
27. A. K. Gupta and A. S. Curtis, *Biomaterials*, 2004, **25**, 3029–3040.
28. L. Hansen, E. K. U. Larsen, E. H. Nielsen, F. Iversen, Z. Liu, K. Thomsen, M. Pedersen, T. Skrydstrup, N. C. Nielsen and M. Ploug, *Nanoscale*, 2013.
29. T. Dunn, K. Gable and T. Beeler, *J. Biol. Chem.*, 1994, **269**, 7273–7278.
30. V. Maheshwari, D. E. Fomenko, G. Singh and R. F. Saraf, *Langmuir*, 2010, **26**, 371–377.
31. L. Y. Wu, B. M. Ross, S. Hong and L. P. Lee, *Small*, 2010, **6**, 503–507.
32. N. Doshi, A. J. Swiston, J. B. Gilbert, M. L. Alcaraz, R. E. Cohen, M. F. Rubner and S. Mitragotri, *Adv. Mater.*, 2011, **23**, H105–H109.
33. J. B. Gilbert, J. S. O'Brien, H. S. Suresh, R. E. Cohen and M. F. Rubner, *Adv. Mater.*, 2013, **25**, 5948–5952.
34. L. H. Reddy, J. L. Arias, J. Nicolas and P. Couvreur, *Chem. Rev.*, 2012, **112**, 5818–5878.
35. D. Zhang, R. F. Fakhrullin, M. Özmen, H. Wang, J. Wang, V. N. Paunov, G. Li and W. E. Huang, *Microbial Biotechnol.*, 2011, **4**, 89–97.
36. M. R. Dzamukova, E. A. Naumenko, N. I. Lannik and R. F. Fakhrullin, *Biomater. Sci.*, 2013, **1**, 810–813.
37. G. I. Däwlätşina, R. T. Minullina and R. F. Fakhrullin, *Nanoscale*, 2013, **5**, 11761–11769.
38. M. C. Leung, P. L. Williams, A. Benedetto, C. Au, K. J. Helmcke, M. Aschner and J. N. Meyer, *Toxicol. Scie.*, 2008, **106**, 5–28.
39. W. Zhang, B. Sun, L. Zhang, B. Zhao, G. Nie and Y. Zhao, *Nanoscale*, 2011, **3**, 2636–2641.
40. Y. Zhao, Q. Wu, Y. Li and D. Wang, *RSC Adv.*, 2013, **3**, 5741–5757.
41. P. R. Hunt, B. J. Marquis, K. M. Tyner, S. Conklin, N. Olejnik, B. C. Nelson and R. L. Sprando, *J. Appl. Toxicol.*, 2013.

CHAPTER 4

Bioinspired Encapsulation of Living Cells within Inorganic Nanoshells

JI HUN PARK,[a] JUNO LEE,[a] BEOM JIN KIM[a] AND
SUNG HO YANG*[b]

[a] Center for Cell-Encapsulation Research and Molecular-Level Interface
Research Center, Department of Chemistry, KAIST, Daejeon 305-701,
Korea; [b] Department of Chemistry Education, Korea National University of
Education, Chungbuk 363-791, Korea
*Email: sunghoyang@knue.ac.kr

4.1 Introduction

For a decade, individual living cells have been encased within polyelectrolytes and nanomaterials.[1,2] Layer-by-layer (LbL) self-assembly is the deposition or assembly of polyelectrolytes/nanomaterials, applied sequentially to the surfaces of living cells, through electrostatic interactions between oppositely charged polyelectrolytes or nanomaterials. LbL films are useful in integrating living cells with various materials for the following reasons.[3] (1) LbL films are formed in a substrate-nonspecific manner, but are only manipulated by the surface charge. (2) Their thickness can be finely controlled on a 1-nm scale by the number of depositions. (3) LbL films are fabricated with extremely simple and low-cost processes, which make them highly accessible. (4) All the processes involved can be performed in biorelevant

RSC Smart Materials No. 9
Cell Surface Engineering: Fabrication of Functional Nanoshells
Edited by Rawil F Fakhrullin, Insung S Choi and Yuri Lvov
© The Royal Society of Chemistry 2014
Published by the Royal Society of Chemistry, www.rsc.org

conditions, such as aqueous media, ambient pressure, and room temperature.

To apply LbL assembly to the encapsulation of living cells, researchers have predominantly focused on the biocompatibility of coating materials and the conditions required for the LbL process, including deposition time, number of LbL steps, and salt concentration. Poly(allylamine hydrochloride) (PAH), poly(dimethyl diallylammonium chloride) (PDDAC), poly(ethyleneimine) (PEI), and chitosan are used as positively charged polyelectrolytes, whereas poly(styrenesulfonate sodium salt) (PSS), poly(acrylate sodium salt) (PAA), and sodium alginate are used as the counterpart for sequential deposition (Figure 4.1).[1–3] In early studies, micro-organisms, including *Saccharomyces cerevisiae*, *Escherichia coli*, and *Trichoderma asperellium*, were chosen to demonstrate the polyelectrolyte coating of cells, and their viability was maintained even after the shells were formed.[1,2] Polyelectrolyte-coated yeast cells can also be deposited spatioselectively onto oppositely charged surfaces to produce cellular micropatterns.[4] Even more interestingly, poly-electrolyte-coated red blood cells displayed suppressed immunogenicity, or "immunocamouflage", which is useful in blood transfusions and cell implantation.[5] The cell type encapsulated has also been extended to include mammalian cells, which are relatively susceptible to their chemical environments. Fibroblast cells were encapsulated with fibronectin and gelatin to generate multilayered tissue accumulations through biospecific interactions in the extracellular matrix.[6]

By combining polyelectrolytes and nanomaterials, the physicochemical properties of living cells have been modulated in diverse ways. Living cells have been decorated with nanomaterials, including gold nanoparticles,[7] magnetic nanoparticles,[8] silica nanoparticles,[9] carbon nanotubes,[10]

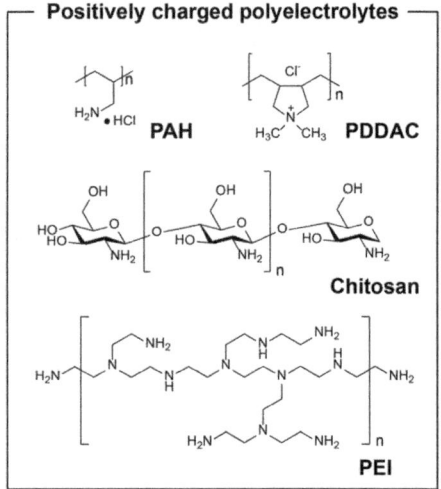
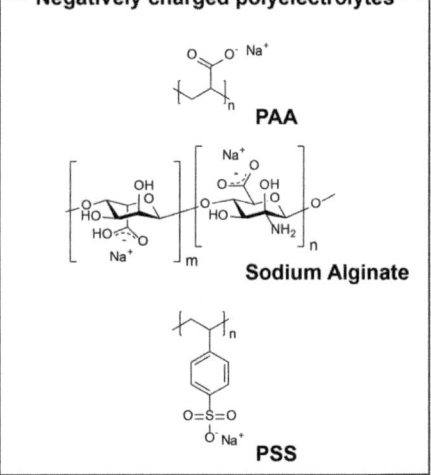

Figure 4.1   Representative polyelectrolytes for LbL assembly.

halloysite clay,[11] and graphene,[12] by replacing certain steps in the deposition of the polyelectrolytes with the deposition of nanomaterials that have inherent functions. Magnetic-nanoparticle-decorated cells, in particular, could be used in biosensors, bioreactors, and microfluidics, which require spatioselective manipulation.[8] Multiwalled carbon nanotubes have also been incorporated into the polyelectrolyte shells on cells to study their electrochemical properties, anticipating their use in the development of microelectronic devices and whole-cell-based biosensors.[10] Interestingly, silica-particle-decorated cells show resistance to thermal stress.[9]

Although polyelectrolyte-based shells have been formed successfully and have shown potential utility in nanobioapplications, this artificial shells could not compete with the protecting layers found in Nature (*e.g.*, cyst wall or spore coat) especially with regards to mechanical strength. Inorganic nanoshells could be the most plausible alternative to polyelectrolyte shells because these materials are both robust and hard.[13,14] However, cell encapsulation within inorganic shells faces serious obstacles compared with their encapsulation within polyelectrolyte shells. Polyelectrolyte coatings are biocompatible predominantly because no chemical reaction is required to establish strong bonding because the LbL process proceeds by the simple deposition of presynthesized materials. In contrast, the formation of inorganic nanoshells generally requires *in situ* chemical reactions that involve toxic inorganic precursors under conditions that are harmful to living cells, such as high temperatures and high pressures.[15]

To overcome these obstacles, researchers have turned to Nature. They have sought solutions from biological systems that generate inorganic materials, including bones, dentins, and shells, as well as intricate and highly organized organic materials, such as DNA, proteins, and peptides, under mild conditions.[16–18] In recent years, there has been a considerable increase in studies of biomimetic or bioinspired mineralization, which generates inorganic materials.[19–21] Based on our knowledge of biomineralization, researchers have attempted to encapsulate living cells within inorganic nanoshells using a bioinspired route under cytocompatible conditions. The strategies used to produce inorganic nanoshells are categorized as direct metal reduction, *in situ* crystallization, and *in situ* polycondensation (Figure 4.2). More specifically, gold shells were formed by the direct reduction of the metal when bacteria were incubated with gold precursors. Calcium-based mineral shells were formed by *in situ* crystallization on cell surfaces through ionic interactions. Shells of silica, titania, and their hybrids were formed by *in situ* polycondensation with the involvement of catalytic templates. The mechanical robustness of inorganic shells has also provided researchers with unexpected tools for controlling cell division, protecting the cells from foreign aggression, and enhancing their long-term viability. In this chapter, we review recent progress in the formation of inorganic nanoshells for cell encapsulation according to the materials used, and will discuss the trends in and perspectives on the encapsulation of cells with inorganic materials.

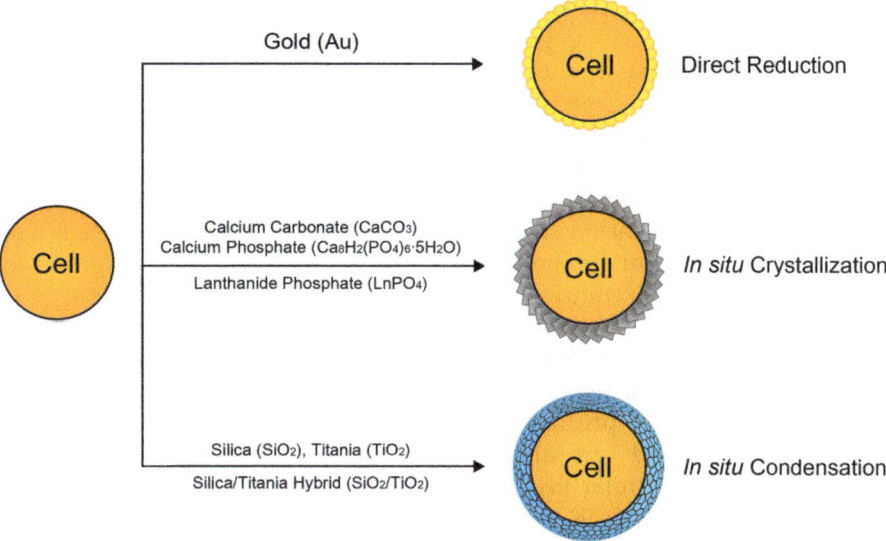

Figure 4.2 Schematic representation of the formation of nanometric inorganic shells for the encapsulation of individual cells: (1) direct reduction, (2) *in situ* crystallization, (3) *in situ* condensation.

4.2 Direct Reduction of Inorganic Precursors on Cell Surfaces

Micro-organisms participate actively in the geochemical cycling of metals because the redox reactions of metals are essential to the microbial catabolic reactions that generate the energy for their survival.[22,23] These phenomena, in particular, have inspired geochemists to utilize micro-organisms for the remediation of toxic-metal contaminants in the soil.[24,25] Researchers have also investigated the cellular responses to various metal ions that influence microbial activities.[26,27] Metal-reducing micro-organisms are of interest to material science because they can generate metals from low concentrations of metal ions, while maintaining their viability. For example, iron and uranium ions are reduced by micro-organisms, generating precipitates and accumulations that are believed to be the sources of mineral ores.[28,29] These microbial activities have recently been extended experimentally to other minerals, including manganese, silver, and gold.[30,31]

4.2.1 *In Vivo* Gold Mineralization

Gold mineralization by micro-organisms is an intriguing issue in the field of biomineralization because gold is a rare, inert, and inessential element in cell metabolism. Free gold ions are known to be unstable in aqueous solution and highly toxic to micro-organisms.[32,33] However, it has been demonstrated that specific sorts of micro-organisms can reduce solubilized gold ions to inert metallic gold *in vivo* in gold-ion-supplemented medium.[34] This ability of micro-organisms is considered to be a highly evolved strategy for protecting their cytoplasm against heavy metal ions. With the aim of condensing gold or purifying water or soils, researchers have incubated micro-organisms with sources of gold ions, such as gold(III) chloride,[35] gold(I) thiosulfate,[36] or gold L-asparagine.[37] Sequential studies of *in vivo* gold mineralization have shown that bacteria are the species that most actively accumulate gold from media. For example, *Cupriavidus metallidurans*, a representative model organism for studying the effects of metal ions on cellular metabolism, is one of a few bacteria that can produce nanosized gold nuggets from small amounts of solubilized gold ions in mines.[38] *Acidithiobacillus thiooxidans*, a thiosulfate-oxidizing bacterium, can also generate gold precipitates from gold(I) thiosulfate on the cell surface when the gold precursor is the only available energy source (Figure 4.3).[39] In a similar fashion, gold reduction also occurs in wild-type *Bacillus subtilis* and *E. coli*, which do not accumulate gold in Nature,[40,41] and in Fe(III)-reducing bacteria and archaea (Figure 4.3).[42] Although the mechanism of gold mineralization is not fully understood, it is noteworthy that gold can be integrated within the cells in a biologically controllable manner.

4.2.2   Gold Shells for Cell Encapsulation

Taking advantage of gold (Au) mineralization *in vivo*, Yeh and coworkers reported the formation of Au-encapsulated *E. coli* (*E. coli*@Au) composites, and investigated their photothermal activities in destroying cancer cells.[43] They chose Gram-negative *E. coli* for the formation of *E. coli*@Au composites, and gold shells were grown on the surfaces of the cells by simply incubating them in a solution of 5 mM HAuCl$_4$ or Au(OH)$_x$Cl$_y^-$ ions (Figures 4.4a and b). Images of microtome-sectioned *E. coli*@Au composites showed that the thickness of the gold shells was about 7–8 nm when both methods were used. The authors suggested that the reduction of gold ions might involve the aldehyde groups in the peptidoglycan of the cell wall of *E. coli*. In other words, the aldehyde groups might act as electron donors to reduce the surface-tethered Au^{3+} ions. Interestingly, *E. coli*@Au exhibited low cytotoxicity against mammalian cells, although *E. coli* is usually a pathogenic bacterium. Cytotoxicity assays were performed by coculturing nonmalignant human keratinocyte cells or malignant human lung carcinoma cells with native *E. coli.* or *E. coli*@Au (Figure 4.4c). The viability of the mammalian cells decreased significantly to 40% after one day in the presence of native *E. coli*, whereas their viability was maintained at > 80% when they were exposed to the *E. coli*@Au composites. These results are presumably attributable to the gold shells, which could inhibit the production of exotoxins and endotoxins or block their physical contact with the mammalian cells. More importantly, malignant cells cultured with the *E. coli*@Au composites

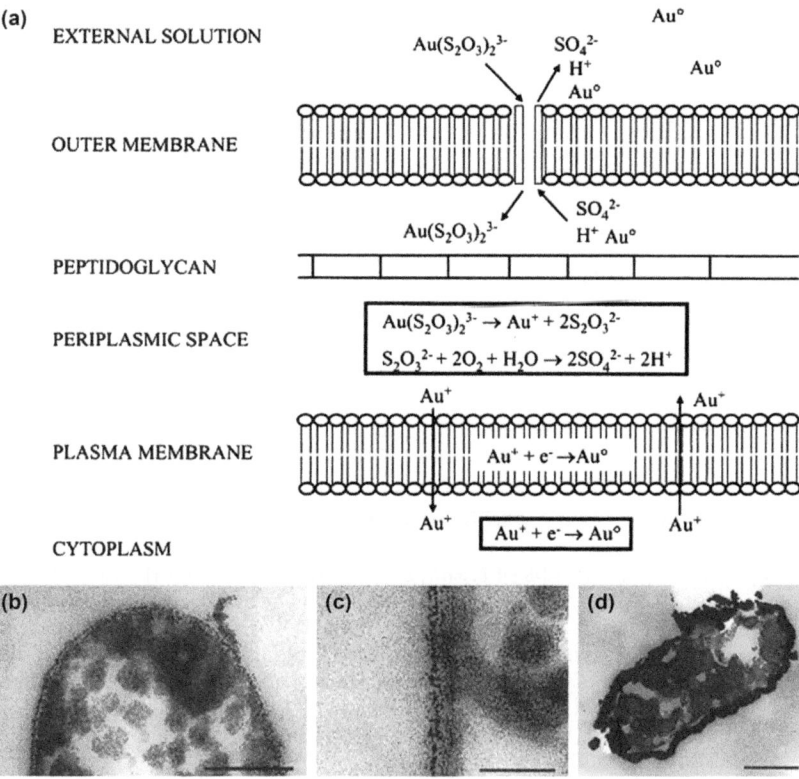

Figure 4.3 (a) Schematic models of cellular utilization of $Au(S_2O_3)_2^{3-}$ in *A. thiooxidans* when $Au(S_2O_3)_2^{3-}$ is the only available source in solutions. The internalized $Au(S_2O_3)_2^{3-}$ was decomplexed into Au^+ and thiosulfate $(S_2O_3^{2-})$, and the produced Au^+ was reduced directly on the plasma membrane or ejected out from the cell. (b, c) Transmission electron microscopy (TEM) micrographs of microtome-sectioned bacteria that were incubated with 0.24 mM of gold in dark reaction system for 75 days. Colloidal gold was deposited on outer wall layer, periplasm, and cytoplasmic membrane. (d) After 7 months, colloidal gold was grown and thickened from the continuous reduction. Scale bars in (b), (c), and (d) are 200 nm, 30 nm, and 300 nm, respectively. Reproduced with permission from Elsevier (Copyright 2005).[39]

were destroyed by irradiation with a near-infrared (NIR) laser (808 nm) because of the photothermal effect of the gold. A significant loss of cancer-cell viability was observed with NIR laser irradiation at 260 mW, and continued with irradiation up to 300 mW. However, irradiation with the NIR laser at 300 mW resulted in nonspecific cell injury, which was not caused by a gold-specific photothermal treatment. The researchers concluded that the range of 260–280 mW is optimal for killing cancer cells. These data imply that direct gold reduction on living cells can be used as an antimicrobial treatment or cell-based cancer therapy.

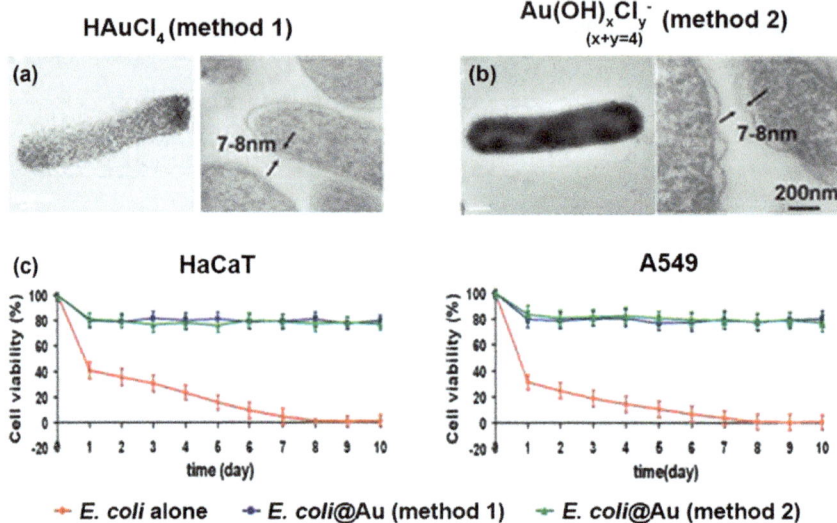

Figure 4.4 TEM micrographs of *E. coli*, DH5α, incubated with (a) 5 mM of HAuCl₄
for 32 h (method 1) and (b) gold growing solution containing
Au(OH)ₓCl$_y^-$ made by mixing HAuCl₄ with K₂CO₃(method 2). Both
methods gave similar thickness of gold nanoshell, measured to be
7–8 nm. (c) Cytotoxicity assay of HaCaT and A549 cells in the presence
of *E. coli* and *E.coli*@Au at a dosage of 100 mg/mL for 10 days.
Reproduced with permission from the Royal Society of Chemistry
(Copyright 2008).[43]

Kaehr and Brinker suggested another approach to the formation of gold
shells on *E. coli* using the enhancement of gold nanoparticles (AuNPs) on the
cell surface (Figure 4.5a).[44] Positively charged AuNPs were seeded onto the
surface of *E. coli* and used as nucleation sites for enhancing the gold when
the AuNP-seeded *E. coli* cells were incubated with metal ions, such as Au^{3+}
and Pt^{2+}. In terms of the growth rates and chemotactic properties of the
cells, AuNPs-seeded *E. coli* exhibited the same biological activities as native
E. coli (Figure 4.5b). Even after the gold was deposited to produce *E. coli*@Au,
the bacteria retained their ability to grow and divide, confirming that this
gold mineralization was conducted in a cytocompatible way. The characteristics of bacterial division, involving the segregation and septation of
murein, meant that the division of *E. coli*@Au caused the gold shell to be
asymmetrically distributed on the progeny cells, and the broken gold
shells were placed at the ends of the *E. coli* progeny cells (Figures 4.5a and c).
It is noteworthy that the asymmetric distribution of the gold shells was
achieved by the combination of the cellular metabolism and anthropogenic
gold mineralization. This interesting strategy could be utilized in micro/
nanoscale devices, including in energy storage, nanopropulsion, and
nanomachines.

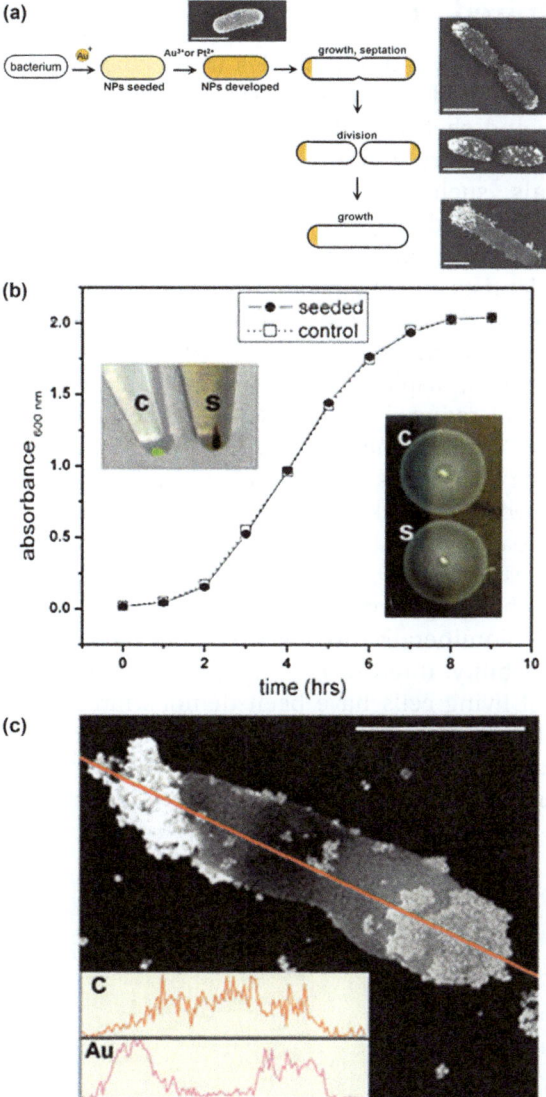

Figure 4.5 (a) Schematic representation of biocompatible AuNPs seeding and metallization on bacterial cell envelope. The gold enhanced bacterial cells were grown and divided, finally resulting in metallic asymmetry. (b) AuNPs seeded *E. coli* (denoted as "S") displayed no difference in growth rates and chemotaxis properties (right inset) compared to controls (denoted as "C"). (c) Backscattered scanning electron (BSE) microscopy image of *E. coli*, poles of which are coated with gold. The presence of gold was confirmed by the elemental profiling of energy-dispersive X-ray spectroscopy across the red line.
Reproduced with permission from the Royal Society of Chemistry (Copyright 2010).[44]

4.3   Bioinspired Shells Formed by Ionic Interactions

Crystalline minerals play important roles in the support and protection of biological systems,[45] and the minerals strengthen and stiffen tissues, such as bones, teeth, and shells.[46,47] Living organisms have developed the capacities to control the crystallization processes and the morphologies of inorganic crystals, such as calcium phosphate and calcium carbonate (Figure 4.6). Understandably, the mineralization processes in biological systems take place under physiologically mild conditions, and are therefore exemplary for scientists and engineers. However, anthropogenic approaches have never achieved the degree of structural and functional complexity of biologically formed inorganic crystals.[48] Biogenic crystals, with highly complex architectures, display superior mechanical properties, such as enhanced rigidity and reduced brittleness, to those of their synthetic counterparts.[17,18,49,50] To utilize the superior properties of biogenic crystals and the biocompatible conditions of their crystallization, biomimetic synthesis has been developed by mimicking the mechanism of biomineralization. The bioinspired approach, which utilizes the mechanistic principles of biomineralization processes for *in vitro* crystallization, is better than conventional chemical approaches in generating cytocompatible inorganic shells for living cells because biomineralization occurs under biocompatible conditions, sustaining cell viability. Interestingly, the properties of inorganic shells required to control living cells have been demonstrated by the synthesis of biomineral shells composed of calcium phosphate or calcium carbonate. This approach confers various positive effects on living cells: long-term cell storage with sustained cell viability, the protection of cells from foreign

Figure 4.6   (a) Scanning electron microscopy (SEM) micrograph of the fracture surfaces of human lamellar bone. (b) BSE micrograph of fracture section of the interface between aragonite nacre (top) and aragonite prisms (bottom) at the innermost shell of *Modiolus modiolus*.
Reproduced with permission from Elsevier (Copyright 1999),[46] and the American Chemical Society (Copyright 2008).[45]

aggression, and the control of cell division. In this section, we discuss several strategies for encapsulating living cells within inorganic shells formed by ionic interactions.

4.3.1 Calcium Phosphate Shells

Calcium and phosphorus are elements widely distributed throughout the earth, and the surface layer of the earth contains approximately 3.4 wt% calcium and 0.1 wt% phosphorus.[51] The nature of calcium phosphate is determined by the contents of calcium, oxygen, and phosphorus, with or without the incorporation of water. Although calcium phosphates are less-abundant minerals in unicellular organisms than calcium carbonate or silica, they are important inorganic compounds in biological hard tissues. For instance, a hierarchically ordered form of carbonated hydroxyapatite affects the stability, hardness, and functions of the teeth and bones of vertebrates.[52] Calcium phosphate is suitable for biological applications because of the biocompatibility and rigidity of the material.

Tang and coworkers reported a bioinspired approach to the encapsulation of individual living yeast cells with a calcium phosphate (CaP) shell.[53] Because the cell wall of yeast is mainly composed of polysaccharides of sugars, glucose, mannose, and *N*-acetylglucosamine,[54] the authors suggested that it would be hard to directly crystallize calcium phosphate on the cell-wall surface because of the relatively low electronic charge density on the surface of yeast (Figures 4.7a and b).[55] They enhanced the charge density on the cell surface by coating the cells alternately with PDDAC and PAA in an LbL self-assembly process. Their strategy was supported by a previous study, in which Mann and coworkers demonstrated that this polyelectrolyte multilayer induced the nucleation of calcium phosphate nanocrystals.[56] The highly dense carboxylate groups in PAA act as active binding sites for calcium ions (Ca^{2+}). The crystallization of calcium phosphate on the surfaces of LbL-coated yeast cells was achieved by the careful mixing of calcium chloride ($CaCl_2$) solution and disodium hydrogen phosphate (Na_2HPO_4) solution, generating yeast@CaP. The calcium phosphate shell formed was composed of many flake-like crystallites (Figures 4.7c and d) and the encapsulated cells remained viable, according to a live/dead fluorescent probe assay (Figures 4.7e–h). Importantly, it was shown that the calcium phosphate shell protected the cells from foreign aggressors, such as zymolyase, which digests the cell wall, mainly because of the robustness of the artificial shell. This result implies that the lifetimes of cells can be extended by robust encapsulation, and that this encapsulation can be used for the long-term storage of cells.

Although this chapter addresses the encapsulation of living cells, it is appropriate to introduce a strategy for generating inorganic microcapsules by scarifying biological templates because this strategy corresponds in many ways to the strategy for cell encapsulation. Although there are many methods for generating calcium phosphate microcapsules,[57–61] it is difficult to obtain uniform and spherical calcium phosphate microcapsules with porous

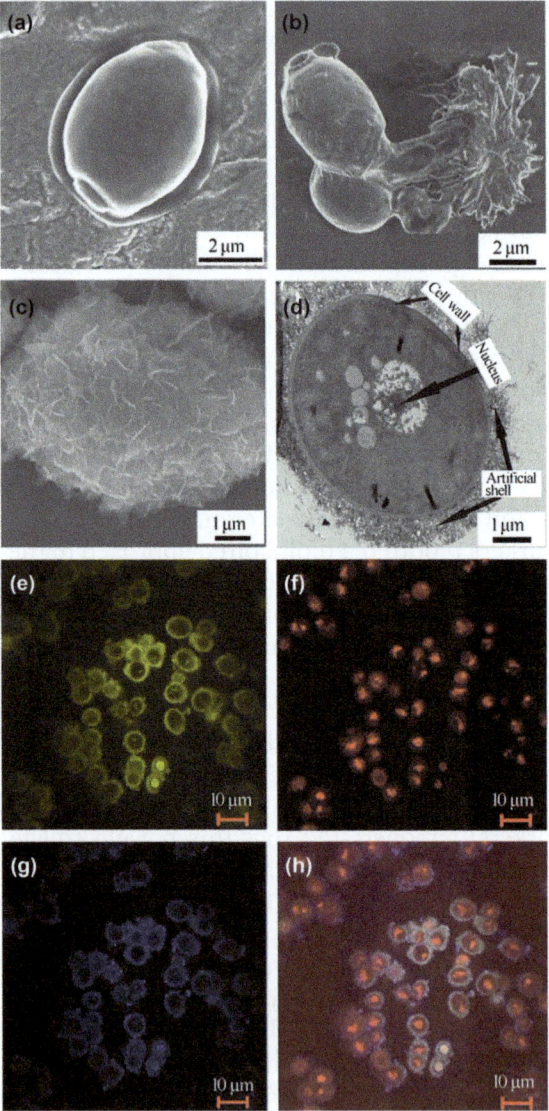

Figure 4.7   SEM micrographs of yeast cells: (a) native cell, (b) cells that were calcified without LbL process, (c) cell with CaP shell after LbL process, (d) encapsulated cell that was sliced. Confocal microscopy images of live/dead fluorescent probe for testing for yeast viability. (e) Mineral shells were stained by tetracycline hydrochloride. (f) Red spots imply that the cells are alive. (g) Cell walls were stained by fluorescent blue. (h) Combination of the above fluorescence results.
Reproduced with permission from Wiley (Copyright 2008).[53]

microstructures. Huang and Wang developed methods for synthesizing calcium phosphate microcapsules using yeast cells as templates.[62] To generate the calcium phosphate microcapsules, the yeast cells were removed by the calcination of yeast@CaP formed by a similar method to that of Tang and coworkers. Both the morphology and structure of the microcapsules were controlled by tuning the sintering temperature.

4.3.2　Calcium Carbonate Shells

Calcium carbonate, which forms the bones and shells of many organisms, is one of the most abundant biominerals in Nature. Anhydrous calcium carbonate crystals have three polymorphs: rhombohedral calcite, needle-shaped aragonite, and spherical polycrystalline vaterite (Figure 4.8).[49,63] Calcite is thermodynamically stable and is the most common polymorph on the surface of the earth. Aragonite and vaterite are metastable and are thermodynamically transformed into the more stable calcite under the appropriate conditions. However, it is noteworthy that these two polymorphs are frequently found in biological systems. There is also a hydrated state, amorphous calcium carbonate (ACC).[64] The crystallization of calcium carbonate is achieved by the intimate cooperation of an insoluble matrix and soluble macromolecules at the organic/inorganic interface.[65–69] Although it is thought that ACC acts as a precursor for the crystallization of $CaCO_3$,[70–81] the detailed process of how ACC nucleates, grows, and transforms into crystals is not yet clearly understood. However, previous reports have shown that the crystallization of calcium carbonate is induced by a negatively charged matrix and that the morphology of the crystal is influenced by negatively charged additives in the Ca^{2+} solution, such as PAA.[65,82,83]

Interestingly, the $CaCO_3$ shells of unicellular micro-organism, such as coccolithophores and foraminifera, are necessary for their survival under adverse environmental conditions.[45,84,85] By mimicking the process of $CaCO_3$ mineralization, Fakhrullin and Minullina encapsulated living yeast cells within $CaCO_3$ shells by the direct deposition of $CaCO_3$ on the cell walls (Figure 4.9a).[86] When yeast cells were placed into supersaturated solutions of calcium ions and carbonate ions, micrometric vaterite shells formed on the surfaces of the cells within several minutes. The successful deposition of $CaCO_3$ on the native cell wall, without LbL coating, can be explained by the inherent negative charges on the yeast cell wall. A negatively charged matrix is known to enhance the nucleation and growth of $CaCO_3$ crystal through

Figure 4.8　SEM micrographs of $CaCO_3$ polymorphs. (a) calcite, (b) aragonite, and (c) vaterite.
Reproduced with permission from Elsevier (Copyright 2013).[63]

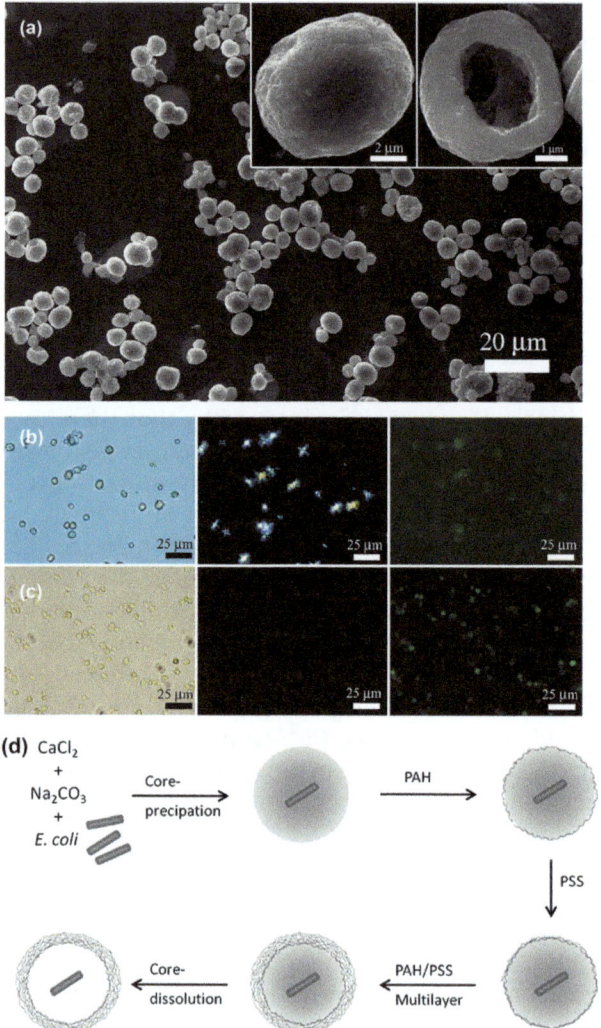

Figure 4.9   (a) SEM micrographs of yeast cells with CaCO$_3$ shells. The left insert is magnified SEM micrograph of single cell with the CaCO$_3$ shell. The right insert shows a hollow microshell that was produced after the removal of the cell. (b, c) Optical micrographs (from left to right in each row: bright-field, polarized, and fluorescent): (b) cells with CaCO$_3$ shells. (c) cells released from CaCO$_3$ shells by adding 0.1 M EDTA. The released cells show no birefringence at the polarized optical micrograph. This means that the released cells are optically isotropic like native cells. At the fluorescent optical micrographs, green fluorescence means that cells are still alive, which are stained by FDA. (d) Schematic illustration of the preparation of hollow polymeric microcapsules that include living *E. coli*.
Reproduced with permission from the American Chemical Society (Copyright 2009, 2013).[86,90]

electrostatic interactions because the surface of ACC presents a positive charge.[87–89] In the case of yeast cells, the negative charges on the cell walls are sufficient to nucleate ACC and initiate crystal growth, leading to the generation of vaterite shells. However, the metastable shells undergo recrystallization during their long-term incubation in the reaction solution, resulting in the aggregation of the yeast cells in calcite microcrystals. To prevent this aggregation and produce individually encapsulated yeast cells (yeast@$CaCO_3$), it is necessary to remove the cells from the reaction solution after several minutes. The yeast@$CaCO_3$ survived for several months and could be released from their shells by dissolving the $CaCO_3$ shell with ethylenediaminctetraacetic acid disodium salt (EDTA), while maintaining the viability of the yeast cells (Figures 4.9b and c).

Sieber and coworkers used the dissolution of the $CaCO_3$ shell with EDTA to capture living *E. coli* in hollow polymeric microcapsules (Figure 4.9d).[90] *E. coli* cells were suspended in a supersaturated Ca^{2+} solution with mild stirring. After incubation for 15 min, an equal amount of $CO_3{}^{2-}$ solution was added to precipitate the $CaCO_3$, producing *E.coli*@$CaCO_3$. PAH and PSS were deposited alternately on top of the $CaCO_3$ shell with LbL self-assembly. The $CaCO_3$ cores were finally removed with EDTA treatment, generating *E. coli*@hollow polymeric microcapsules. Although only approximately one-third of the hollow polymeric microcapsules carried a single *E. coli* cell, these cells showed a prolonged lag phase under the culture conditions used, compared with that of native *E. coli*.

4.3.3   Lanthanide Phosphate Shells

Tang and coworkers demonstrated that the encapsulation of living cells could be achieved by nonbiogenic crystals, as well as by biominerals. By introducing the appropriate templates for mineralization onto the surfaces of living cells, the cells were encapsulated within lanthanide phosphate (Figures 4.10a–d).[91] Lanthanide compounds are well-recognized functional materials because of their broad applicability in magnets, phosphors, catalysts, biochemical probes, and medical diagnostics.[92,93] Among the various lanthanide compounds available, lanthanide phosphate ($LnPO_4$) can absorb UV–visible light because the lanthanide (III) ion displays many *f–f* transitions in the visible range.[94] The *Danio rerio* (zebrafish) egg was chosen as the cellular model to test lanthanide applications because UVB has a greater negative effect on the hatching success of these fish than on that of other fish.[95] The method used to generate calcium phosphate shells was followed to achieve $LnPO_4$ encapsulation. Coated eggs were produced by the alternate deposition of chitosan and PAA, and $LnPO_4$ shells were precipitated onto the coated eggs when they were suspended in a lanthanide solution titrated with phosphate solution. The protective effect of the $LnPO_4$ shell on *Danio rerio* was investigated by examining embryonic development under increased UVB radiation (Figure 4.10e). The $LnPO_4$ shell reduced the malformation of the embryos and increased their viability and hatchability during UVB exposure.

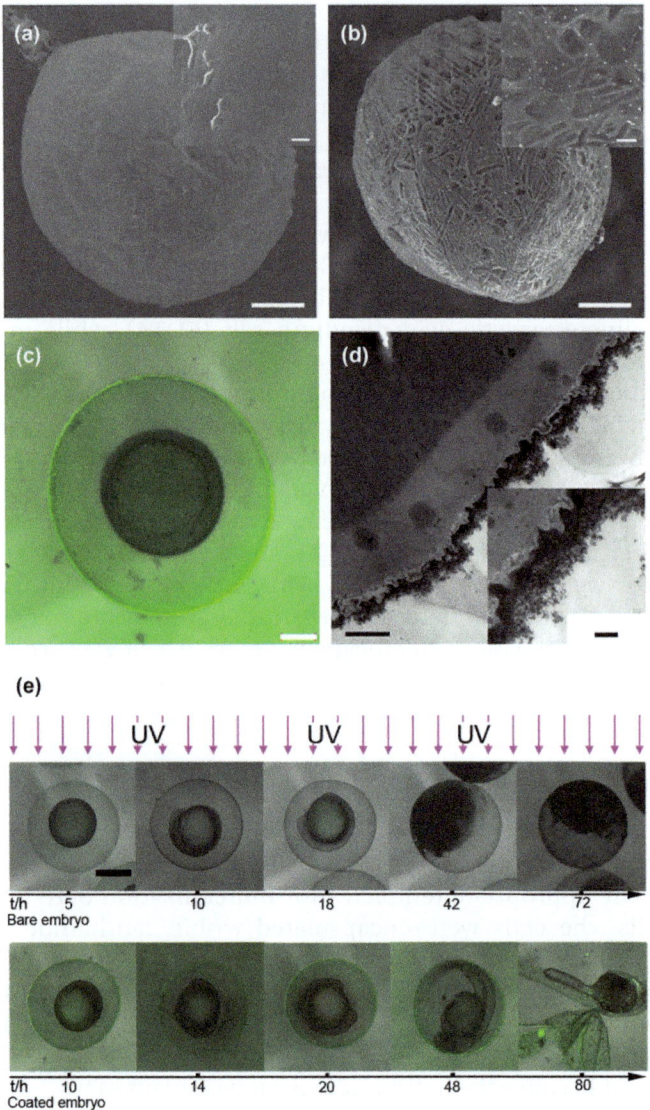

Figure 4.10   (a, b) SEM micrographs: (a) native embryo (insert: surface of native embryo), (b) embryo with LaPO$_4$ shell (insert: surface of the shell). (c) Laser scanning micrograph of embryo with the green-emission fluorescent LaPO$_4$ shell. (d) TEM micrographs of encapsulated embryo which was sliced. (e) Parallel development stages of the native embryo and the coated embryo under UVB radiation. At 42 h, the death of native embryos is distinctly observed. At 80 h, the LaPO$_4$ shell is broken by the larvae. Scale bars: 200 μm (a, b, c), 2 μm (a, b; the inset), 1 μm (d), 200 nm (d; the inset).
Reproduced with permission from PLOS (Copyright 2010).[91]

4.4 Artificial Shells Inspired by Biosilicification

Certain unicellular organisms, such as diatoms, radiolaria, and synurophytes, and multicellular glass sponges produce siliceous exoskeletons, which have exquisite hierarchical structures and superior mechanical properties (Figures 4.11a and b),[96–98] whereas most cells in Nature do not have siliceous shells to protect their cytoplasm from the external environment. In biological systems, peptides or proteins play a pivotal role as structure-directing agents in the accumulation of mineral ions and in the

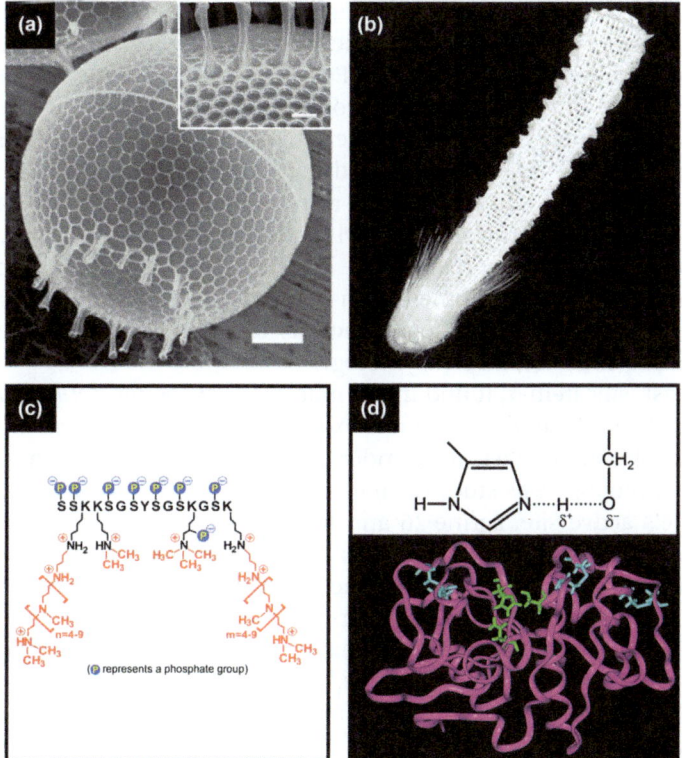

Figure 4.11   SEM micrographs of (a) *Stephanopyxis turris* cell walls (scale bar: 10 μm, 5 μm for the inset) and (b) The glass sponge, *Euplectella*, showing the basket-like cage structure. (c) Chemical structure of silaffin-1A$_1$ from *Cylindrotheca fusiformis*. (d) The catalytically essential moieties in a serine-hydrolase active site (top), and Ribbon model of silicatein-a from an energy minimization program (INSIGHT II) with green highlights for the catalytic site (bottom).
Reproduced with permission from Wiley (Copyright 2006),[97] the Nature Publishing Group (Copyright 2003),[98] the Royal Society of Chemistry (Copyright 2004),[102] the National Academy of Sciences, USA (Copyright 2006).[126]

morphogenesis of exoskeletons. With much work in the fields of biology and biochemistry, silica-forming peptides, called "silaffins", have been extracted from silica cell walls. Silaffins are post-translationally modified peptides in which many lysine residues are modified to ε-N-dimethyllysines or oligo-N-methylpropyleneimine-linked lysine residues, and serine residues are modified to phosphorylated serine residues (Figure 4.11c).[99,100]

The biosilicification in diatoms is achieved by specific interactions between silicic acid derivatives and cationic biopolymers, and the self-assembled structure of cationic polypeptides is thought to act as a template for the *in vivo* polycondensation of the silicic acid derivatives.[101] Inspired by the *in vivo* silicification of diatoms, materials scientists have attempted to mimic the catalytic activity of silaffin in *in vitro* silicification using a synthetic counterpart of silaffin, usually polyamines.[102,103] For instance, silica nanospheres have been synthesized *in vitro* under mild conditions with various polyamines, including poly L-lysine,[104] poly L-histidine,[105] poly L-arginine,[106] poly(allylamine hydrochloride),[107–109] amine-terminated dendrimers (PPI and PAMAM),[110,111] and others.[112,113] Thin silica films have also been fabricated on the nanometer scale using poly(2-(dimethylamino)ethyl methacrylate),[114–119] PDDAC,[120–122] or PEI[123] as a synthetic counterpart of the silaffins. The biomimetic approach to silicification has been combined with synthetic macromolecules, and subsequently applied in various fields, such as in biosensors, microelectronics, heterogeneous catalysis, and enzymology.[124]

In contrast, silicatein-α, found in the marine demosponge *Tethya aurantia*, is a representative protein that produces exquisite silica exoskeletons with a small amount of solubilized silicic acid under marine environmental conditions.[20] A site-directed mutagenesis study of silicatein-α revealed that two amino acids in the enzyme's active site, serine-26 and histidine-165, positioned at an appropriate distance for hydrogen bonding, play a key role not only in the hydrolysis of the alkoxide precursor, but also in the polycondensation of monosilicic acid to silica (Figure 4.11d).[125,126] Researchers have attempted to emulate this active site, and thereby suggest synthetic analogs for use in *in vitro* silicification, including polymers,[127–129] polypeptides,[130] and small molecules.[131–133]

4.4.1   Silica Shells

Biomimetic silica is formed under cytocompatible conditions, including at near-neutral pH, in aqueous solution, under ambient pressure, and at low concentrations of silicic acids or catalytic polymers.[20,134–136] By mimicking the silica-forming process that occurs in diatoms, living cells were artificially encapsulated within silica shells. These shells should expand the areas in which living-cell-based systems can be used and enhance the viability of the encapsulated cells under harsh conditions.

Choi and coworkers successfully encapsulated living yeast cells within silica shells using two consecutive, biocompatible processes, LbL self-assembly and biomimetic silicification, without disturbing cell viability (Figure 4.12a). Typically, the introduction of a catalytic template to the cell

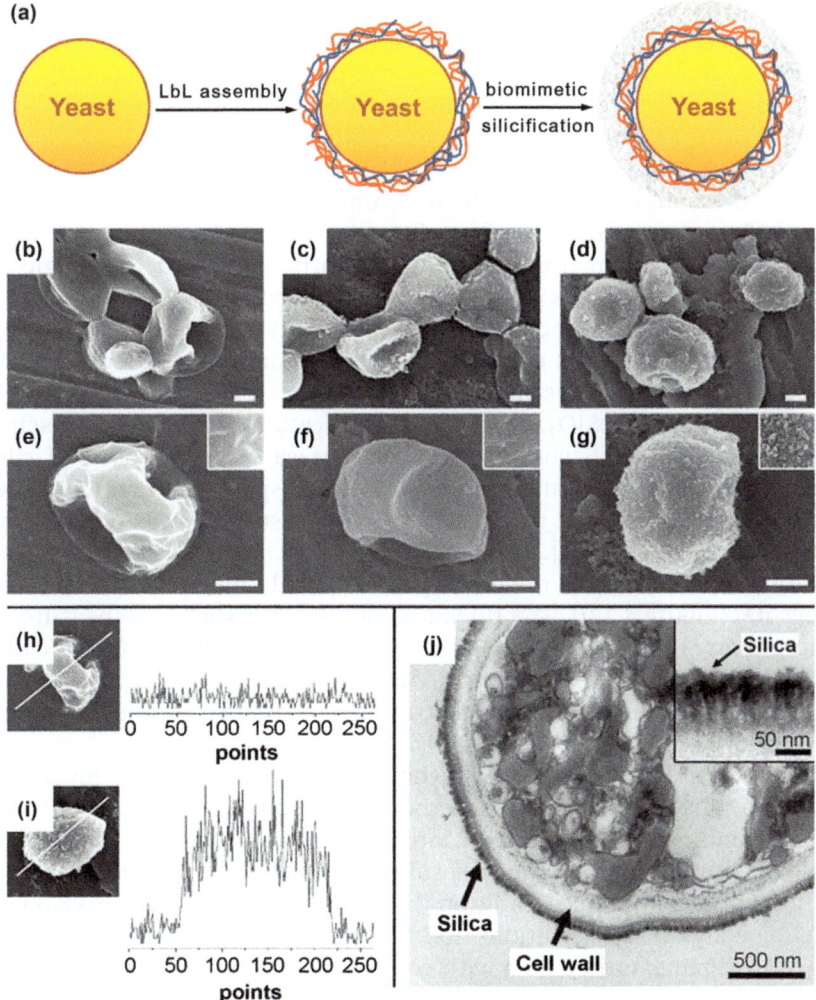

Figure 4.12 SEM micrographs of (a, b) native yeast, (c, d) PDADMA/PSS multilayer-coated yeast, and (e, f) yeast@SiO$_2$ at different magnifications. The scale bars are 1 μm. Insets in (b), (d), and (f) show the surface morphologies of yeast cells at each step. EDX spectroscopy line profiles for silicon of (g) native yeast and (h) yeast@SiO$_2$. (i) TEM micrographs of microtome-sliced yeast@SiO$_2$, and the magnified micrograph (inset).
Reproduced with permission from Wiley (Copyright 2009).[137]

surface for biosilicification is considered the first step in encapsulating living cells within silica.[137] Because all the processes must be mild enough to maintain the viability of the cells and because the LbL process has already been used to encapsulate individual living cells within polyelectrolyte multilayers, these researchers selected the LbL self-assembly of polyelectrolytes to introduce catalytic templates onto the surfaces of living cells. Synthetic polymers containing quaternary amines were found to chemically catalyze biomimetic silica formation under physiologically mild conditions,[114–117,120–122] and PDDAC was deposited onto the surfaces of yeast cells with an LbL process by pairing it with PSS, a negatively charged counterpart of PDDAC. After coating individual yeast cells with the catalytic template, the template-coated cells were immersed in a solution of 50 mM silicic acid to apply silica onto the cells.

Scanning electron micrographs (SEM) clearly confirmed the single-cell encapsulation of yeast cells within silica shells (Figures 4.12a–f). Yeast@SiO$_2$ maintained its original shape even after 24 h of drying, whereas native yeast noticeably shrunk because of dehydration. The multilayer-coated yeast also showed some ability to resist dehydration, but this effect was not as pronounced as for yeast@SiO$_2$. The high-magnification micrographs (insets of Figures 4.12b, d and f) showed that the surface was composed of silica nanoparticles. The line-scan analysis of energy-dispersive X-ray (EDX) spectroscopy clearly confirmed that the surface of yeast@SiO$_2$ was covered with silica (Figures 4.12g and h). The presence of silica shells was directly shown by transmission electron microscopy (TEM) with microtomous slices of yeast@SiO$_2$ (Figure 4.12i). The thickness of the silica shell was measured to be more than 50 nm. Furthermore, it was shown that encapsulation within silica shells increased the long-term viability of the yeast cells and delayed cell division. These results demonstrate that physical hardness was successfully conferred on the silica shells.

Silica encapsulation is plausible when the chemical bonding involved is considered. It is also an excellent method for generating artificial shells with covalent bonding. Compared with ionic bonds, which can be formed with simple precipitation in an aqueous solution, covalent bonds are difficult to form under biocompatible conditions. However, the researchers developed a set of biocompatible conditions for the formation of covalent bonds in biological systems, rather than with conventional chemical reactions.

Artificial inorganic shells have also been formed by genetically modifying microbial cells. The catalytic templates for biosilicification were produced by the metabolism of the living host cells, and the metabolically produced proteins diffused to the cell wall and reacted with a silica precursor on the surface of the cell. For example, Tan and coworkers achieved the siliceous bioencapsulation of yeast cells by expressing lysozyme. The polycondensation of silica precursors was triggered by the positively charged lysozyme, presumably through silaffin-mimicking mechanisms (Figure 4.13a).[138] Muller *et al.* also demonstrated the formation of poly(silicate) on the membranes of *E. coli* with self-expressed silicatein-α (Figure 4.13b).[139] Silicatein-α produces silica using the mechanism of glass sponges and the simplicity and cytocompatibility of these processes make them very useful. They also circumvent the exposure of cells to additional chemicals and processes because the bioengineered cells themselves produce the catalytic templates.

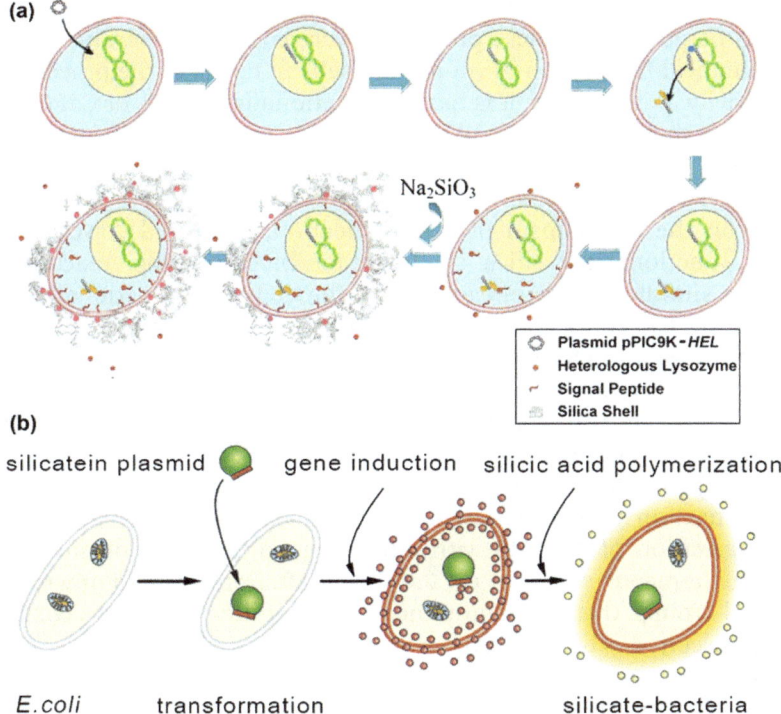

Figure 4.13   Schematic description of silica shell formation by bioengineered cells. (a) *P. Pastoris* transformed with HEL gene, and (b) *E. coli* are transformed with the silicatein gene.
Reproduced with permission from Springer (Copyright 2008),[138] and Elsevier (Copyright 2008).[139]

4.4.2   Functionalized Silica Shells

Cell surface modifications have largely been achieved using complicated methods, such as the introduction of functional groups with metabolic or genetic engineering.[140–145] Although bioengineering methods are progressing in more biocompatible ways, it is still possible that the direct insertion of functional groups will cause harmful effects on cell membranes. Encapsulation methods have been investigated enthusiastically to develop an indirect approach to cell-surface modification, as an alternative to the direct modification of cell membranes. It is presumed that cell integrity will not be compromised by these encapsulation methods, because the functional groups can be introduced onto an artificial shell with no direct contact with the cell membrane. For a decade, the noncovalent adsorption of macromolecules was achieved with LbL self-assembly, which introduced various functionalities onto cell surfaces.[1,2,146,147] The recent formation of inorganic shells on living cells presented researchers with the challenge of functionalizing these inorganic shells. The clear advantages of inorganic shells derive mainly from their mechanical robustness and chemical inertia. However, paradoxically, these properties were considered serious obstacles to the chemical functionalization of the shells in terms of the processability and reactivity. For example, it is hard to functionalize $CaCO_3$ or $CaPO_4$ shells because the shells are chemically inert.[148] It is impossible to functionalize silica shells without harsh conditions, such as extreme pH, high temperatures, and harmful solvents, which are fatal to cells.[149]

Based on the mechanism of biosilicification, Choi and coworkers successfully functionalized silica shells under biocompatible reaction conditions by simultaneously encapsulating the cells and functionalizing the shells during the course of silicification (Figures 4.14a and b).[150] In their experiment, PEI and PSS were introduced onto the surfaces of yeast cells by LbL self-assembly as the catalytic templates for biomimetic silicification. After the template was coated with the polyelectrolytes, thiol functional groups were incorporated directly onto the silica shell by adding (3-mercaptopropyl)trimethoxysilane (MPTMS) during the course of the biomimetic polycondensation of silicic acid derivatives, leading to the simultaneous polycondensation of MPTMS with silicic acid under physiologically mild conditions (aqueous solution, pH 7.4). Even after encapsulation with these functional groups, the cells maintained their viability, which was confirmed with fluorescein diacetate (FDA).

The authors suggested that the thiol functional groups on silica shells can be utilized in versatile ways to further functionalize the shells (Figure 4.14c). Various chemical or biological functions can be introduced onto silica shells by preparing maleimide derivatives linked to the desired moieties, because the specific coupling reaction between the thiol functional group and maleimide occurs in aqueous solution at pH 7.4, which is compatible with living cells. As a proof of concept, fluorescein-linked or biotin-linked maleimide was introduced onto the surface of yeast@SiO_2^{SH}, and rhodamine-linked streptavidin was conjugated onto the biotin-functionalized yeast@SiO_2^{SH}. Thus, the authors developed a versatile method for the chemical functionalization of silica shells with covalent bonding and the simultaneous encapsulation and functionalization of cells.

4.4.3   Titania Shells

In the previous sections, we have described the successful formation of inorganic shells using biocompatible materials and processes inspired by biomaterials and biomineralization. Compared with biogenic materials, including $CaCO_3$, $CaPO_4$, and silica, which are inherently compatible with biological systems, nonbiogenic materials are much more difficult to interface with living cells. In addition to the toxicity of nonbiogenic materials, the reactions required to synthesize them take place under harsh conditions, including high temperatures, high pressures, extreme pHs, and with caustic reagents, which usually compromise cell viability.[151] However, encapsulating living cells with nonbiogenic materials would equip artificial shells

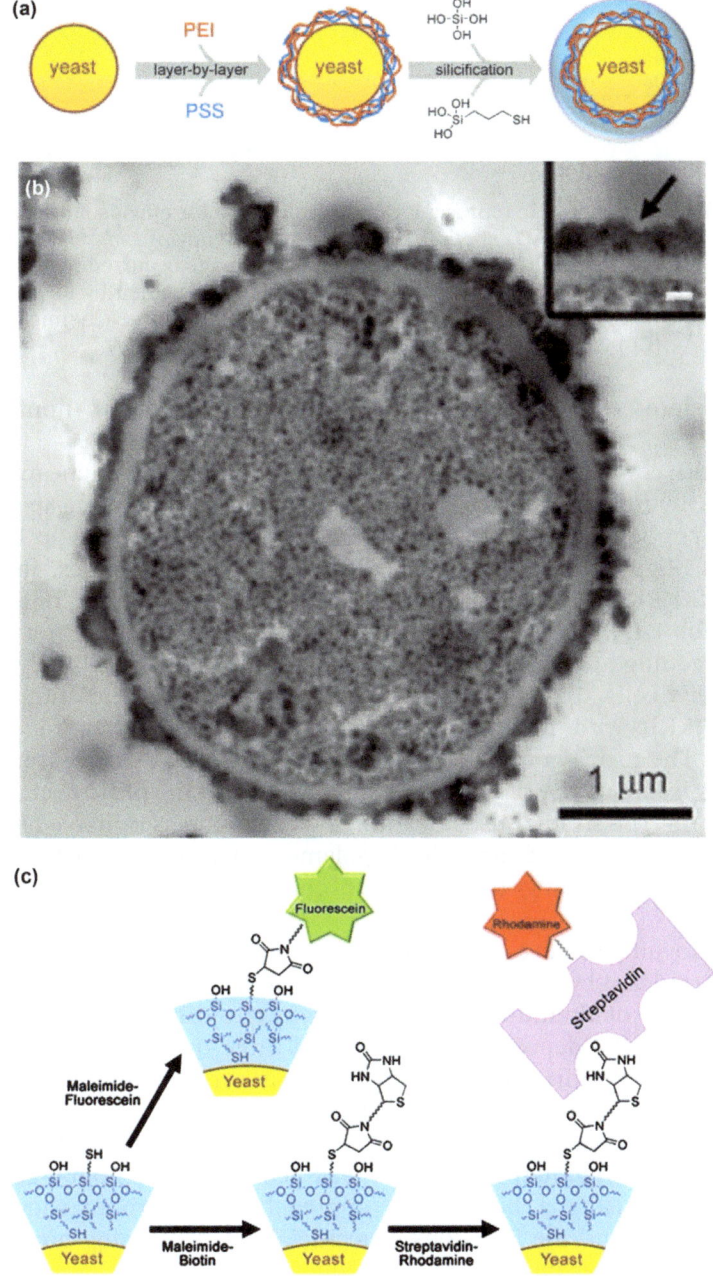

Figure 4.14 (a) Schematic representation of procedure for functionalization of silica shell. (b) TEM micrograph of microtome-sectioned yeast@SiO$_2$^{SH}. (c) chematic representation of procedure for postfunctionalization of SiO$_2$^{SH} shell with fluorescein, biotin, and streptavidin. Reproduced with permission from Wiley (Copyright 2011).[150]

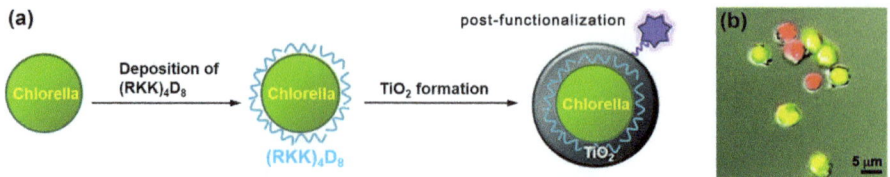

Figure 4.15   (a) Schematic representation of procedure for encapsulating *Chlorella* cells with TiO$_2$ shell. (b) Confocal microscopy of *Chlorella*@TiO$_2$. The bright red fluorescence indicated no damage occurred to chloroplasts and cells in green were considered alive, examined by FDA.
Reproduced with permission from the American Chemical Society (Copyright 2012).[152]

with the interesting properties and useful functions of nonbiogenic materials.

Choi and coworkers reported a bioinspired method of encapsulating individual living *Chlorella* cells within titania (TiO$_2$) using a specifically designed peptide, which was both cytocompatible and catalytic for TiO$_2$ formation (Figure 4.15a).[152] Arginine/lysine (R/K)-rich peptides, including RKKRKKRKKRKK ((RKK)$_4$), were particularly selected as the bioinspired catalysts for TiO$_2$ formation after they had been identified with a peptide library screening method.[153] However, (RKK)$_4$ was found to be highly toxic to the *Chlorella* cells. To reduce the toxicity of the peptide, the authors derivatized (RKK)$_4$ with additional units of aspartic acid (D), to produce (RKK)$_4$D$_8$. The toxicity of polyelectrolytes during their LbL self-assembly on living cells is known to originate predominantly from the positive charges on the polymers. The strategy used by Choi and coworkers is valuable because it can be applied to many toxic cationic polymers, the toxicity of which can be reduced by neutralizing their positive charges with negatively charged moieties.

After cytocompatible catalytic templates were achieved, *Chlorella* cells were encapsulated by immersing them alternately in a solution of (RKK)$_4$D$_8$, and a solution of titanium bis(ammoniumlactato)dihydroxide (TiBALDH). (RKK)$_4$D$_8$/TiBALDH deposition was repeated three times, generating *Chlorella*@TiO$_2$. The viability of *Chlorella*@TiO$_2$ was confirmed by FDA staining and chlorophyll autofluorescence.[154] The viability of the cells was about 69% after 3 by 3 LbL processes (Figure 4.15b). It is noteworthy that the encapsulation of individual living cells was achieved with nonbiogenic, inorganic shells using a bioinspired approach. The authors showed the possibility to extend the functions of artificial shells by using nonbiogenic materials for cell encapsulation.

4.4.4   Titania–Silica Hybrid Shells

In the previous sections, we reviewed artificial shells made of biogenic silica and nonbiogenic titania. After their successes with biogenic materials,

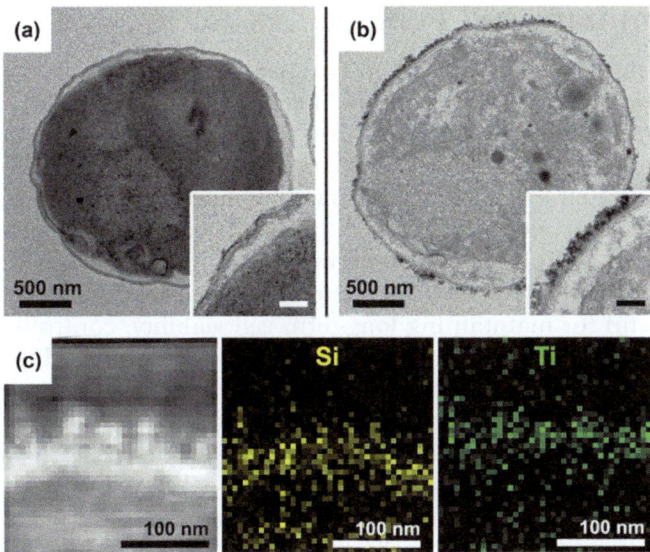

Figure 4.16 TEM micrograph of (a) native Chlorella and (b) Chlorella@SiO₂/TiO₂ (scale bar in the inset: 100 nm). (c) EDX mapping of microtomed Chlorella@SiO₂/TiO₂ by TEM.
Reproduced with permission from Wiley (Copyright 2013).[155]

researchers investigated how to equip artificial shells with expanded functions using nonbiogenic materials. By extension, the next step should be the hybridization of biogenic and nonbiogenic materials. As a proof of concept, Choi and coworkers used a nanocomposite of SiO_2 and TiO_2 to encapsulate *Chlorella* cells.[155] The cells were successfully encapsulated within the SiO_2/TiO_2 nanocomposite with a method similar to that established for TiO_2 encapsulation, in which a mixture of silicic acid and TiBALDH was added to (RKK)₄D₈-coated cells (Figures 4.16a and b). The authors characterized the nature of the SiO_2/TiO_2 nanocomposite with energy-dispersive X-ray spectroscopy and TEM (Figure 4.16c). The encasing layer was composed of SiO_2 and TiO_2, and the outermost particulate structures were mainly composed of amorphous titania, in which 3-nm anatase titania was embedded. Even after their encapsulation, the viability of *Chlorella*@SiO_2-TiO_2 remained above ∼ 87%, and retardation of cell division was controlled by tuning the thickness of the SiO_2/TiO_2 shell. Furthermore, the thermal tolerance of *Chlorella*@SiO_2/TiO_2 was greater than that of native cells, presumably because the SiO_2/TiO_2 nanocomposite effectively dissipated the heat energy.[156]

Their work demonstrates that both biological and abiological materials can be hybridized *in situ* onto the cell surface to encapsulate the cell. Furthermore, it should be feasible to utilize the newly obtained or synergistically combined properties of the hybrid materials to introduce novel functions to living cells through cell encapsulation.

4.5   Summary and Perspectives

In the last decade, the field of cell encapsulation has emerged and grown rapidly. However, the focus of the field is shifting from coating single cells with polymeric materials, with the aim of immunocamouflage, to encapsulating single cells with hard shells to control cellular behavior. In the course of this conceptual shift, inorganic materials have been widely used to establish the mechanical stability of the artificial shells, inspired by biomineralization. Cellular viability and metabolic activities are ensured by the careful selection of biocompatible materials and processes. Inorganic shells are also useful for maintaining long-term cell viability, controlling cell division, protecting cells from foreign aggression, and providing cells with chemical functionalization. They also offer a platform for understanding cellular metabolism at the single-cell level and for the development of many cell-based applications, including cell-based sensors, cell therapies, and regenerative medicine. To maximize the controllability of cellular behavior, researchers are actively trying to enhance the mechanical stability and cytocompatibility of artificial shells, to realize the dream of *artificial spores*[13] that imitate simulate the inherent functions of bacterial endospores.

Although inorganic shells have been successfully adapted to living cells and support many novel functions, our future goals include artificial shells that are more versatile, more compatible, and more protective. Until now, the novel functions of inorganic shells and the control of cellular behaviors have largely derived from the mechanical stability of the inorganic shells, which are strong enough to resist internal or external stressors, such as osmotic pressure, dehydration, and lytic enzymes, and to suppress cell division. However, this mechanical robustness could be an obstacle to equipping inorganic shells with more advanced functions.

Several properties of inorganic shells have not yet been fully demonstrated or remain challenging. (1) Tunable permeability: a series of studies has shown that inorganic shells are permeable to small molecules, such as water, nutrients, and fluorescent dyes, but block, at least to some extent, the penetration of macromolecules, such as lytic enzymes. The porosity of inorganic shell has not been systematically tuned as yet, although the control of porosity will be the starting point for accurately controlling cell division and the fundamental study of cell–cell communication. Because tuning the permeability of inorganic shells is important to the bioapplications of single cells, the systematic control of their porosity will allow the modification of their properties.

(2) Controlled degradability: although calcium phosphate and calcium carbonate shells can be removed by a chemical cue (*e.g.*, EDTA), the removal of silica and titania shells has not yet been demonstrated. In shells made with covalent bonds, the robustness of the shells is useful for retarding cell division, but the removal of the shells without disturbing the viability of the living encapsulated cells is difficult because breaking strong chemical bonds usually requires extreme conditions. In contrast to the passive germination

of natural spores, which occurs uncontrolledly in the presence of nutrients and other factors, the stimulus-responsive degradation of shells will allow the chemical control of cellular metabolism and cell division. In terms of controllability, the *artificial spores* formed by inorganic shells would surpass natural spores.

(3) Higher biocompatibility: to date, encapsulation within inorganic shells has been limited to microbial cells, which are useful models for demonstrating the control of cellular behaviors. In fact, the success of previous demonstrations depended largely on the strong viability of micro-organisms. In contrast to micro-organisms, the membranes of mammalian cells are not reinforced by a polysaccharide layer, and this is considered to be a major obstacle to encapsulating mammalian cells within inorganic shells. Because of the susceptibility of mammalian cells to environmental changes, whether these involve chemicals, pH, or osmotic pressure, more biocompatible materials and processes must be developed if we are to apply inorganic shells to mammalian cells. Although the construction of hard shells on mammalian cells requires further development, the studies reviewed here indicate that it would be feasible to encapsulate mammalian cells within inorganic shells by carefully selecting nontoxic inorganic precursors and organic templates that catalyze biomineralization. Because the manipulation of mammalian cells is in high demand for biomedical purposes, establishing more biocompatible materials and processes will be an unavoidable task in the near future.

In this chapter, we have only reviewed inorganic shells for cell encapsulation, which have clear advantages, mainly deriving from the mechanical strength of the materials. However, recent reports have demonstrated that organic shells can also be used to enhance the long-term viability of cells, control their division, and protect their cytoplasm against foreign aggression.[157–159] These organic shells are strengthened by mussel-inspired polymerization/crosslinking and a coupling reaction between the polyelectrolyte layers. This has confirmed that the essential properties of artificial shells are conferred by the mechanical robustness of the materials used, rather than by the sort of material (organic or inorganic) used. However, it must be recognized that the field of cell encapsulation took a major step forward with the development of inorganic shells, which have been the starting point for fabricating robust shells inspired by biomineralization. In the future, materials and processes will be developed to produce advanced functional shells that are more versatile, more cytocompatible, and more protective.

References

1. R. F. Fakhrullin, A. I. Zamaleeva, R. T. Minullina, S. A. Konnova and V. N. Paunov, *Chem. Soc. Rev.*, 2012, **1**, 4189.
2. R. F. Fakhrullin and Y. M. Lvov, *ACS Nano*, 2012, **6**, 4557.
3. Z. Tang, Y. Wang, P. Podsiadlo and N. A. Kotov, *Adv. Mater.*, 2006, **18**, 3203.

4. S. Krol, M. Nolte, A. Diaspro, D. Mazza, R. Magrassi, A. Gliozzi and A. Fery, *Langmuir*, 2005, **21**, 705.
5. S. Mansouri, Y. Merhi, F. M. Winnik and M. Tabrizian, *Biomacromolecules*, 2011, **12**, 585.
6. A. Nishiguchi, H. Yoshida, M. Matsusaki and M. Akashi, *Adv. Mater.*, 2011, **23**, 3506.
7. R. F. Fakhrullin, A. Zamaleeva, M. V. Morozov, D. I. Tazetdinova, F. K. Alimova, A. K. Hilmutdinov, R. I. Zhdanov, M. Kahraman and M. Culha, *Langmuir*, 2009, **25**, 4628.
8. I. Safarik, K. Pospiskova, K. Horska and M. Safarikova, *Soft Matter*, 2012, **8**, 5407.
9. G. Wang, L. Wang, P. Liu, Y. Yan, X. Xu and R. Tang, *ChemBioChem*, 2010, **11**, 2368.
10. A. Zamaleeva, I. R. Sharipova, A. V. Porfireva, G. A. Evtugyn and R. F. Fakhrullin, *Langmuir*, 2010, **26**, 2671.
11. S. A. Konnova, I. R. Sharipova, T. A. Demina, Y. N. Osin, D. R. Yarullina, O. N. Ilinskaya, Y. M. Lvov and R. F. Fakhrullin, *Chem. Commun.*, 2013, **49**, 4208.
12. S. H. Yang, T. Lee, E. Seo, E. H. Ko, I. S. Choi and B.-S. Choi, *Macromol. Biosci.*, 2012, **12**, 61.
13. D. Hong, M. Park, S. H. Yang, J. Lee, Y.-G. Kim and I. S. Choi, *Trends Biotechnol.*, 2013, **31**, 442.
14. S. H. Yang, D. Hong, J. Lee, E. H. Ko and I. S. Choi, *Small*, 2013, **9**, 178.
15. C. N. R. Rao and B. Raveau, *Transition Metal Oxides: Structure, Properties, and Synthesis of Ceramic Oxides*, 2nd edn, Wiley, New York, 1998.
16. H. A. Lowenstam, *Science*, 1981, **211**, 1126.
17. H. A. Lowenstam and S. Weiner, *On Biomineralization*, Oxford University Press, Oxford, 1989.
18. S. Mann, J. Webb and R. J. P. Williams, *Biomineralization: Chemical and Biochemical Perspectives*, Wiley-VCH, New York, 1989.
19. L. C. Palmer, C. J. Newcomb, S. R. Kaltz, E. D. Spoerke and S. I. Stupp, *Chem. Rev.*, 2008, **108**, 4754.
20. R. L. Brutchey and D. E. Morse, *Chem. Rev.*, 2008, **108**, 4915.
21. M. B. Dickerson, K. H. Sandhage and R. R. Naik, *Chem. Rev.*, 2008, **108**, 4935.
22. H. L. Enrlich, *Geomicrobiology*, Marcel Dekker, Inc., New York, 2002.
23. D. R. Lovley and E. R. Phillips, *Appl. Environ. Microbiol.*, 1988, **54**, 1472.
24. J. R. Stephen and S. J. Macnaughton, *Curr. Opin. Biotechnol.*, 1999, **10**, 230.
25. G. M. Gadd, *Geoderma*, 2004, **122**, 109.
26. R. K. Poole and G. M. Gadd, *Metal-Microbe Interaction*, IRL Press, Oxford, 1989.
27. S. Silver and L. T. Phung, *Microbiol. Biotechnol.*, 2005, **32**, 587.
28. J. C. G. Walker, *Nature*, 1987, **329**, 710.
29. D. R. Lovley and E. J. P. Phillips, *Appl. Environ. Microbiol.*, 1992, **58**, 850.
30. D. R. Lovley, *Adv. Agron.*, 1995, **54**, 175.

31. T. Klaus, R. Joerger, E. Olsson and C.-G. Granqvis, *Proc. Natl. Acad. Sci.*, 1999, **96**, 13611.
32. R. W. Boyle, *Geol. Surv. Can. Bull.*, 1979, **280**, 583.
33. S. Karthikeyan and T. J. Beveridge, *Environ. Microbiol.*, 2002, **4**, 667.
34. F. Reith, M. F. Lengke, D. Falconer, D. Craw and G. Southam, *ISME J.*, 2007, **1**, 567.
35. T. J. Beveridge and R. G. E. Murray, *J. Bacteriol.*, 1976, **127**, 1502.
36. M. F. Lengke, M. E. Fleet and G. Southam, *Langmuir*, 2006, **22**, 2780.
37. G. Southam, W. S. Fyfe and T. J. Beveridge, *Mineral. Metallurg. Proc.*, 2000, **17**, 129.
38. L. Fairbrother, B. Etschmann, J. Brugger, J. Shapter, G. Southam and F. Reith, *Environ. Sci. Technol.*, 2013, **47**, 2628.
39. M. F. Lengke and G. Southam, *Geochim. Cosmochim. Acta.*, 2005, **69**, 3759.
40. T. J. Beveridge and R. G. E. Murray, *J. Bacteriol.*, 1976, **127**, 1502.
41. J. V. Stoyanov and N. L. Brown, *J. Biolog. Chem.*, 2003, **278**, 1407.
42. K. Kashefi, J. M. Tor, K. P. Nevin and D. R. Lovley, *Appl. Environ. Microbiol.*, 2001, **67**, 3275.
43. W.-S. Kuo, C.-M. Wu, Z.-S. Yang, S.-Y. Chen, C.-Y. Chen, C.-C. Huang, W.-M. Li, C.-K. Sun and C.-S. Yeh, *Chem. Commun.*, 2008, 4430.
44. B. Kaehr and C. J. Brinker, *Chem. Commun.*, 2010, **46**, 5268.
45. M. Cusack and A. Freer, *Chem. Rev.*, 2008, **108**, 4433.
46. S. Weiner, W. Traub and H. D. Wagner, *J. Struct. Biol.*, 1999, **126**, 241.
47. L. Addadi, S. Raz and S. Weiner, *Adv. Mater.*, 2003, **15**, 959.
48. S. Mann, *Angew. Chem. Int. Ed.*, 2000, **39**, 3392.
49. F. C. Meldrum, *Int. Mater. Rev.*, 2003, **48**, 187.
50. S. Weiner and L. Addadi, *J. Mater. Chem.*, 1997, 7, 689.
51. R. C. Weast, *The CRC Handbook of Chemistry and Physics*, 66th edn, CRC Press, Boca Raton, FL, 1985–1986.
52. L. C. Palmer, C. J. Newcomb, S. R. Kaltz, E. D. Spoerke and S. I. Stupp, *Chem. Rev.*, 2008, **108**, 4754.
53. B. Wang, P. Liu, W. Jiang, H. Pan, X. Xu and R. Tang, *Angew. Chem. Int. Ed.*, 2008, **47**, 3560.
54. E. Cabib, D. H. Roh, Schmidt, L. B. Crotti and A. Varma, *J. Biol. Chem.*, 2001, **276**, 19679.
55. B. Wang, P. Liu, Z. Liu, H. Pan, X. Xu and R. Tang, *Biotechnol. Bioeng.*, 2013, **111**, 386.
56. A. Bigi, E. Boanini, D. Walsh and S. Mann, *Angew. Chem. Int. Ed.*, 2002, **41**, 2163.
57. L. L. Hench, *J. Am. Ceram. Soc.*, 1991, **74**, 1487.
58. S. C. Liou, S. Y. Chen and D. M. Liu, *Biomaterials*, 2003, **24**, 3981.
59. M. Wei, A. J. Ruys, B. K. Milthorpe and C. C. Sorrell, *J. Biomed. Mater. Res.*, 1999, **45**, 11.
60. D. M. Liu, Q. Yang, T. Troczynski and W. J. Tseng, *Biomaterials*, 2002, **23**, 1679.
61. I. S. Neira, Y. V. Kolen'ko, O. I. Lebedev, G. V. Tendeloo, H. S. Gupta, F. Guitian and M. Yoshimura, *Cryst. Growth Des.*, 2009, **9**, 466.

62. M. Huang and Y. Wang, *J. Mater. Chem.*, 2012, **22**, 626.
63. J. Küther, R. Seshadri, W. Knoll and W. Tremel, *J. Mater. Chem.*, 1998, **8**, 641.
64. D. Gebauer, A. Vclkel and H. Cölfen, *Science*, 2008, **322**, 1819.
65. S. H. Yang and I. S. Choi, *Chem. Asian. J.*, 2010, **5**, 1586.
66. M. Balz, E. Barriau, V. Istratov, H. Frey and W. Tremel, *Langmuir*, 2005, **21**, 3987.
67. T. Kato, A. Sugawara and N. Hosoda, *Adv. Mater.*, 2002, **14**, 869.
68. J. Aizenberg, J. Hanson, T. F. Koetzle, S. Weiner and L. Addadi, *J. Am. Chem. Soc.*, 1997, **119**, 881.
69. S. Albeck, J. Aizenberg, L. Addadi and S. Weiner, *J. Am. Chem. Soc.*, 1993, **115**, 11691.
70. Y. Chen, J. Xiao, Z. Wang and S. Yang, *Langmuir*, 2009, **25**, 1054.
71. X. R. Xu, A. H. Cai, R. Liu, H. H. Pan, R. K. Tang and K. Cho, *J. Cryst. Growth*, 2008, **310**, 3779.
72. F. M. Michel, J. MacDonald, J. Feng, B. L. Phillips, L. Ehm, C. Tarabrella, J. B. Parise and R. J. Reeder, *Chem. Mater.*, 2008, **20**, 4720.
73. M. Faatz, F. Grohn and G. Wegner, *Adv. Mater.*, 2004, **16**, 996.
74. A. Becker, U. Bismayer, M. Epple, H. Fabritius, B. Hasse, J. M. Shi and A. Ziegler, *Dalton Trans.*, 2003, 551.
75. M. J. Olszta, D. J. Odom, E. P. Douglas and L. B. Gower, *Connect. Tissue Res.*, 2003, **44**, 326.
76. S. Weiner, Y. Levi-Kalisman and S. Raz, *Connect. Tissue Res.*, 2003, **44**, 214.
77. S. Raz, P. C. Hamilton, F. H. Wilt, S. Weiner and L. Addadi, *Adv. Funct. Mater.*, 2003, **13**, 480.
78. J. Aizenberg, G. Lambert, S. Weiner and L. Addadi, *J. Am. Chem. Soc.*, 2002, **124**, 32.
79. L. B. Gower and D. J. Odom, *J. Cryst. Growth*, 2000, **210**, 719.
80. S. Raz, S. Weiner and L. Addadi, *Adv. Mater.*, 2000, **12**, 38.
81. J. Aizenberg, G. Lambert, L. Addadi and S Weiner, *Adv. Mater.*, 1996, **8**, 222.
82. X. Xu, J. T. Han and K. Cho, *Langmuir*, 2005, **21**, 4801.
83. J. T. Han, X. Xu, D. H. Kim and K. Cho, *Adv. Funct. Mater.*, 2005, **15**, 475.
84. J. Henderiks, *Mar. Micropaleontol.*, 2008, **67**, 143.
85. M. R. Langer, M. T. Silk and J. H. Lipps, *J. Foraminiferal Res.*, 1997, **27**, 271.
86. R. F. Fakhrullin and R. T. Minullina, *Langmuir*, 2009, **25**, 6617.
87. S. C. Huang, K. Naka and Y. Chujo, *Langmuir*, 2007, **23**, 12086.
88. E. Chibowski, L. Hotysz and A. Szczes, *Colloids Surf. A*, 2003, **222**, 41.
89. A. Jada and A. Verraes, *Colloids Surf. A*, 2003, **219**, 7.
90. J. Flemke, M. Maywald and V. Sieber, *Biomacromolecules*, 2013, **14**, 207.
91. B. Wang, P. Liu, Y. Y. Tang, H. H. Pan, X. R. Xu and R. Tang, *PLoS One*, 2010, **5**, e9963.
92. C. F. Wu, W. P. Qin, G. S. Qin, D. Zhao, J. S. Zhang, S. H. Huang, S. Z. Lu, H. Q. Liu and H. Y. Lin, *Appl. Phys. Lett.*, 2003, **82**, 520.

93. M. Yada, M. Mihara, S. Mouri, M. Kuroki and T. Kijima, *Adv. Mater.*, 2000, **12**, 309.
94. K. Hickmann, V. John, A. Oertel, K. Koempe and M. Haase, *J. Phys. Chem. C*, 2009, **113**, 4763.
95. T. Imai and T. Ohno, *Appl. Environ. Microbiol.*, 1995, **61**, 3604.
96. C. J. Weaver, J. Aizenberg, E. G. Fantner, D. Kisailus, A. Woesz, P. Allen, K. Fields, J. M. Porter, W. F. Zok, K. P. Hansma, P. Fratzl and D. E. Morse, *J. Struct. Biol.*, 2007, **158**, 93.
97. M. Sumper and E. Brunner, *Adv. Funct. Mater.*, 2006, **16**, 17.
98. V. C. Sundar, A. D. Yablon, J. L. Grazul, M. Ilan and J. Aizenberg, *Nature*, 2003, **424**, 899.
99. N. Kröger, S. Lorenz, E. Brunner and M. Sumper, *Science*, 2002, **298**, 584.
100. M. Sumper, *Science*, 2002, **295**, 2430.
101. M. Sumper, S. Lerenz and E. Brunner, *Angew. Chem. Int. Ed.*, 2003, **42**, 5192.
102. M. Sumper and N. Kröger, *J. Mater. Chem.*, 2004, **14**, 2059.
103. M. B. Dickerson, K. H. Sandhage and R. R. Naik, *Chem. Rev.*, 2008, **108**, 4935.
104. S. V. Patwardhan, N. Mukherjee and S. J. Clarson, *J. Inorg. Organomet. Polym.*, 2001, **11**, 19.
105. S. V. Patwardhan and S. J. Clarson, *J. Inorg. Organomet. Polym.*, 2003, **13**, 49.
106. S. V. Patwardhan and S. J. Clarson, *J. Inorg. Organomet. Polym.*, 2003, **13**, 193.
107. F. Noll, M. Sumper and N. Hampp, *Nano Lett.*, 2002, **2**, 91.
108. S. V. Patwardhan, N. Mukherjee and S. J. Clarson, *Silicon Chem.*, 2002, **1**, 47.
109. S. V. Patwardhan, N. Mukherjee and S. J. Clarson, *J. Inorg. Organomet. Polym.*, 2001, **11**, 117.
110. E. Brunner, K. Lutz and M. Sumper, *Phys. Chem. Chem. Phys.*, 2004, **6**, 854.
111. S. V. Patwardhan and S. J. Clarson, *J. Mater. Sci. Eng.*, 2003, **23**, 495.
112. M. R. Knecht and D. W. Wright, *Langmuir*, 2004, **20**, 4728.
113. S. V. Patwardhan and S. J. Clarson, *Silicon Chem.*, 2002, **1**, 207.
114. S. H. Yang, *Solid State Sci.*, 2013, **23**, 1.
115. S. H. Yang and I. S. Choi, *Bull. Korean Chem. Soc.*, 2010, **31**, 753.
116. S. H. Yang, J. H. Park, W. K. Cho, H.-S. Lee and I. S. Choi, *Small*, 2009, **5**, 1947.
117. W. K. Cho, S. M. Kang, D. J. Kim, S. H. Yang and I. S. Choi, *Langmuir*, 2006, **22**, 11208.
118. D. J. Kim, K.-B. Lee, T. G. Lee, H. K. Shon, W.-J. Kim, H.-j. Paik and I. S. Choi, *Small*, 2005, **1**, 992.
119. D. J. Kim, K.-B. Lee, Y. S. Chi, W. J. Kim, H.-j. Paik and I. S. Choi, *Langmuir*, 2004, **20**, 79.
120. N. Laugel, J. Hemmerle, C. Porcel, J.-C. Voegel, P. Schaaf and V. Ball, *Langmuir*, 2007, **23**, 3706.

121. S. H. Yang and I. S. Choi, *Chem. Asian J.*, 2009, **4**, 382.
122. S. H. Yang, J. H. Park and I. S. Choi, *Bull. Korean Chem. Soc.*, 2009, **30**, 2165.
123. S. H. Yang, E. H. Ko and I. S. Choi, *Macromol. Res.*, 2011, **19**, 511.
124. H. R. Luckarift, J. C. Spain, R. R. Naik and M. O. Stone, *Nat. Biotechnol.*, 2004, **22**, 211.
125. Y. Zhou, K. Shimizu, J. N. Cha, G. D. Stucky and D. E. Morse, *Angew. Chem. Int. Ed.*, 1999, **6**, 799.
126. D. Kisailus, Q. Truong, Y. Amemiya, J. C. Weaver and D. Morse, *Proc. Natl. Acad. Sci. USA*, 2006, **103**, 5652.
127. D. H. Adamson, D. M. Dabbs, C. R. Pacheco, M. V. Giotto, D. E. Morse and I. A. Aksay, *Macromolecules*, 2007, **40**, 5710.
128. L. Zhang, K. Yu and A. Eisenberg, *Science*, 1996, **272**, 1777.
129. M. I. Liff and M. N. Zimmerman, *Polym. Int.*, 1998, **47**, 375.
130. J. N. Cha, G. D. Stucky, D. E. Morse and T. J. Deming, *Nature*, 2000, **403**, 289.
131. J. H. Park, J. Y. Choi, T. Park, S. H. Yang, S. Kwon, H.-S. Lee and I. S. Choi, *Chem. Asian J.*, 2011, **6**, 1939.
132. J. H. Park and I. S. Choi, *Bull. Korean Chem. Soc.*, 2010, **31**, 1831.
133. K. M. Roth, Y. Zhou, W. Yang and D. E. Morse, *J. Am. Chem. Soc.*, 2005, **127**, 325.
134. O. Helmecke, A. Hirsch, P. Behrens and H. J. Menzel, *Coll. Interface Sci.*, 2008, **321**, 44.
135. E. A. Coffman, A. V. Melechko, D. P. Allison, M. L. Simpson and M. J. Doktycz, *Langmuir*, 2004, **20**, 8431.
136. L. L. Brott, R. R. Naik, D. J. Pikas, S. M. Kirkpatrick, D. W. Tomlin, P. W. Whitlock, S. J. Clarson and M. O. Stone, *Nature*, 2001, **413**, 291.
137. S. H. Yang, K.-B. Lee, B. Kong, J. H. Kim, H. S. Kim and I. S. Choi, *Angew. Chem. Int. Ed.*, 2009, **48**, 9160.
138. C. Guan, G. Wang, J. Ji, J. Wang, H. Wang and M. Tan, *J. Sol-Gel Sci. Technol.*, 2008, **48**, 369.
139. W. E. G. Müller, S. Engel, X. Wang, S. E. Wolf, W Tremel, N. L. Thakur, A. Krasko, M. Divekar and H. C. Schröder, *Biomaterials*, 2008, **29**, 771.
140. S. T. Laughlin, J. M. Baskin, S. L. Amacher and C. R. Bertozzi, *Science*, 2008, **320**, 664.
141. L. K. Mahal, K. J. Yarema and C. R. Bertozzi, *Science*, 1997, **276**, 1125.
142. W. Liu, A. Brock, S. Chen, S. Chen and P. G. Schultz, *Nat. Methods*, 2007, **4**, 239.
143. S. Boonyarattanakalin, S. E. Martin, Q. Sun and B. R. Peterson, *J. Am. Chem. Soc.*, 2006, **128**, 11463.
144. I. Chen, M. Howarth, W. Lin and A. Y. Ting, *Nat. Methods*, 2005, **2**, 99.
145. B. Kellam, P. A. De Bank and K. M. Shakesheff, *Chem. Soc. Rev.*, 2003, **32**, 327.
146. R. F. Fakhrullin, L. V. Shlykova, A. I. Zamaleeva, D. K. Nurgaliev, Y. N. Osin, J. García-Alonso and V. N. Paunov, *Macromol. Biosci.*, 2010, **10**, 1257.

147. J. T. Wilson, V. R. Krishnamurthy, W. Cui, Z. Qu and E. L. Chaikof, *J. Am. Chem. Soc.*, 2009, **131**, 18228.
148. J. W. Morse, R. S. Arvidson and A. Lüttge, *Chem. Rev.*, 2007, **107**, 342.
149. S. Onclin, B. Jan Ravoo and D. N. Reinhoudt, *Angew. Chem. Int. Ed.*, 2005, **44**, 6286.
150. S. H. Yang, E. H. Ko and I. S. Choi, *Angew. Chem. Int. Ed.*, 2011, **50**, 6115.
151. X. Chen and S. S. Mao, *Chem. Rev.*, 2007, **107**, 2891.
152. S. H. Yang, E. H. Ko and I. S. Choi, *Langmuir*, 2012, **28**, 2151.
153. M. B. Dickerson, S. E. Jones, Y. Cai, G. Ahmad, R. R. Naik, N. Kröger and K. H. Sandhage, *Chem. Mater.*, 2008, **20**, 1578.
154. Y. V. Nancharaiah, M. Rajadurai and V. P. Venugopalan, *Environ. Sci. Technol.*, 2007, **41**, 2617.
155. E. H. Ko, Y. Yoon, J. H. Park, S. H. Yang, D. Hong, K.-B. Lee, H. K. Shon, T. G. Lee and I. S. Choi, *Angew. Chem. Int. Ed.*, 2013, **52**, 12279.
156. V. Zeleňák, V. Hornebecq, S. Mornet, O. Schäf and P. Llewellyn, *Chem. Mater.*, 2006, **18**, 3184.
157. S. H. Yang, S. M. Kang, K.-B. Lee, T. D. Chung, H. Lee and I. S. Choi, *J. Am. Chem. Soc.*, 2011, **113**, 2795.
158. J. Lee, S. H. Yang, S.-P. Hong, D. Hong, H. Lee, H.-Y. Lee, Y.-G. Kim and I. S. Choi, *Macromol. Rapid Commun.*, 2013, **34**, 1351.
159. I. Drachuk, O. Shchepelina, M. Lisunova, S. Harbaugh, N. Kelley-Loughnane, M. Stone and V. V. Tsukruk, *ACS Nano*, 2012, **6**, 4266.

CHAPTER 5

Characterization Techniques of Living Cells Encapsulated with Nanomaterials

MUSTAFA ÇULHA

Department of Genetics and Bioengineering, Faculty of Engineering and
Architecture, Yeditepe University, Ataşehir, Istanbul, 34755, Turkey
Email: mculha@yeditepe.edu.tr; mculha2@gmail.com

5.1 Introduction

The use of molecules and polymers of biological origins to construct novel
biomolecular structures with improved properties or to replace existing en-
vironmentally unfriendly materials is a major area of research. Although a
tremendous amount of success in the construction of such materials has
been achieved in recent years, reaching the level of complexity and function
of living systems is far from where we stand at the moment. Another approach is to use existing living systems to construct novel tools and devices.
In this context, the modification of living cells of micro-organisms and eukaryotic cells is pursued.[1–3]

Since a foreign material is to be placed onto the cell surface, it is critical to
understand the nature of interactions between the living system and the
material. A microscopic technique can help to visualize the localization of
the materials in the cell or on the cell surface. However, this information
alone may not be enough to understand the molecular nature of the interactions between molecular structures and materials placed on the cell surface. Therefore, the use of a spectroscopic technique is necessary. In

RSC Smart Materials No. 9
Cell Surface Engineering: Fabrication of Functional Nanoshells
Edited by Rawil F Fakhrullin, Insung S Choi and Yuri Lvov
© The Royal Society of Chemistry 2014
Published by the Royal Society of Chemistry, www.rsc.org

particular, a molecular spectroscopic technique can quickly provide very valuable information about the molecular environment of the material placed on or in the cell. In this chapter, the characterization techniques used in living-cell modifications are discussed. The techniques are grouped into three categories as imaging, spectroscopic, and other techniques and summarized by giving examples from the literature.

5.2 Imaging Techniques

Since the aim is to alter cell surface functionality, the foreign materials, such as nanoparticles and polymers, are targeted to be placed onto the cell surface. Therefore, the imaging is performed to visualize the location of the materials placed on the surface of a cell. There are two important parameters for decision making for which imaging technique should be used for characterization. These parameters are the resolution limit of the technique and the nature of the materials such as a polymer or a nanomaterial attached to the cell surface. In a microscopic technique, a light beam, electron beam or physical probe is used for imaging. When light is used as the probe, it is called optical microscopy, when an electron beam is used, it is called electron microscopy (EM) and when a tip or tapered fiber optic is used, it is called scanning-probe microscopy (SPM).

A light microscope can be used to gather spectroscopic information. For example, fluorescence or Raman scattering are commonly collected using an optical microscope. In conventional light microscopy, all scattered light originating, regardless whether from the focal plane and out-of-focus plane, is collected. However, it is possible to collect scattered light for imaging from the focal plane. In confocal microscopy, the image is collected from the focal plane. Only light collected from the focal plane is used in imaging and out-of-plane light is removed. Therefore, most of the collected image originates from the focal plane in confocal microscopy. Since light is used as a probe, the spatial resolution is mostly defined by the diffraction of the light. The spatial resolution can given with a formula of $S = 0.61\ \lambda/\text{NA}$, where λ is the wavelength of the light used for imaging and NA is the numerical aperture of the objective used. Although this formula gives the theoretical limits of how a light beam can be focused on an area, the diameter of the light beam impinging onto the sample from the objective could be greater than 1 μm. Therefore, it is important to realize the size of the material placed onto the cell surface. If it is a nanometer-sized material without fluorescing or plasmonic property, it cannot be visualized with an optical microscope.

Fluorescence spectroscopy is a molecular spectroscopic technique based on the excitation of a molecule at a certain wavelength and recording the emitted light with a detector. Not every molecule has a fluorescence property and it is strongly related to the chemical structure of a molecule varying from molecule to molecule. The florescence property of molecules is utilized in fluorescence and confocal microscopies. Fluorescence microscopy can simply be defined as optical microscopy with a proper light source for excitation and a detector to record the fluorescence emission but confocal micro-scopy has a confocality property and a laser is used as a monochromatic light source. It is also possible to collect Raman scattering in the confocal mode.

Plasmonics is a term used to define the light-metal interactions.[4,5] Upon interaction of light with metallic NPs, two fundamental processes take place: absorption and scattering of a portion of impinging light simultaneously. While the absorbed light causes the formation of surface plasmons, a magnetic field, on the surface of the metal structure, the scattered light can be used for imaging.[5] Since noble metals, such as gold, and silver absorb the light in the visible region of the spectrum, they are widely used in sensing, imaging and therapy benefiting from their extraordinary plasmonic prop-erties.[6,7] When NPs with plasmonic properties such as gold and silver are used, they and their aggregates can be visualized using an ordinary light microscope or confocal microscope.

In order to overcome the diffraction-limited resolution problem in light-based microscopic techniques, electron- and probe-based microscopic techniques were developed. For example, with SEM, down to subnanometer resolution is possible and with TEM atomic-level resolution can be easily achieved.

In probe-based imaging techniques, a probe such as a tip is used to scan a defined surface area. Atomic force microscopy (AFM) and scanning tunnel-ing microscopy (STM) are the most commonly employed modes of this group of microscopic techniques. AFM provides topographic information down to a few nanometer resolution and the obtained image offers similar infor-mation to the images obtained with SEM. However, SEM is much faster to acquire the image compared to AFM but it may not be suitable to image biological samples due to the destructive nature of the electron beam. The STM is based on quantum tunneling, and a conductive tip is scanned across the surface by applying a bias voltage. Although the technique is mentioned here, it is a difficult technique for the applications addressed here since it requires an extremely clean environment and it is performed in a vacuum.

Another mode of scanning-probe microscopy called near-field scanning optical microscopy (NSOM) combining optical microscopy and scanning probe mode is also available.[8] In NSOM, a fiber optic tip tapered down to about 30 nm coated with aluminum is used as a scanning probe. Since light cannot escape from the tapered end because the diameter of a fiber optic is much smaller than the wavelength of the light, an electromagnetic field is formed at the end of the fiber optic tip. This magnetic field is used to excite the molecules either for fluorescence Raman scattering measurements. However, NSOM is a difficult technique for probing the surfaces with high roughness due to the fact that the fiber optic tip should be located as close as a few nm to the sample surface but a spatial resolution down to 20 nm can be achieved.[9] Tip-enhanced Raman scattering (TERS) is also becoming popular in recent years for characterization of not only polymeric surfaces but also biological structures.[10] In this technique, AFM is combined with a confocal Raman microscopy, where laser light is focused onto the AFM tip coated with gold. Molecular level information from a sub-100 nm region can be obtained with the technique.

The type of material used is the second parameter for decision making for which spectroscopic technique should be used for characterization. For example, if it is a polymer with a fluorescing property, fluorescence based microscopic techniques such as fluorescence and confocal microscopy should be used to confirm the coating. Upon attachment of a nanomaterial with a high density, either SEM or TEM should be used for the localization. The visualization of AuNPs and AgNPs attached onto *S. cerevisiae* cells and *T. asperellum* conidia with TEM can be given as an example.[11] For clarity, the steps of the coating procedure of yeast cells are provided in Figure 5.1. During the first step (1), the intact cells are coated with a triple layer of either PAH/PSS/PAH or BSA/DNA/BSA. In the second step (2), the polyelectrolyte-coated cells are further coated with either AuNPs or AgNPs and finally in the last step (3) the polyelectrolyte and metal NP-coated cells were treated with a double layer of PAH/PSS or BSA/DNA. Although a color change upon coating the cells with AuNPs or AgNPs due to the scattered light from the noble metal NPs is observed, indicating the presence of metal NPs in the suspension containing cells, it is necessary to confirm the metal NPs in this particular case, and nanomaterials in general, are on the cell surface and cannot be released in the suspension or leaked into the cells. Figure 5.2 shows the TEM images of a *S. cerevisiae* cell coated with AuNPs and AgNPs. As seen from the TEM images, AuNPs or AgNPs are clearly attached onto the

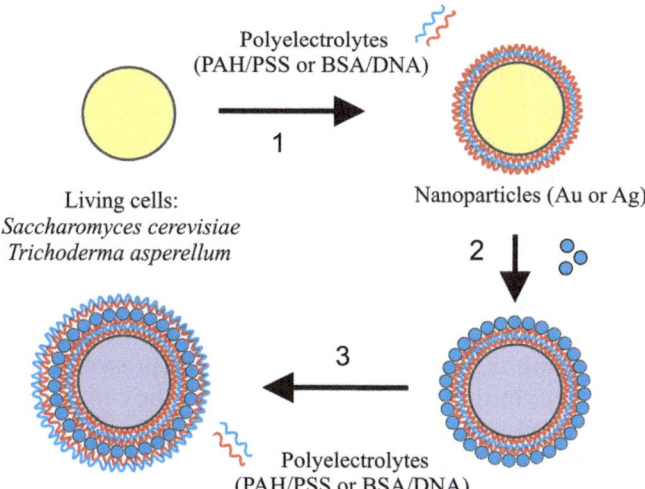

Figure 5.1   Schematic representation of the coating of *S. cerevisiae* cells and *T. asperellum conidia* with the polyelectrolyte shells and AuNPs or AgNPs. (Reprinted with permission from ref. 11. Copyright 2009 the American Chemical Society.)

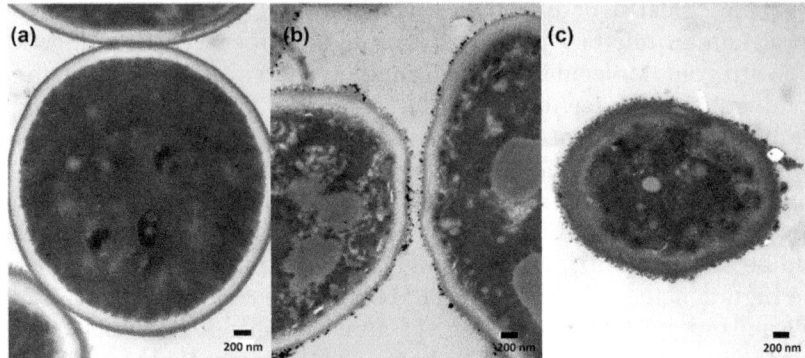

Figure 5.2   TEM images of bare *S. cerevisiae* cells (a), *S. cerevisiae* cells coated with
                AuNPs (b) and AgNPs (c).
                (Reprinted with permission from ref. 11. Copyright 2009 the American
                Chemical Society.)

cell surface. Further, energy-dispersive X-ray spectroscopy (EDS or EDX)
confirming the presence of NPs can be performed. Most SEM and TEM
microscopic systems are usually equipped with an EDS or EDX detector. For
example, the presence of magnetic NPs immobilized onto unicellular algae
and yeast cells was verified with EDX analysis during TEM imaging.[12,13] EDX
analysis can be performed as a line analysis or an area on the cell surface can
be mapped to obtain information about the elemental distribution, and thus
the material distribution. Figure 5.6c shows the line analysis of a yeast cell
coated with SiO_2SH showing the presence of silicon and sulfur on the cell
surface.

In electron-based scanning microscopies such as SEM and TEM, the
density of the material to be imaged is critical. For example, most nanoparticles such as noble metals and CNTs can be visualized with TEM or SEM
but polymers and biological materials with low density may not be imaged
with these techniques since they are transparent to electrons. In such cases,
atomic force microscopy (AFM) can be used to visualize surface topography
down to nanometer resolution. Figure 5.3 shows the comparison of the AFM
images of before and after coating E. coli cells with AgNPs and AuNPs as an
example of AFM imaging.[14] A line analysis (green and red lines and corresponding graphs on the right side of the Figure 5.3) is performed to
measure the topographic roughness to evidence the presence of noble metal
NPs. As can be seen, in the absence of the NPs on the cell surface, both the
lines are smoother while it is coarser upon attachment of NPs.

Other materials such as proteins have also been used to coat the living
cells. In a recent report by Drachuk *et al.*, *Saccharomyces cerevisiae* cells were
coated with fibroin (silk) using a layer-by layer approach.[15] It was claimed that
the cells were still viable after the coating process. The surface modification of
the cells was also verified with AFM. Another technique used in that study for
the characterization of the cell coating was fluorescence-spectroscopy-based

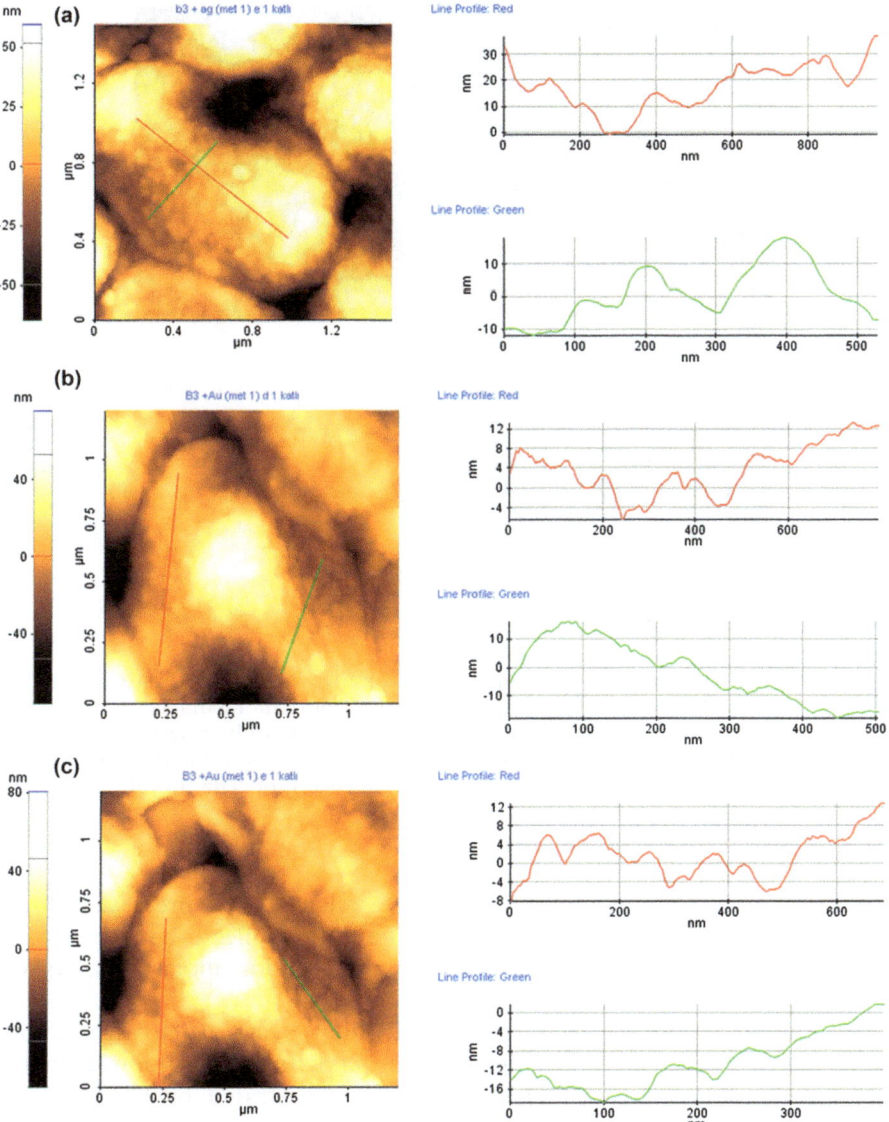

Figure 5.3 AFM image of *E. coli* before (a) and after coating with PAH/AgNPs/PAH (b) and PAH/AuNPs/PAH (c).
(Reprinted from ref. 14 by permission of Springer).

confocal microscopy. Although the same resolution level achieved with electron and probe based microscopies cannot be achieved, confocal microscopy may provide very useful information. Figure 5.4 shows the comparison of the confocal microscopy images of the cells for the viability verification. For this, the fibroin was labeled with a fluorescence dye (Alexa Fluor 532) to visualize the coating on the cells. Then, the cells coated with

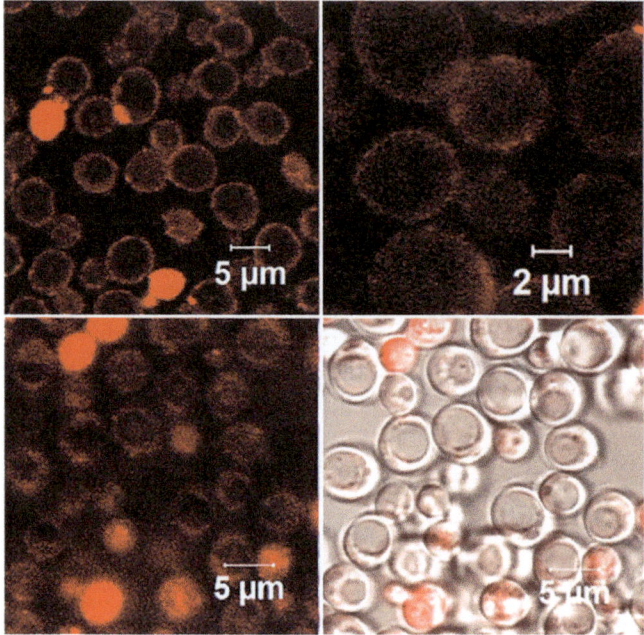

Figure 5.4   Confocal microscopy images of fibrion encapsulated yeast cells before
incubation (top) and after 12 h (bottom) in media containing 2%
galactose to for yEGFP expression. Bottom right images show over-
lapped image of fluorescent and optical images to identify the cells.
(Reproduced from ref. 15 by permission of The Royal Society of
Chemistry).

fibroin labeled with fluorescence dye were allowed to express yEGFP to
confirm the viability of the coated cells. The fluorescence originating from
the periphery or center of the cells provides information about the status of
the coating and the cell viability.

In another example, coating of individual yeast cells with thiol-
functionalized silica (SiO_2) shells and the attachment of fluorescein or
streptavidin onto the SiO_2 SH shells was demonstrated by Yang *et al.*[3]
Figure 5.5 shows the coating process of yeast cells followed by attachment
of fluorescein or streptavidin onto the SiO_2 SH coating. Figure 5.6 shows the
SEM, TEM and confocal microscopy images of the coated cells and is a good
example of combining both electron microscopies and fluorescence confocal
microscopy.

In another interesting study, hemagglutinating virus was coated with
the biocompatible polymer hyaluronic acid (HA). In that study, TEM was
used to visualize the nanometer size layer of HA.[16] There are several
examples in the literature that use the above-mentioned techniques for the
characterization of living cells; and the reader may consult to these reports
for details.[17–19]

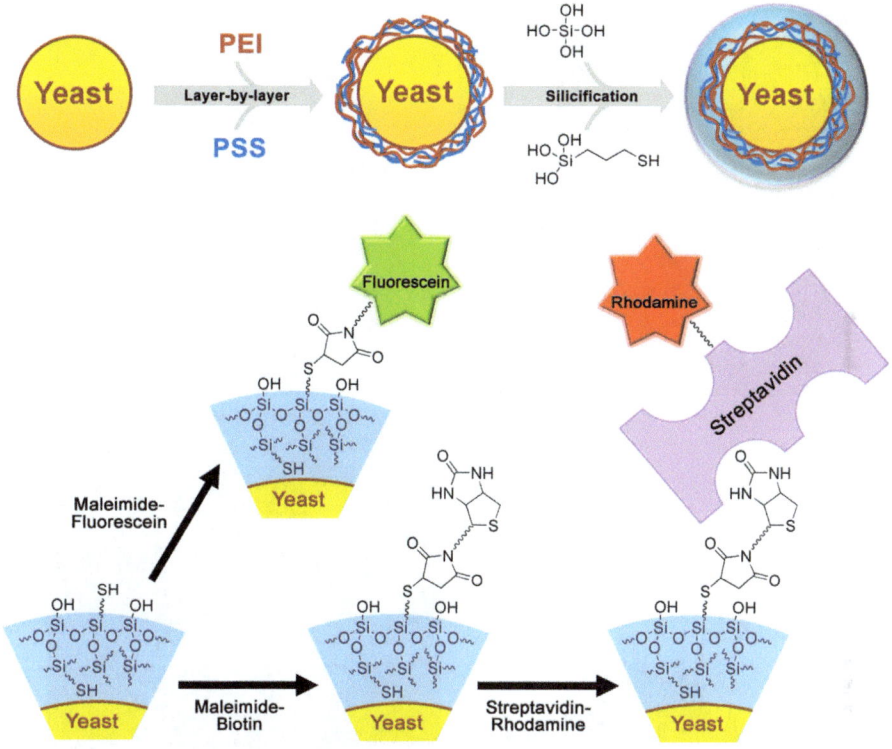

Figure 5.5 Coating of yeast cells with a silica layer before attachment of fluorescein or streptavidin onto SiO_2 SH coating.
(Reprinted from ref. 3 by permission of John Wiley & Sons, Inc.)

5.3   Spectroscopic Techniques

Spectroscopic techniques can provide very specific information about the chemical environment of materials used to coat the cells. The choice of spectroscopic techniques is mostly driven by the type of the material used to coat the cell surface. Among these techniques, Raman Spectroscopy and its modes such as SERS are the most convenient ones. For example, when AgNPs or AuNPs are used for the coating, SERS is the technique of choice. This is due to the fact that a nanostructured noble metal surface for the enhanced Raman scattering is an obligatory condition.[20–21] The use of SERS for a variety of applications is enthusiastically pursued in several fields.[22] The reader can find a wealth of information about the applications of SERS from medicine to materials science in the literature.[23–25]

Raman spectroscopy is a vibrational spectroscopy and considered as complementary to IR spectroscopy since both arise from the same fundamental molecular processes. Although both are very useful characterization techniques, IR spectroscopy is traditionally more established compared

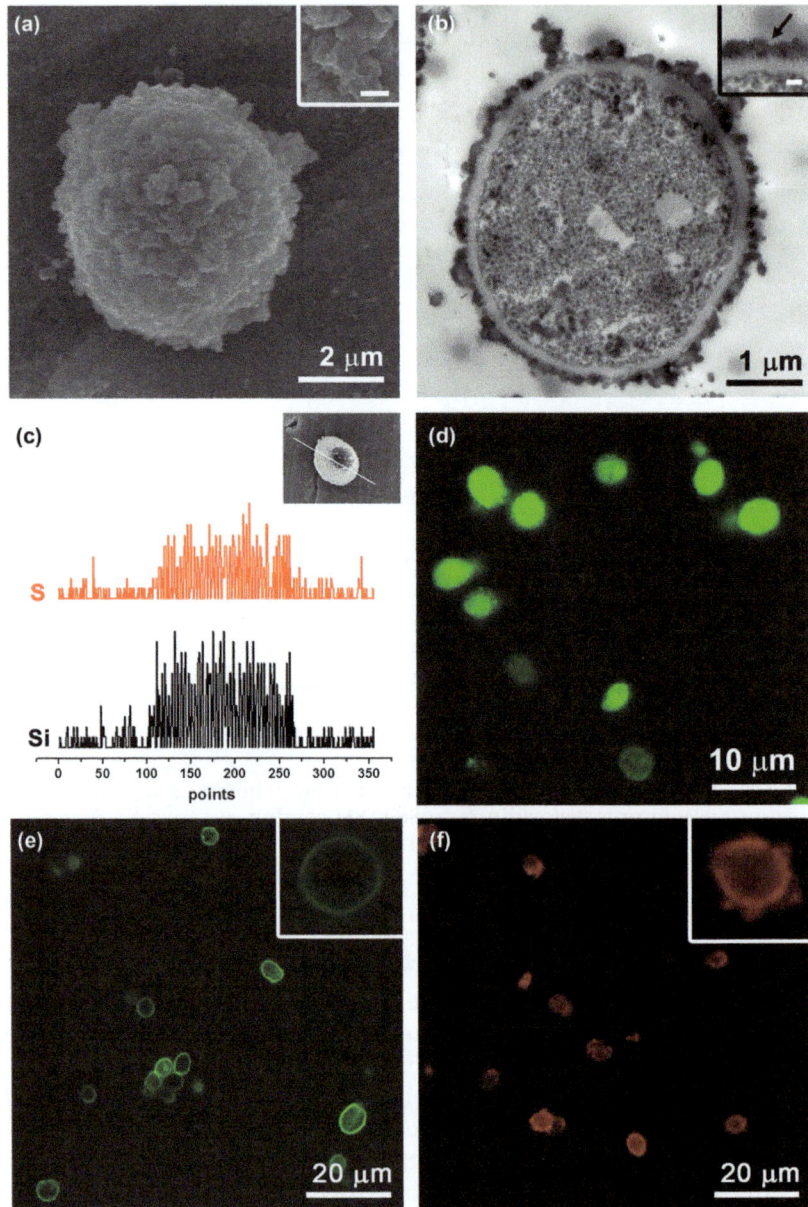

Figure 5.6 (a) SEM image of a yeast cell coated with SiO$_2$ SH, (b) TEM image of microtome-sliced yeast coated with SiO$_2$ SH, (c) EDX line profiles of a coated yeast cell indicating the presence of silicon and sulfur, (d) confocal microscopy image of coated yeast cells stained with fluorescein diacetate for viability test. Confocal microscopy images of (e) fluorescein attached and (f) rhodamine-streptavidin attached yeast cells. Insets show magnified images.
(Reprinted from ref. 3 by permission of John Wiley & Sons, Inc.)

to Raman spectroscopy. Since IR spectroscopy is an absorption-based technique, significantly high absorption of light in the IR region by water limits its use for biological applications. Although the developments in instrumentation have helped to reduce the effect of water in IR applications, it is still not often preferred for biological applications. However, the presence of water in a sample has a limited effect on Raman scattering. However, it is inherently very weak, its use has been delayed until the significant development of laser and detector technology in the last two decades. It is now possible to use Raman spectroscopy for routine applications without significantly sacrificing the sensitivity.

In the early 1970s, enhanced Raman scattering from pyridine molecules on a silver electrode was observed and it was explained with the increased concentration effect on the electrode surface.[26] In the mid-1970s, the discovery that enhanced Raman scattering was due to two possible mechanisms: chemical and electromagnetic effects for the molecules brought to the surface of the noble metals such as gold and silver, was the birth of so-called surface-enhanced Raman scattering (SERS).[21,22] In the following years, tremendous effort was devoted to better understanding the mechanism and the potential of this new technique. The fact that the increased sensitivity of SERS with an enhancement up to 10^{14}, even competing with fluorescence, excited the scientific community to extensively research the potential of the technique from single molecule to living cells.[27-29]

SERS has also been used to investigate the molecular environment of AgNPs and AuNPs placed onto yeast cell surface.[11] Figure 5.7 shows the comparison of the SERS spectra of single *T. asperellum* conidia coated

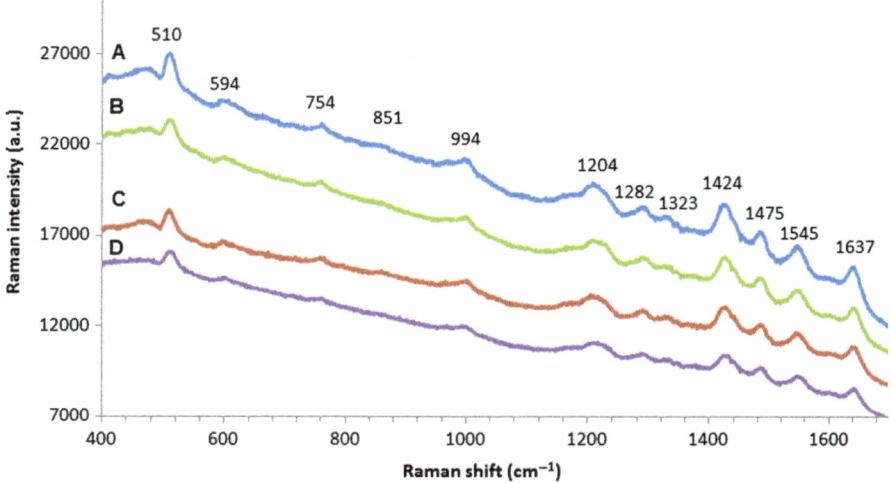

Figure 5.7 SERS spectra of the single fungi cell (A) PAH/PSS/AuNPs, (B) BSA/DNA/AuNPs, (C) PAH/PSS/AgNPs, (D) BSA/DNA/AgNPs. Conditions: laser power: 30 mW; exposure time 10 s; accumulation: 1.

with PAH/PSS/AuNPs (A), BSA/DNA/AuNPs (B), PAH/PSS/AgNPs (C), and BSA/DNA/AgNPs (D). Although the polymeric material varies, the SERS spectra obtained under four different conditions are surprisingly similar. This indicates that the AuNPs and AgNPs (or their aggregates are sharing similar molecular environments. This can be explained with the affinity of the noble metal NPs to the similar functional groups such as $-SH$, $-NH_2$ and $-COOH$. In order to identify the possible molecular structures used to coat the cells into contact with the noble metal NPs, the SERS spectra from individual molecules used for cell coating and cells were obtained. Figures 5.8A and B show the comparison of the components of two polymer/cell groups. The inspection of all these spectra indicate that the polyelectrolytes and noble-metal NPs are interacting from different locations on the living cell.

The attachment of AuNPs and AgNPs on bacterial cells with a layer-by-layer method was also investigated.[14] In that study, *E. coli* and *S. cohnii*, a Gram-negative and Gram-positive bacterium, respectively, are used as model micro-organisms to coat them layer-by-layer employing poly (allylamine hydrochloride) (PAH), a positively charged polymer, and citrate reduced AuNPs or AgNPs. Figure 5.9 shows the coating procedure of bacteria cells (A), color of the suspension-coated bacteria (B) and fluorescence microscopy image of *E. coli cell* coated with fluorescein-isothiocyanate-labeled PAH (FITC-PAH) (C). As can be seen, the color of the suspension of AuNP or AgNP coated is different since the AuNPs and AgNPs scatter light at different wavelengths. The color change may give an indication of the presence of plasmonic AuNPs or AgNPs but it is not enough for a solid conclusion if the NPs are attached to the cell or are free in the suspension, and SEM, TEM or AFM analyses should be performed to confirm the presence of the NPs on the cell surface. Figures 5.10 and 5.11 show the SEM images AuNP and AgNP coated *E. coli* cells. As seen from the SEM images, AuNPs or AgNPs and/or their aggregates are visible on the cell surface.

The understanding of molecular moieties that in interaction with the AuNPs/AgNPs or their aggregates on the bacterial cell surface, can help to develop novel applications and tools of the future. In order to understand this, the SERS spectra of *S. cohnii* upon coating with AgNPs and AuNPs were compared. As seen in Figure 5.12, there is great similarity in both spectra. The only difference is the higher intensity of the bands on AgNP-coated bacteria cells. The similarity of the spectral pattern between two spectra indicate that the AuNPs and AgNPs are interacting with the similar molecular moieties, perhaps carrying similar functional groups such as $-SH$ and $-NH_2$. The difference in intensity of the bands is due to the difference in their plasmonic properties, where AgNPs show better plasmonic properties due to their larger size (60 nm compared to 13-nm AuNPs) and higher surface roughness.

The comparison of the SERS spectra AgNP coated *E. coli* and *S. cohnii* also provides valuable information about the possible interaction points of the NPs on the cell surface. Although Gram + and − bacteria cell wall structures

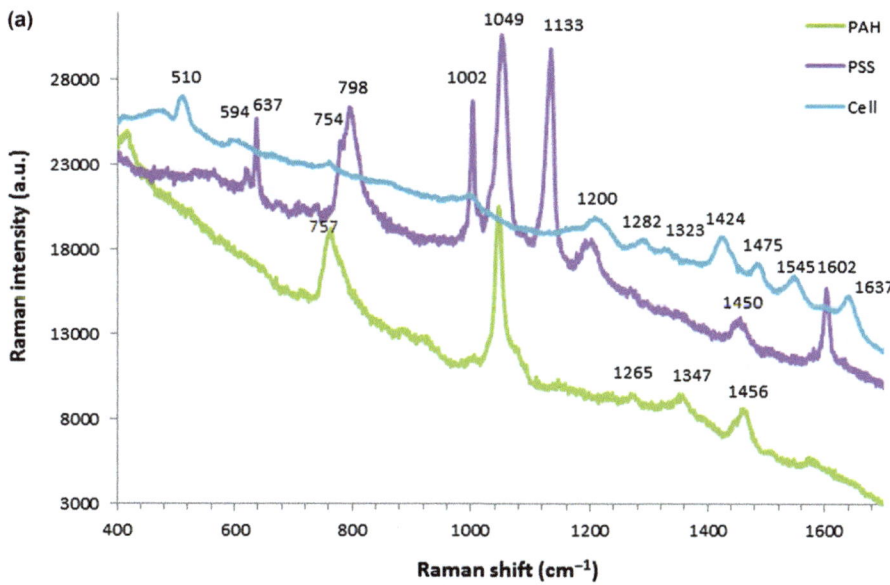

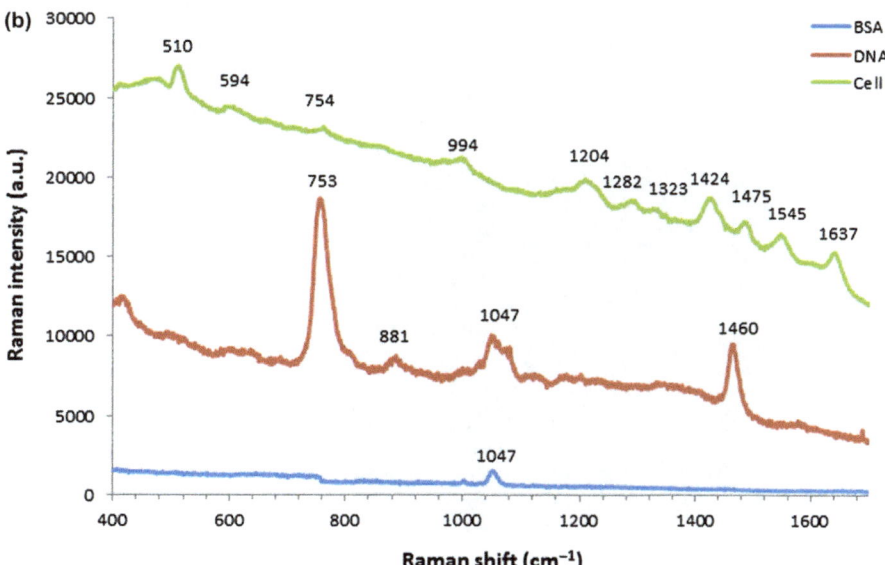

Figure 5.8   SERS spectra of two groups of polyelectrolytes and cells; (A) PAH/PSS/Cell and (B) BSA/DNA/Cell.

are rather different, the molecular components forming the structure are very similar and the only variation is in the combination of biomacromolecules. As can be seen from Figure 5.13, the SERS spectra of two different

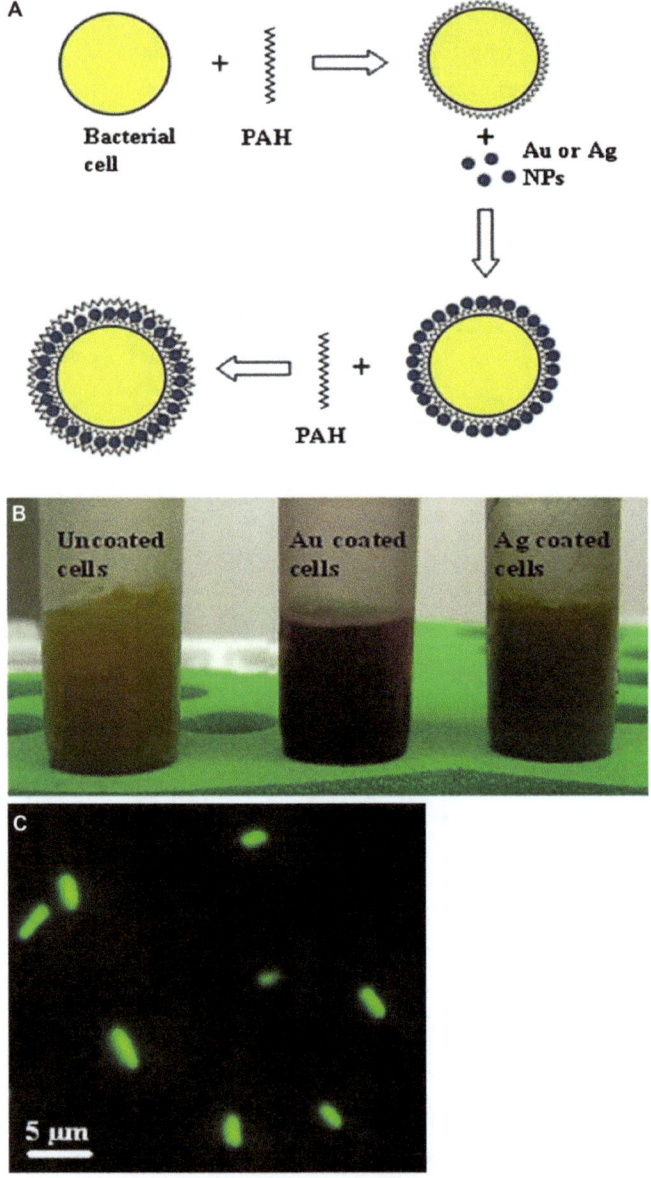

Figure 5.9   Schematic illustrations of coating process (a), color change of the coated cells with AuNPs and AgNPs (b), and fluorescence microscopy image of FITC-PAH-coated E. coli cells (c).
(Reprinted from ref. 14 by permission of Springer).

bacteria coated with AgNPs, the spectral pattern is rather similar, suggesting that the AgNPs are interacting with the similar biomacromolecular moieties on the cell surface.

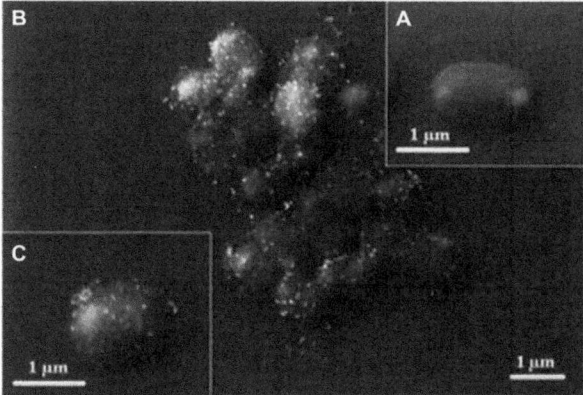

Figure 5.10 SEM images of uncoated *E. coli* (A) *E. coli* coated with PAH/AuNPs/PAH (B) and single *E. coli* cell coated with PAH/AuNPs/ PAH at high magnification (C).
(Reprinted from ref. 14 by permission of Springer).

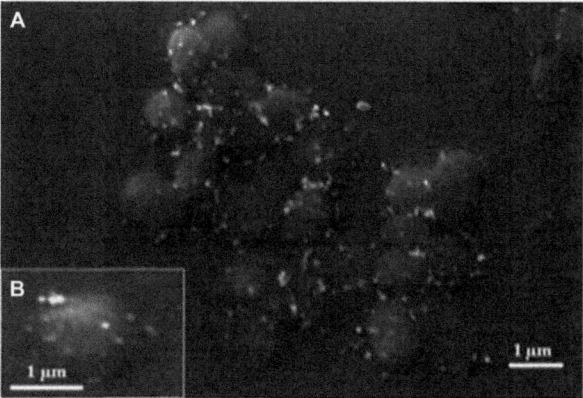

Figure 5.11 SEM images of coated *E. coli* with PAH/AgNPs/ PAH (A) and coated *E. coli* with PAH/AgNPs/PAH at high magnification (B).
(Reprinted from ref. 14 by permission of Springer).

5.4 Other Techniques

The other two important approaches worth mentioning to monitor the change on the micro-organism cell surface are the measurements of cell surface charge and mass change. The first can easily be monitored using a zetasizer analyzer. The latter can be estimated by using a quartz microbalance (QCM).

As an electrical phenomenon, the zeta potential (ζ) forms at an interface due to formation of an electrical double layer composed of ions and counterions. The nature of the formed double layer strongly depends on solvent and the surface properties. The zeta potential cannot be measured directly

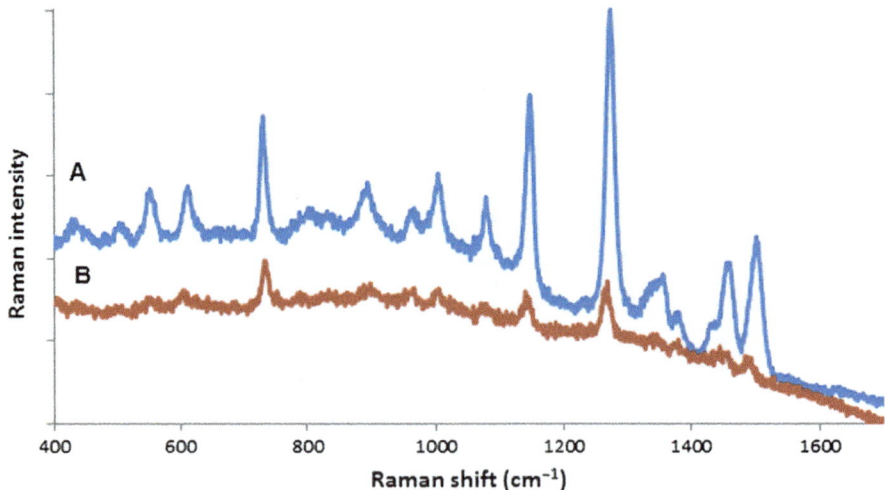

Figure 5.12   SERS spectra of the *S. cohnii* coated with PAH/Ag/PAH (A), and PAH/
Au/PAH (B).
(Reprinted from ref. 14 by permission of Springer).

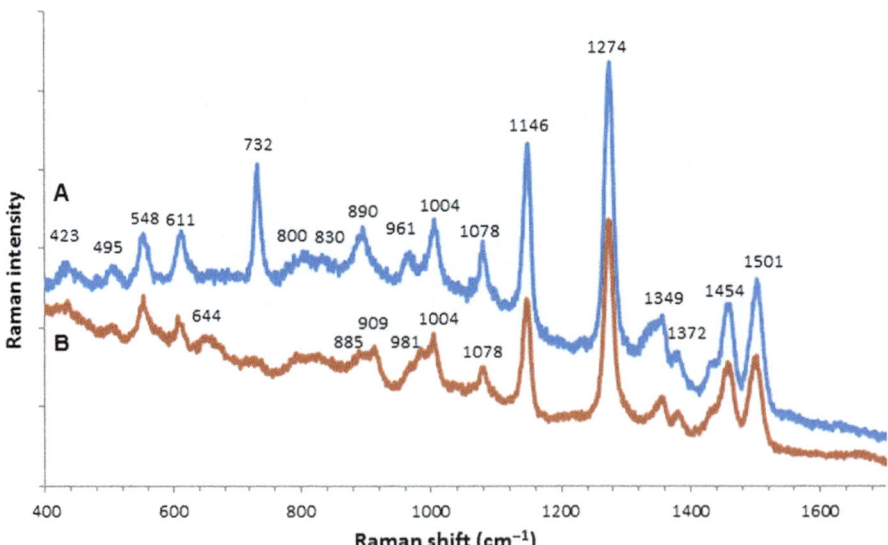

Figure 5.13   Comparison of SERS spectra obtained from *S.cohnii* (A), and *E.coli*
(B) coated with PAH/AgNPs/PAH.
(Reprinted from ref. 14 by permission of Springer).

and it can only be calculated from experimentally predicted electrophoretic
mobility or dynamic electrophoretic mobility. In a colloidal suspension, a
double electrical layer composed of ions on the surface of the colloidal
particles and the counterions are attracted to the surface from the liquid

media. The nature of this electrical double determines the stability of the colloidal particles in the liquid media. The higher the zeta potential the higher the stability of the particles in the suspension. The zeta potential can be measured from the mobility of particles by monitoring the frequency or phase shift of a laser beam impinging onto colloidal suspension as particles move around using Smoluchowski theory.[30] The solvent viscosity used to prepare the suspension and dielectric permittivity should be known for the calculation of zeta potential.

The change in surface potential of *B. subtilis* spores was monitored as their surface was modified with a cationic and anionic polyelectrolyte.[31] The native *B. substilis* spores had a negative surface potential (~ -17 mV) in water at pH 6.8. When the spore surface was coated with poly(dimethyldiallyl ammonium chloride) (PDDA, a positively charged polyelectrolyte, the zeta potential shifted to about 20 mV. When these PDDA-coated spores were coated with sodium poly(styrene sulfonate) (PSS a negatively charged polyelectrolyte), the zeta potential shifted to about -65 mV. The periodic shift between negative and positive layers continues as the number of polyelectrolyte added onto the top of each other.

A quartz crystal microbalance (QCM) is an instrument that is composed of an electrode sensitive to small mass changes. The working principle of a QCM is explained with the piezoelectric effect. Certain materials such as quartz crystal can generate potential upon application of a physical strain on them. In a similar way, an applied voltage can cause physical strain on the material. The alternative potential application results with the oscillation of the quartz with a certain frequency. When a material adheres to the quartz electrode surface, it causes a shift in the frequency, which can be monitored. Note that similar to other transducers used in sensing, a receptor phase or molecular structure should be placed for selective capturing target molecule or molecular structure since QCM responds unselectively to all mass changes.

The use of QCM was first demonstrated by Sauerbrey.[32] The earlier applications of QCMs were to monitor the thickness of the material deposited onto the quartz surface in air. However, it was later realized that the QCMs can be used in liquid media.[33] In the following years, several reports demonstrated the applicability of QCMs in biological sensing applications and even in nanomedicine more recently.[34-36] In living-cell modification applications, positively or negatively charged polymeric materials are simultaneously deposited onto the cell surface. When integration of a nanomaterial into the cell surface is desired, it is sandwiched in between the polymeric layers. The thickness of the layer during a layer-by-layer deposition process can be monitored with a QCM. There are several reports employing QCM for the estimation of film thickness on a cell surface.[37-39] Note that the material is deposited onto the QCM surface not to the cell surface and it is used to estimate the material thickness onto the cell surface. However, mimicking the cell surface may help to obtain more realistic results.

5.5   Conclusions

In this chapter, a summary of techniques used for characterization of en-capsulated cells is provided. The rationale behind the utility of the techniques is discussed and some examples from the literature are given. While em-ploying the imaging and spectroscopic techniques, it is important to under-stand both the techniques and material used for coating. The right choice of the characterization techniques will not only help to obtain more reliable information but also will enhance the value of the study. Although all imaging and characterization techniques used in the current studies are conventional techniques, it is also possible to employ unconventional techniques such as NSOM and TERS, which may provide additional valuable information to understand the nature of the interaction of the materials placed onto the living cell surface.

Acknowledgement

The author acknowledges the financial support of The Scientific and Tech-nological Counsel of Turkey (TUBITAK) and Yeditepe University.

References

1. R. F. Fakhrullin and Y. M. Lvov, *ACS Nano*, 2012, **6**, 4557.
2. A. I. Zamaleeva, I. R. Sharipova, A. V. Porfireva, G. A. Evtugyn and R. F. Fakhrullin, *Langmuir*, 2010, **26**(4), 2671–2679.
3. S. H. Yang, E. H. Ko, Y. H. Jung and I. S. Choi, *Angew. Chem.*, 2011, **123**(1), 6115.
4. E. Ozbay, *Science*, 2006, **311**, 189.
5. S. A. Maier, *Plasmonics: Fundamentals and Applications*, Springer, New York, 2007.
6. S. Link and M. A. El-Sayed, *Annu. Rev. Phys. Chem.*, 2003, **54**, 331.
7. I. H. El-Sayed, X. Huang and M. A. El-Sayed, *Cancer Lett.*, 2006, **239**, 129.
8. E. Betzig, A. Lewis, A. Harootunian, M. Isaacson and E. Kratschmer, *Biophys. J.*, 1986, **49**, 269.
9. D. W. Pohl, W. Denk and M. Lanz, *Appl. Phys. Rev. Lett.*, 1984, **44**, 651.
10. T. Deckert-Gaudig and V. Deckert, *Chem. Chem. Phys.*, 2010, **12**, 12040.
11. R. F. Fakhrullin, A. I. Zamaleeva, M. V. Morozov, D. I. Tazetdinova, F. K. Alimova, A. K. Hilmutdinov, R. I. Zhdanov, M. Kahraman and M. Culha, *Langmuir*, 2009, **25**, 4628.
12. R. F. Fakhrullin, L. V. Shlykova, A. I. Zamaleeva, D. K. Nurgaliev, Y. N. Osin, J. García-Alonso and V. N. Paunov, *Macromol. Biosci.*, 2010, **10**, 1257–1264.
13. J. García-Alonsoa, R. F. Fakhrullin and V. N. Paunov, *Biosens. Bioelec-tron.*, 2010, **25**, 1816.
14. M. Kahraman, A. I. Zamaleeva, R. F. Fakhrullin and M. Culha, *Anal. Bioanal. Chem.*, 2009, **395**, 2559.

15. I. Drachuk, O. Shchepelina, S. Harbaugh, N. Kelley-Loughnane, M. Stone and V. V. Tsukruk, *Small*, 2013, **9**(18), 3128.
16. T. Okada, K. Uto, M. Sasai, C. M. Lee, M. Ebara and T. Aoyagi, *Langmuir*, 2013, **29**, 7384.
17. S. Mansouri, Y. Merhi, F. M. Winnik and M. Tabrizian, *Biomacromolecules*, 2011, **12**, 585.
18. R. Kempaiah, S. Salgado, W. L. Chung and V. Maheshwari, *Chem. Commun.*, 2011, **47**, 11480.
19. S. A. Konnova, I. R. Sharipova, T. A. Demina, Y. N. Osin, D. R. Yarullina, O. N. Ilinskaya, Y. M. Lvov and R. F. Fakhrullin, *Chem. Commun.*, 2013, **49**, 4208.
20. D. L. Jeanmaire and R. P. Van Duyne, *J. Electroanal. Chem.*, 1977, **84**, 1.
21. M. G. Albrecht and J. A. Creighton, *J. Am. Chem. Soc.*, 1977, **99**, 5215.
22. K. Kneipp, A. S. Haka, H. Kneipp, K. Badizadegan, N. Yoshizawa, C. Boone, K. E. Shafer-Peltier, J. T. Motz, R. R. Dasari and M. S. Feld, *Appl. Spectrosc.*, 2002, **56**, 150.
23. Mustafa Culha, Brian Cullum, Nickolay Lavrik and Charles K. Klutse, *J. Nanotechnol.*, 2012, Article ID 971380.
24. K. Hering, D. Cialla, K. Ackermann, T. Dörfer, R. Möller, H. Schneidewind, R. Mattheis, W. Fritzsche, P. Rösch and J. Popp, *Anal. Bioanal. Chem.*, 2008, **390**, 113.
25. S. D. Hudson and G. Chumanov, *Anal. Bioanal. Chem.*, 2009, **394**, 679.
26. M. Fleischman, P. J. Hendra and A. J. McQuillan, *Chem. Phys. Lett.*, 1974, **26**, 163.
27. K. Kneipp, Y. Wang, H. Kneipp, L. T. Perelman, I. Itzkan, R. R. Dasari and M. S. Feld, *Phys. Rev. Lett.*, 1997, **78**, 1667.
28. K. Kneipp, A. S. Haka, H. Kneipp, K. Badizadegan, N. Yoshizawa, C. Boone, K. E. Shafer-Peltier, J. T. Motz, R. R. Dasari and M. S. Feld, *Appl. Spectrosc.*, 2002, **56**, 150.
29. J. P. Scaffidi, M. K. Gregas, V. Seewaldt and T. Vo-Dinh, *Anal. Bioanal. Chem.*, 2009, **39**(Gram-3), 1135.
30. M. von Smoluchowski, *Bull. Int. Acad. Sci. Cracovie*, 1903, 184.
31. S. B. Shantanu, G. V. Nalinkanth, D. M. Eby, G. R. Johnson and Y. M. Lvov, *Langmuir*, 2009, **25**(24), 14011.
32. G. Sauerbrey, *Z. Phys.*, 1959, **155**, 206.
33. D. Johannsmann, *Phys. Chem. Chem. Phys.*, 2008, **10**(31), 4516–34.
34. G. N. M. Ferreira, A.-C. da-Silva and B. Tomé, *Trends Biotechnol.*, 2009, **27**(12), 689.
35. A. C. Hunter, *J. Biomed. Nanotechnol.*, 2009, **5**(6), 669.
36. M. Nirschl, F. Reuter and J. Vörös, *Biosensors*, 2011, **1**(3), 70.
37. H. Ai, M. Fang, S. A. Jones and Y. M. Lvov, *Biomacromolecules*, 2002, **3**, 560.
38. N. G. Veerabadran, P. L. Goli, S. S. Stewart-Clar, Y. M. Lvov and D. K. Mills, *Macromol. Biosci.*, 2007, 7, 877.
39. A. I. Zamaleeva, I. R. Sharipova, A. V. Porfireva, G. A. Evtugyn and R. F. Fakhrullin, *Langmuir*, 2010, **26**(4), 2671.

CHAPTER 6

Cytocompatibility and Toxicity of Functional Coatings Engineered at Cell Surfaces

EUGENIA KHARLAMPIEVA* AND VERONIKA KOZLOVSKAYA

University of Alabama at Birmingham, Chemistry Department,
901 14th Street South, Birmingham, Alabama USA, 35294
*Email: ekharlam@uab.edu

The cell surface is very complex and regulates diverse physical and chemical processes. Engineering the cell features by modifying cell surfaces is a powerful tool for manipulating cell properties and controlling interactions between cells and the environment. In this way, a protective environment can be created to prolong cell lifetime and functions; and biological hybrids based on synergistic properties of living cells and nanomaterials can be developed. To be modified, cells are exposed to various types of nano-materials including natural and synthetic polymers, nanoparticles, and small reactive molecules. Modification conditions may require exposure of modified cells to chemical and physical stresses provided through changes in pH, temperature, ionic shocks, or toxic media. Depending on the procedure, various types of cells may react in a different manner. The cell response to the modification processes or their short- or long-term effects are evaluated by measuring cytotoxicity, which is determined by cell death, or the inhibition of cell proliferation or its metabolic activity.

In this chapter we will discuss various methods used in evaluation of the cell toxicity; cytotoxic effects of various nanomaterials on different cells; and

RSC Smart Materials No. 9
Cell Surface Engineering: Fabrication of Functional Nanoshells
Edited by Rawil F Fakhrullin, Insung S Choi and Yuri Lvov
© The Royal Society of Chemistry 2014
Published by the Royal Society of Chemistry, www.rsc.org

cytocompatibility issues of various approaches used for cell surface functionalization. The impact of various physicochemical properties of nanomaterials including the chemical compositions, surface charge and chemistry, dimensions and shape on cytotoxicity will be reviewed. The effect of cell type on their capability to tolerate chemical and physical stresses will be presented.

6.1 Methods for Evaluation of Cell Toxicity

Cell viability is an important parameter in cell and tissue engineering as well as in culture studies to evaluate survival of cells. At the same time, viability may not be a sufficient indicator of biosynthetic capacity of modified cells; a cell may tolerate the treatment but still develop a high stress affecting cell productive functions. Currently, there are a number of assays available to determine cell viability and its functional capacities in culture or in three-dimensional scaffolds. Thus, cell viability and function may be inferred from (1) their ability to divide and proliferate, (2) their morphological integrity and metabolic activity, or (3) the cell response to a certain physicochemical trigger.

6.1.1 Colorimetric Assays

In vitro colorimetric assays provide a simple, reproducible, sensitive and affordable method to assess cell survival or proliferation in the process of cell surface modification. Dyes are often used to assess the viability of functionalized cells using untreated cells as a control group. Fluorescence-based techniques provide alternative methods for viability assays and allow rapid and direct measurements. Here, we will describe the advantages and limitations of several colorimetric probes that are routinely used to assess the viability (cell counting) and the proliferation of cells in order to determine the cytotoxicity potential of a cell treatment and its effect on cell viability, growth, and response.[1] The colorimetric assays can be divided into those that usually probe (1) changes in membrane integrity based on the exclusion of certain dyes or the uptake and retention of others, or (2) metabolic activity of cells.

6.1.1.1 Probing Cell Membrane Integrity

Viable cells have intact membranes and are distinguished by their ability to exclude dyes that easily penetrate dead or damaged cells. Staining of dead cells with propidium iodide has been performed on most cell types. Propidium iodide (PI) is a small molecule of ~ 1 nm size and is normally excluded from cells with an intact plasma membrane; however, it can penetrate cells in which the plasma membrane has been disrupted. The PI stains nucleic acids resulting in bright red fluorescence from the dead cells and is commonly used as a marker of dead cells. Trypan Blue is another widely used assay to probe the cell membrane integrity and identify dead cells. In the

Trypan Blue exclusion method, cell viability is determined by counting the unstained cells using a hemacytometer and a microscope or automatic cell counter, which determines the number of dead cells, and calculates percent cell viability in a treatment group. Viable cells with intact membranes exclude the probe, while nonviable cells become labeled with the dye and can be observed in the bright-field mode. This method is simple and useful in assessing not only functional integrity, but also as an objective method of determining overall viable cell count prior to cell treatment. However, this staining can only determine live or dead cells without distinguishing the live cells that are already losing cell functions.

6.1.1.2   Probing Metabolic Cell Activity

The viability of cell cultures is routinely assessed by utilizing the metabolic capacity of cells that biochemically convert chemicals that can then be conveniently measured at specific wavelengths using a multiwell plate reader. These assays should be rapid, sensitive and inexpensive in order to measure the effects of cell treatment on survival and proliferation of cells. In this approach, cytoplasmic esterases present in the metabolically active cells can cleave moieties from a lipid-soluble nonfluorescent probe to result in a fluorescent product. The product should be retained within the cell if membrane function is intact. Therefore, the bright fluorescence from live cells can be registered in contrast to nonviable cells. The most widely used probes and assays include calcein AM, fluorescein diacetate, the 3-(4,5-dimethylthiazol-2yl)-2,5-diphenyl etrazolium bromide (MTT) and the resazurin assays. The efficiency of particular probes may depend on uptake or retention of the dye among individual cells or on different conditions. Since dead cells can be detected using nucleic acid staining with propidium iodide, the live–dead cell calcein AM/propidium iodide staining assays can be used to simultaneously stain living and dead cells.[2] The calcein AM is enzymatically hydrolyzed into calcein in living cells, coloring them with a green fluorescence. The commonly used live–dead cell-staining kits are commercially available and easy to use.

One of the well-established cell viability assays is the MTT assay that is based on the capacity of the cellular mitochondrial dehydrogenase enzyme in living cells to reduce the yellow water-soluble MTT into a purple formazan[3,4] therefore not evaluating the cell but its mitochondrial activity. The MTT assay can be highly useful in the case of metabolically active cells that do not divide.[4] The assay can be used for cell monolayers or suspensions.[5,6] The MTT assay is highly reproducible and is widely used in both cell viability and cytotoxicity tests. However, since a produced formazan is insoluble in water and forms crystals in the cells needed to be dissolved in organic solvent prior to absorbance reading, cell culture follow-up is impossible.[7]

Fluorescein diacetate (FDA) can serve as a viability probe that measures both enzymatic activity, required to activate its fluorescence, and cell-membrane integrity required for intracellular retention of the fluorescent product. Upon hydrolysis by endogenous esterases, FDA yields green fluorescent fluorescein. However, hydrolysis of FDA to fluorescein by common media components such as tryptone, peptone, as well as by Tris-HCl and sodium phosphate buffers in the absence of live cells and the quenching of fluorescein fluorescence by assay solutions might cause problems in cell-viability determination.[8] Moreover, because FDA can leak from cells, the analysis should be performed immediately after incubation.[9]

Resazurin (7-hydroxy-10-oxido-phenoxazin-10-ium-3-one) is used to measure cytotoxicity in different types of cells.[10,11] Resazurin is a redox dye exhibiting changes in absorbance and fluorescence due to cellular metabolic activity.[11,12] The resazurin assay is based on the ability of viable, metabolically active cells to reduce resazurin (the oxidized, blue (549 nm), nonfluorescent (530 nm)) to resorufin (the reduced, red (630 nm), fluorescent (590 nm)). This conversion occurs intracellularly, where the oxidized form of the resazurin enters the cytosol and is converted to the reduced form by mitochondrial enzyme activity.[7] Since resazurin is nontoxic to cells and is stable in culture media, continuous measurement of cell activity in culture can be achieved.[13,14]

The release of cell-death markers upon cell functionalization can also be used to evaluate inhibition in cell viability (cytotoxicity). Among those, lactate dehydrogenase (LDH) is a stable oxidative cytoplasmic enzyme that changes lactate into pyruvate during glycolysis. LDH exists in cell membranes and cytoplasm in all cells and preserves its activity during cell-death assays.[15] When the plasma membrane is damaged, LDH is immediately released into the culture supernatants and can be measured photospectrometrically, *e.g.*, by the conversion of tetrazolium salt to soluble formazan. The medium from cell cultures is assayed, and the amount of LDH activity is measured as an indicator of relative cell viability as well as a function of membrane integrity.[16]

6.1.2   Cell-Growth Measurements

Monitoring cell growth has traditionally been done by colony counts on agar plates[17] or with light scattering or turbidity measurements.[18,19] In the latter case, the optical density of cell suspensions as an indicator of cell density is measured in the visible region (~ 600 nm). Since this approach is based on measuring the total amount of scattered light, the measured values include scattering from the living cells, dead cells, and cell debris. However, direct observation of cell growth using optical and confocal microscopy can be more indicative of the cell viability such as, for instance, as budding of yeast cells enclosed in nanothin multilayer polymer shells when the growing daughter cell from the encapsulated mother yeast cell could be observed.[20,21]

6.1.3   Activation of Cell Functionality

The preservation of appropriate cell functional capacity is critical for successful development of a functional coating for biochemical sensing devices or immunomodulating strategies. There exists a possibility when the cell coating, while not adversely influencing cell viability, may hamper cell functional capacity. For example, one of the directions of cell surface modification is to create optical sensing biochemical assays based on living cells that can respond to the presence of a specific small molecule by giving a fluorescence signal.[22,23] Thus, for instance, *E. coli* cells or yeast cells modified to express a fluorescent protein (*e.g.*, green fluorescent protein, GFP) can significantly raise fluorescence from the protein in response to a certain molecular inducer, such as theophylline[22] or aminoglycoside neomycin.[23] The functional capacity of *Saccharomyces cerevisiae* yeast cells (YPH501) with incorporated yEGFP (yeast-enhanced green fluorescence protein) reporters after their surfaces were coated with thin multilayers of hydrogen-bonded tannic acid/poly(N-vinylpyrrolidone) (TA/PVPON) was evaluated.[20,24] The possible adverse effects of the coating treatment on the ability of the engineered yeast cells to express yEGFP after induction of the reporter protein expression by galactose were checked by continuous monitoring of the yEGFP fluorescence from the coated yeast cells after the induction. In this example, the gradually increased fluorescence could be detected from the cells coated with 3, 4, 5, and 6 bilayers of TA/PVPON multilayer. Similarly, the viability of GFP-expressing yeast cells coated with ionically paired multilayers of $(PAH/PSS)_{2.5}$ was confirmed by monitoring the rising fluorescence from GFP expressed by the yeast cells in response to galactose inducer.[21]

The ability of the metabolically active *Allochromatium vinosum* bacterial cells to oxidize reduced inorganic sulfur compounds into sulfate was used to demonstrate the viability of the microbial cells encapsulated into ionically pared polyelectrolyte multilayers containing poly(allylamine hydrochloride) or poly(diallyldimethyl ammonium chloride) as polycations and poly(styrene sulfonate) sodium salt, poly(glutamic acid), or poly(acrylic acid) as polyanions.[25] The sulfide (or sulfur) supplied to the modified cells was oxidized by the living cells to intermediate sulfur globules or to sulfate products that were determined either colorimetrically (sulfur) by cyanolysis or using the turbidimetric method (sulfates).[25]

The functional capacity of living red blood cells to uptake oxygen from solutions was utilized to determine erythrocytes viability after their surfaces were camouflaged with multilayers comprised of alginate, chitosan-graft-phosphorylcholine and polylysine-graft-poly(ethylene glycol).[26] The decrease in oxygen dissolved in PBS solution was measured with a micro-oxygen electrode and served as a means to confirm that the functioning of hemoglobin, oxygen-transporting iron-containing protein inside the erythrocytes, and therefore RBC activity was not affected by the cell surface modifications.

The viability of *Bacillus subtilis* bacterial spores encased in polyelectrolyte multilayers of biocompatible polymers such as poly(glutamic acid), poly(lysine), albumin, lysozyme, gelatin A, protamine sulfate, and chondroitin sulfate, and silica nanoparticles (22 and 72 nm) was quantified by measuring the amounts of pyridine-2,6-dicarboxylic acid (dipicolinic acid,

DPA) released by the spores exposed to a solution of nutrients to germinate.[27] During germination, the growth and metabolism of dormant bacterial spores is triggered by nutrients such as amino acids, sugars, *etc.* The changes in the inner membrane of the germinating spores upon interaction of the membrane receptors with certain nutrients lead to DPA release from the spores, and, eventually to a fully functioning viable cell.[28] The DPA release from germinating nanocoated *Bacillus subtilis* was monitored by measuring fluorescence from complexes of released DPA with terbium ions after the germination was induced in the presence of 1.1 mM L-alanine at 37 °C.[27]

The induction of a colorimetric signal from fluorescent species within the metabolically active cells as a reporter of functional capability of cells unaffected by cell modification was also utilized for *Acinetobacter baylai* bacterial cells[29] and *Chlorella pyrenoidosa* green algae cells[30] entrapped into shells of magnetic particles of iron oxide. In the latter case, photosynthetic activity of the green algae cells after encasing with 15-nm iron oxide nanoparticles modified with PAH (M_w 15 000 g mol^{-1}) was tested. The fluorescence from Chlorophyll B was recorded as the reporter of functionally unaffected algae cells.[30]

To evaluate functional capacity of insulin-producing β-cells within the modified pancreatic islets, the glucose-stimulated insulin release in static incubation or in a cell perifusion system is usually carried out for noncoated and coated pancreatic islets as a function of time in response to variations in glucose concentration. The glucose concentration of ~ 3 mM refers to a low glucose (basal glucose) level, while ~ 16 mM glucose concentration is used as a high glucose level. The stimulation of β-cells in pancreatic islets with a high glucose concentration should result in a larger amount of released insulin. As a result, a stimulation index of the islets, defined as the ratio of the released insulin at high glucose level to that at basal level, can be calculated and utilized as a measure of functional capacity of the pancreatic cell clusters.

6.2 Cytocompatibility of Functionalizing Species

Here, we will discuss possible cytotoxic effects of synthetic and natural polymers, metal and metal-oxide nanoparticles and carbon-based materials on different types of cells under modification. There exists a seeming discrepancy in toxicity data in the reports on encapsulating cells with various functional species. For example, very often, functional species used for functionalization of yeasts and bacterial cells that are found to be cytocompatible demonstrate acute toxicity when applied for engineering of mammalian cells. The differences of cyto-architecture in various cell types might be critical in understanding the reported variability of toxicity.

6.2.1 Architectural Differences of Cells

Among individual living cells that have been used for cell functionalization are fungi,[31,32] bacteria,[17,27] and mammalian cells.[18,26,33–37] The cellular envelope of bacterial and fungal (yeast) cells is comprised of the inner plasma membrane and an outer cell wall that protects the cell and maintains its shape. The bacterial cell wall is composed of layers of peptidoglycan, a complex of proteins and oligosaccharides, and is quite thin for some bacteria, *e.g.*, in *Escherichia coli*, while others possess a thicker cell wall.[38] The rigid cell wall of plant cells is composed of cellulose and other polymers. In *Saccharomyces cerevisiae* (yeast), the plasma lipid bilayer is about 7 nm thick and contains proteins of various functions. The space between the plasma membrane and the cell wall (periplasm) in yeast is a thin layer of 35–45 Å containing mostly mannoproteins. The wall of a yeast cell is 100 to 200 nm thick and contains polysaccharides (80–90%) and a small percentage of chitin providing strength to the cell wall and forming a network of microfibrils[39,40]

Eukaryotic (*e.g.*, mammalian) cells are surrounded by a plasma membrane without a reinforcing cell wall. Also, for surface modification of pancreatic islets, it is important to know that the islet is a multicellular cluster of secretory endocrine cells, *i.e.* insulin-producing β-cells, glucagon-containing α-cells, somatostatin-containing δ-cells, and the pancreatic polypeptide-producing cells. Although different types of mammalian islets contain similar types of cells, the islet architectures are different. For instance, insulin-producing β-cells of rodent islets form an inner core of the islet with other secretory cells on the periphery, while in human islets β-cells are interspersed with other cells more on the islet periphery.[41,42]

Because of these structural and morphological differences between various types of cells, there might be a different tolerance of the cells towards various cell surface modifications resulting in either cytocompatibility or cytotoxicity of the treatments. Because of the higher vulnerability of mammalian cells, harsh conditions, such as pH variations, salt concentrations, presence of specific ions, and abnormal temperature, a range of materials and/or processes available for encapsulation of individual mammalian cells under specific conditions might be limited.

6.2.2   Cytotoxicity of Polymers

Polymers can induce cell death *via* various routes including plasma membrane permeabilization,[43] formation of harmful polymer degradation products including reactive oxygen species,[44] and permeabilization of a nuclear membrane.[45,46] The cytotoxicity of positively charged polyelectrolytes for mammalian cells was demonstrated in many reports.[47–50] The viability of MELN cells derived from MCF-7 human breast cancer cells transfected with an estrogen-regulated luciferase gene decreased to less than 40% when exposed to poly(L-lysine) (PLL) and protamine sulfate and to less than 10% in the case of poly(allylamine hydrochloride) (PAH), poly(phosphoric acid), and poly(ethylene imine) (PEI). The lowest polycation cytotoxicity, yet that decreased cell viability by ∼30%, was achieved in the presence of poly(diallyldimethyl ammonium chloride) (PDDA) with MELN cells.[47] The slightly lower cytotoxicity of quaternary ammonium groups compared to primary amine groups has also been reported for red blood cells[51] and porcine brain capillary endothelial cells.[52] The toxic effects of the cationic polyelectrolytes were attributed to their direct influence on the plasma membrane fluidity. The toxicity of branched PEI towards HEK293 (human endoderm kidney cells) and CCRF-CEM (acute lymphoblastic leukemia, T cell) cells was found to be concentration dependent and PEI solutions with concentrations larger than 1 µg mL^{-1} exerted a cell viability decrease from 85% for [PEI]> 1 µg mL^{-1}, to less than 20% at [PEI]>10 µg mL^{-1}.[48] As suggested by Hong *et al.*, polycationic PLL and PEI induce nanoscale hole formation and enhance the preexisting defects in the lipid bilayers causing cell death.[53] The polymers became highly toxic to cancerous KB and Rat-2 cells at larger concentrations (12 µg mL^{-1}), however, inducing membrane permeability and release of cytosolic enzymes (cell-death marker LDH) at even lower concentrations of 6-12 µg mL^{-1}. The degree of lipid membrane disruption was found to be strongly dependent on the charge density of the polymer chain and on polymer architecture. Thus, brunched PEI induced the highest cytotoxicity due to the largest density of charged groups on the chain, while positively charged poly(amidoamine) PAMAM G5 dendrimer was more efficient in the membrane hole formation than a linear PLL.[53,54] The higher-generation PAMAM dendrimers (G7) removed the lipid layer from the cell membrane by the formation of dendrimer-filled lipid vesicles leading into the leaky cell membrane.[55] The more toxic effect of a positive charge compared to a negative one has been demonstrated on porcine brain capillary endothelial cells,[52] Cos-1 cells, red blood cells, and on *E. coli* bacteria cultures.[51] The PAH was found to be much more toxic to the brain capillary endothelial cells than poly(styrene sulfonate) sodium salt when they were exposed to the polymer solutions (2 mg mL^{-1}) within 1000 min.[52] The cytotoxic effects for PAH became prominent within 10 min after exposure to cells, and within 40 min for PDDA, while 800 min were required for PSS to induce cell toxicity.[52] The gold nanoparticles with positively charged side chains due to quaternary ammonium groups were found to be toxic to Cos-1 cells, red blood cells, and on *E. coli* bacteria cultures compared to that of negatively charged carboxylate groups that were nontoxic to the studied cells.[51]

Higher molecular weight polymers generally are more toxic to cells.[56,57] The effect of the molecular weight and the architecture of a polycation on cell viability was studied on Chinese hamster ovary cells CHO DG44 upon exposure to PLL analogs.[58] For low molecular weight polycations, osmotic shock was found to be critical in inducing cell death. For polycations of high molecular weight direct interactions between the polymer and cell membrane resulting in membrane disruption were more important. The acute (within 1 h of cell exposure) cytotoxicities of the hyperbranched and dendritic PLL analogs were found to be 5–250 times higher as compared to linear PLL of similar molecular weight.[58] Delayed cytotoxicity (within 3 h of the cell exposure) was more pronounced for dendritic and hyperbranched

PLL of comparable molecular weights as compared to the linear. This difference was attributed to partial enzymatic degradability of the hyperbranched and dendritic PLL molecules during longer exposure times when they still could exert their cytotoxicity unlike the linear PLL which was completely enzymatically degradable.[58]

Because the toxicity of polycationic polymers is dependent on the polymer charge density, it can be attenuated by conjugating neutral molecules, such as poly(ethylene glycol) (PEG), to the critical number of amino groups along the polycation chain. Modification of PLL with PEG is carried out by grafting N-hydroxysuccinimide-PEG (NHS-PEG) chains to amine groups on PLL chain to lead to PLL-g-PEG graft copolymer. The grafted PEG molecules are unbranched and hydrophilic in the form of methyl-(PEG)$_n$-NHS ester ("n" denotes a number of the ethylene glycol monomers). The N-hydroxysuccinimide ester end group reacts with primary amines. The methoxy(ethylene glycol) grafts were conjugated to PLL backbone through a covalent attachment to lysine residues.[59] The PEG substitutes (40%) on the PLL chain led to attenuation of the positive charges without any deleterious effect on the pancreatic islet viability (Figure 6.1).

Similarly, the attenuation of cytotoxicity was achieved with PAMAM dendrimers upon their size reduction (from G7 to G3), and therefore, charge density. A change in the dendrimer surface chemistry from amine to acetamide decreases the cytotoxicity of the PAMAM dendrimers, thus, for example, G5-acetamide PAMAM was no longer able to cause the formation or expansion of defects in the bilayer lipid membrane.[54] The release of LDH from Rat-2 cells decreased from 40% to 7% when G5-acetamide PAMAM was used instead of G5-amine PAMAM (500 nM).[54] Similarly, a 1-fold reduction in cytotoxicity of PAMAM dendrimers towards Caco-2 cells was achieved as the number of surface acetyl groups increased.[60] However, this change in surface chemistry did not affect the dendrimers permeability through the cell membranes.[60] PEGylation of PAMAM G4 dendrimers rendered them nontoxic towards Chinese hamster ovary CHO-K1 cells at concentrations of up to 20 μM after 4 h exposure and assessed by LDH release from the cytoplasm.[61] The 75% release of LDH observed in the case of the cell exposure to G4 and G5 PAMAM dendrimers was prevented from the cells when G4-PEG and G5-PEG were used instead.[61]

6.2.3 Cytotoxicity of Nanoparticles

Cell surface modification involving nanoparticles has become important for development novel biofunctionalities and devices for microelectronics and biosensing. Carbon nanotubes,[62,63] graphene oxide nanosheaths,[64] metal nanoparticles,[65,66] inorganic and polymer nanoparticles[67-69] have been utilized in those efforts. The toxicity of nanoparticles exposed to cells may be originating from the core material[70,71] due to its corrosion and/or from the nanoparticle surface coating.[72] The cytotoxic effects from nanoparticle coating can be associated with particle charge originating from polymer

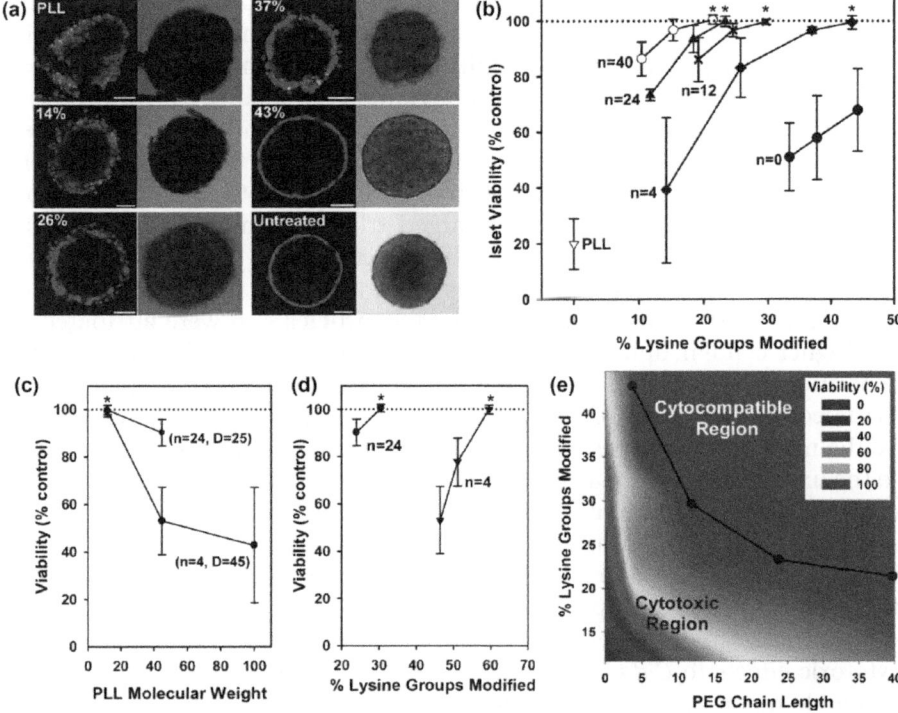

Figure 6.1 Polycations with enhanced cytocompatibility can be designed by tailoring the structure of PLL-*g*-PEG copolymers. (a) Confocal and bright-field micrographs of pancreatic islets stained with calcein AM (viable) and ethidium homodimer (nonviable) after incubation with 80 μM PLL and P12P4[D] copolymers with variable degrees (*D*) of PEG grafting (scale bar, 50 μm). Polycation toxicity is predominantly exerted toward cells on the periphery of the islet, and the absence of fluorescent emission from the islet core is a consequence of the limited tissue penetration depth of confocal microscopy. (b) Islet viability (mean ±SD) after exposure to 80 μMPLL and P12P*n*[D] copolymers (*n* = 0, 4, 12, 24, and 40) with variable degrees of PEG grafting. A unique critical degree (*D*$_c$) of PEGylation (*$p > 0.05$ *vs.* untreated controls) was observed for each PEG chain length explored (*i.e.* *n* = 4, *D*$_c$ = 43%; *n* = 12, *D*$_c$ = 30%; *n* = 24, *D*$_c$ = 23%; *n* = 40, *D*$_c$ = 21%). (c) For fixed *D* and *n*, increasing the PLL backbone molecular weight was found to significantly increase PLL-*g*-PEG copolymer toxicity. (d) Cytocompatible P45P*n* (*n* = 4 and 24) copolymers can be generated by increasing the degree of PEG grafting relative to P12P*n* variants (*n* = 4, *D*$_c$ = 60%; *n* = 24, *D*$_c$ = 30%). (e) Contour plot generated from data in part b demonstrates operative copolymer structurecytotoxicity relationships, with *D*$_c$ and *n* defining a border between cytotoxic and cytocompatible regions in copolymer structure.

Reprinted with permission from J. T. Wilson, W. Cui, V. Kozlovskaya, E. Kharlampieva, D. Pan, Z. Qu, V. R. Krishnamurthy, J. Mets, V. Kumar, J. Wen, Y. Song, V. Tsukruk, E. L. Chaikof, *J. Am. Chem. Soc.*, 2011, **133**, 7054. Copyright 2011 the American Chemical Society.

coating (*e.g.*, polyelectrolyte brushes) introduced to maintain colloidal stability of nanoparticle solutions.[73] Positively charged nanoparticles can easily interact with negatively charged lipid bilayer cell membranes. Similarly to cationic polyelectrolytes, interaction of cationic nanoparticles with the cell membrane causes thinning, disruption of or hole formation in the cell membrane.[74] For instance, when 3T3 fibroblasts were incubated with gold nanoparticles (10–15 nm in size) carrying either negatively charged phosphonate ($-PO(OH)_2$) or positively charged trimethylammonium ($-N(CH_3)_3$) groups as blocks on the amphiphilic diblock copolymer chains, the cytotoxic effects were observed at concentrations above 20 nM for the anionic particles and above 5 nM for the cationic gold nanoparticles and were attributed to the higher cellular uptake of the cationic particles rather than to the concentration of the nanoparticles to which cells were exposed.[73] The cationic PDDA-coated Ag nanoparticles with the average size of 5 nm were reported to be the most toxic to mouse macrophage and lung epithelial cells compared to the uncoated Ag nanoparticles.[75] The use of positively charged surfactants such as hexadecyltrimethylammonium bromide (CTAB) used for colloidal stabilization of nanoparticles may also lead to cytotoxicity.[72] On the other hand, the inorganic nanoparticle cores may induce cytotoxicity by releasing toxic ions.[76,77] For example, the leaching of silver ions from 25-nm silver nanoparticles in the presence of algae due to H_2O_2 production caused cytotoxic effects to *Chlamydomonas reinhardtii*.[78] In the case of soluble and insoluble metal oxides, the exposure of cells to nanoparticles of zinc oxide resulted in death of all mesothelioma MSTO-211H human or rodent 3T3 fibroblast cells at nanoparticle concentrations above 15 ppm and led to a fast drop in cell functionality at concentrations as low as 3.75 ppm.[79] The toxic effect was attributed to the release of Zn^{2+} ions. However, the cytotoxic effect of free ions was cell type sensitive. Thus, for example, human MSTO cells were highly sensitive to Fe_2O_3 while rodent 3T3 cells were not greatly affected and remained viable. The toxicity in this case was from the catalytic production of free radicals through Fenton- and Haber–Weiss-type reactions. Unlike soluble metal oxides, insoluble oxides including SiO_2 and TiO_2 were found to be not exceedingly toxic up to 30 ppm.[79] The cytotoxicity of nanoparticles is generally increased with the particle size decrease that is attributed mainly to increased reactive surface area for smaller nanoparticles.[80] The cytotoxicity of carbon nanotubes has been a contradictory issue in the literature[81] and has been reported[82,83] to be dependent on a variety of factors such as impurities[82,83] and surface functionalization.[84] Carbon nanotubes may induce cytotoxic effects to living cells mainly by physical membrane damage and/or by oxidative stress through generation of reactive oxygen species.[81]

6.3   Cytocompatibility of Cell Functionalization Approaches

Cell surface functionalization can be realized through a conformal protective layer on the cell surfaces using various surface techniques. Natural and

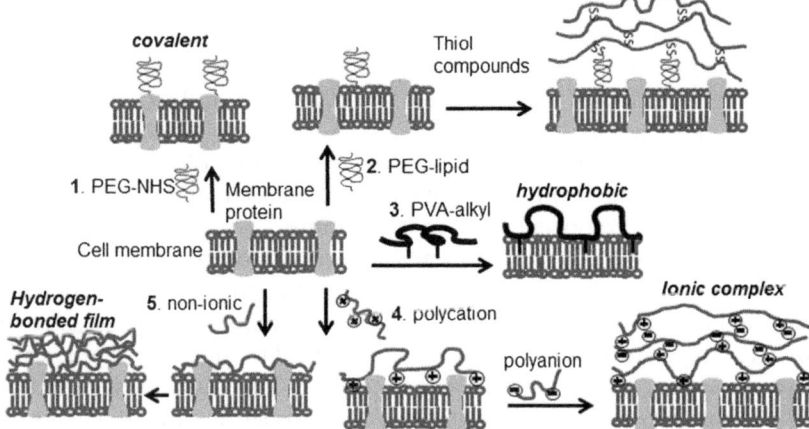

Figure 6.2 Modification of cell surfaces by covalent conjugation to cell surfaces (1), by incorporation of molecules into the cell membrane *via* hydrophobic interactions (2, 3), *via* ionic interactions of polyions with the cell membrane (4), and through hydrogen-bonded interactions with cell membrane (5).

synthetic polymers are often utilized to bring various functional groups to the cell surfaces. Covalent binding to functional groups of cell membrane proteins, insertion of amphiphilic copolymers into the plasma membrane *via* hydrophobic interactions between a lipid bilayer of the membrane are often employed in cell surface modification[85] (Figure 6.2). These covalent and noncovalent modification approaches may be of potential harm to the cell depending on the chemical and physical stresses involved in the modification procedures. Herein, cytocompatibility of various cell modification approaches will be discussed.

6.3.1 Layer-by-layer Assembly of Polymers

The layer-by-layer (LBL) assembly of polymers based on sequential adsorption of oppositely charged components is one of the established methods for the preparation of thin polyelectrolyte multilayer films with controlled properties. The technique is a universal surface-modification approach that allows for building films of controlled thickness, permeability, chemistry and mechanical properties. The technique has been widely explored in modification of various types of cells including pancreatic islets.[49,86] The LbL modification of pancreatic islet surfaces is based on alternating deposition of water-soluble polymers on surfaces from aqueous solutions that results in nanothin coatings of controllable thickness and composition.[87]

To promote a multilayer film formation on the cell surfaces, the negatively charged cell surface is first exposed to a cationic polymer solution followed by an anionic polymer solution to form an ionically paired polyelectrolyte complex film (Figure 6.2). The effect of molecular weight of polyelectrolytes on the cytocompatibility of the animal cells was investigated in the case of the LBL encapsulation of human islets into PAH/PSS and PDDA/PSS multi-layers. Although the islet viability was satisfactory, islets encapsulated into PAH/PSS and PDDA/PSS multilayers using a higher polycation molecular weight demonstrated a limited insulin release due to a lowered permeability of insulin through the polyelectrolyte membrane.[86]

While generally well tolerable by bacterial, fungal or plant cells due to the thick cell wall protecting the plasma membrane, most cationic polymers used in the LbL modification of surfaces such as PLL and PEI are extremely cytotoxic to fragile animal cells that, when exposed to the polycations, can be severely damaged. The cytotoxic effect though has been observed to be dependent on polycation concentration and exposure time.[88] The overall cytotoxicity of the polyelectrolytes originates from the positive charge of polycations that can induce pore formation within the cell membrane causing its damage and, eventually, cell death.[89] The high toxicity of the PAH/PSS LbL film to murine islets was confirmed by Wilson *et al.*[49] The coating of the murine islets with only 3 layers of PAH/PSS/PAH led to the reduction of islet viability by 70%. A similar effect was found for islets coated with 3 layers of PLL/alginate LbL film. Even 15 min of islets incubation with low concentration of PLL results in ~60% decrease in cell viability. Menger *et al.* showed that PLL was able to pass through the lipid bilayer if it was previously allowed to form a complex with anionic lipids.[90] PEI was found to be extremely toxic to the islets. This polycation destroys the cell membrane immediately after its interactions with the membrane surface.[48] The overall charge arrangement of a polycation and its interaction with the cell membrane strongly depends on the three-dimensional structure and flexibility of the polymer chains. It has been shown that polymers with highly flexible chains and a high cationic density will exert tremendous cytotoxicity to animal cells. Thus, the polycations with globular structures demonstrated good biocompatibility, whereas polymers with more linear and flexible structure such as PLL and PEI showed higher cytotoxicity.[48]

A decrease in PLL toxicity was achieved by conjugating a neutral molecule PEG to PLL backbone.[59] Forty percent of PEG substitutes on the PLL chain allowed for attenuation of the PLL positive charges without any disturbing effect on the islet viability (Figure 6.1). A similar approach to attenuate a polycation cytotoxic effect was applied when PEI was functionalized with catechol derivative, 3-(3,4-dihydroxyphenyl)propionic acid. After multilayer assembly of that PEI graft copolymer with hyaluronic acid on the surface of yeast cells, the viability was around 90% for 5-bilayer coated yeast cells and around 87% for those coated with 15 bilayers.[91] Although the cytotoxic effect of PLL and PEI can be reduced by varying polycation concentration, exposure time, and introducing nonionic grafts, polycation-based systems are still challenging for islet modification due to a narrow window of the cyto-compatible graft-copolymers.[35]

The high toxicity inherent to most polyelectrolyte polycations may limit their use in biomedical applications that require cell-modified structures. However, the natural biopolymers chitosan and alginate have more similarities with the extracellular matrix, are chemically versatile and have a good biocompatibility. These linear polysaccharides carry an opposite charge and can be ionically bound in a polyelectrolyte coating. The 3-bilayer coating of chitosan/alginate on pancreatic islet cells was formed *via* alternate deposition starting from positively charged chitosan.[92] The deposition conditions had been shown to greatly influence the islets viability, which well correlates with the difference in charge density and toxicity of the polycation at high and neutral pH values. An increase in the coating thickness up to 5 bilayers did not adversely affect the islet viability or insulin release, and the coated islets were viable up to 5 weeks postencapsulation.

The LbL approach based on hydrogen-bonded LbL suggests new opportunities for cytocompatible cell surface engineering.[20,93] The LbL assembly driven by hydrogen bonds allows for inclusion of polymers that carry no positive charge (Figure 6.2). The nonionic hydrogen-bonded LbL approach was applied for coating individual, living pancreatic islets derived from rat, NHP, and human.[93] The approach can be used as it is more effective than the ionic LbL method. The hydrogen-bonded method allows for conformal coating of the islet surface with multilayers of poly(vinylpyrrolidone) and tannic acid under physiological conditions in a rapid and efficient manner without interfering with the viability and function of the insulin-producing β-cells because no cationic components are used and interactions of the multilayer with the islet surfaces are based on hydrogen bonding between either tannic acid or poly(vinylpyrrolidone) (Figure 6.3).

A multilayer poly(vinyl alcohol) (PVA) thin coating formed layer-by-layer was investigated for islet immunoprotective capabilities[94] (Figure 6.2). In this approach maleimide-PEG-conjugated phospholipids were used to first modify the cell membrane surface to promote further interactions with PVA derivatives. PEG-conjugated phospholipids can be immobilized on the cell membrane through incorporation of the lipid chains into the cell membrane due to their hydrophobic interactions with the lipid bilayers of the membrane.[95] Moreover, PEG-phospholipids are more compatible with cells compared to polycations used in the ionic LbL surface modification of islets. A layer of PVA with introduced thiol groups (PVA-SH) was covalently attached to the maleimide-PEG anchors *via* a thiol–maleimide reaction. The LbL multilayer of PVA was then deposited by alternating immersion of the islets into PVA-SH and PVA-pyridyl disulfide (PVA-PD). The driving force for the multilayer formation was a thiol/disulfide exchange reaction between the PVA derivatives. This ultrathin PVA membrane affected neither cell viability nor insulin release function. However, in this case, despite the advantages, chemical modification of the cell membrane using covalent conjugation of molecules, for example, covalent PEG binding *via* NHS–ester conjugation to the cell-membrane proteins may lead to damaging of the membrane proteins and, consequently, disrupting the cell's physiology.

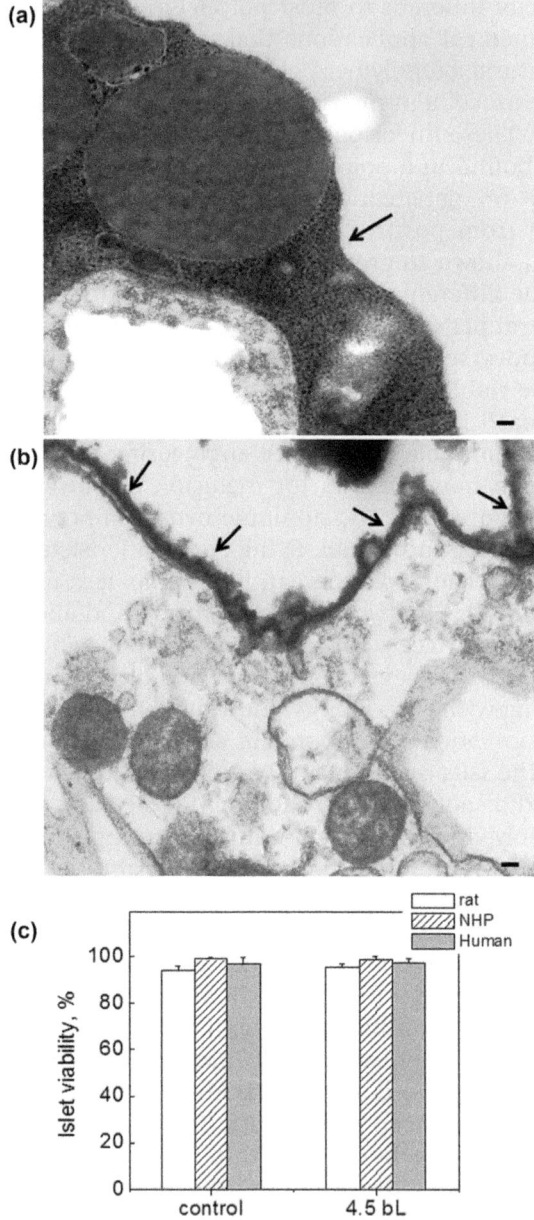

Figure 6.3 TEM images of noncoated (a) and (PVPON/TA)$_4$PVPON-coated (b) NHP islets. The arrows point to the edges of the islet. The scale bars are 100 nm in both images. (c) Viability of noncoated (control) and (PVPON/TA)$_4$PVPON-coated rat (blank), NHP (dashed), and human (filled) islets assayed on the day of coating (within 6 h after film deposition).
Reprinted with permission from V. Kozlovskaya, O. Zavgorodnya, Y. Chen, K. Ellis, H. M. Tse, W. Cui, J. A. Thompson, E. Kharlampieva, *Adv. Funct. Mater.*, 2012, **22**, 3389. Copyright 2012, Wiley-VCH.

6.3.2 Nanothin Organic Shells

Thin conformal coatings based on PEG are highly attractive for cell surface modification due to their biocompatibility and protein-resistant properties based on PEG low interfacial free energy with water, their unique properties in aqueous solutions, their high surface mobility, and their substantial steric stabilization effects.[96] To achieve grafting of the end-functionalized PEG polymer onto the islet surfaces, islets are generally cultured in media in the presence of various concentrations of the polymer.[97–99] The grafting of methoxy-PEG (mPEG) onto the islets surface is usually carried out *via* the succinimidyl ester end groups introduced in mPEG. These groups couple to amino groups present in the collagen matrix around islets. Controlling the grafting time is important since longer grafting times can lead to diffusion of PEG inside the islets that may increase the chance of swelling and exposure of islets to the outside environment. Furthermore, long grafting time increases the probability of islet damage.[98] The molecular weight of the grafted PEG is another important parameter for islet surface modification. Barani *et al.*, showed that grafted mPEG of 5 and 10 kDa onto islets isolated from Wistar rats had a different effect on cell functioning and viability.[100] Insulin secretion for the islets grafted with 5 kDa mPEG was at the same level as for unmodified islets, while overall insulin secretion from islets modified with 10 kDa mPEG decreased.

By dopamine polymerization in the presence of TRIS buffer, an organic 30-nm shell of poly(dopamine) was realized around yeast cells.[101] Dopamine is a catecholamine that functions as a neurotransmitter in the brain. The polymerized coating is believed to be anchored to the yeast cell surface by covalent bonding between poly(dopamine) and amine or thiol moieties of glycoproteins in the yeast cell wall. The FDA viability test revealed a slight cytotoxic effect of the procedure when yeast viability decreased to 70% after the first polymerization step and to 50% after the next polymerization step. The viability decrease is, most probably, because of a basic condition (pH = 8.5) at which dopamine spontaneous polymerization occurs and that is slightly above the tolerance pH range for yeast cells (pH 3–8). The viable modified cells were, however, able to undergo budding when exposed to culture media.

6.3.3 Formation of Thin Solid Shells

Thin solid shells around cells can be achieved through mineralization-based routes or by coating cell surfaces with a layer(s) of nanoparticles. The key factors for preserving cell viability and functioning after introduction of such nano/biointerfaces are maintaining oxygen and nutrient accessibility and removal of cell metabolic waste through the solid shell. One should also keep in mind that possible compressive stresses associated with formation of a solid inorganic shell may result in damage of the cell/inorganic shell interfaces and lead to a cell lysis and a rapid cell death. Herein, toxic and nontoxic effects of the approaches used in building artificial rigid coatings around cells will be examined.

6.3.3.1   Mineralization

Encasing living cells in thin but tough shells is desirable for certain applications such as build-up of biosensing devices or reactors for biosynthesis of proteins. Immobilization of yeast (*Sacharomyces cerevisiae*),[102–106] bacteria[107,108] (*Escherichia coli* and *Bacillus subtilis*) and plant cells[109,110] within inorganic shells has been developed under mild conditions. Entrapment of yeast cells into thin silica shells can be realized by (a) two- or one-step encapsulation routes using silica precursors (sol-gel syntheses),[102,111] or (b) by biomimetic silicification using catalytic molecules.[103,108] In the sol-gel approach, alkoxide precursors, such as tetraetoxysilane, triethoxymethylsilane, and diethoxydimethylsilane are hydrolyzed to alcohols (ethanol) and result in condensation of silicon alkoxides leading to formation of a thin silica layer. Yeast cells are able to convert sugars to alcohol and therefore can withstand the alcoholic byproducts of alkoxide-based sol-gel reactions.[112] Though not detrimental to yeast-cell viability, such an approach is damaging to bacteria cell membranes.[113] The viability of encapsulated cells in the sol-gel silica matrices can also decrease due to the acidity of the silanol groups. For instance, *E. coli* cells encapsulated in silica matrices in the presence of alcohol generated a very low GFP fluorescence signal in response to the inducer, yet the alcohol-free approach, in which harmful byproducts such as methanol or ethanol are removed by controlled distillation, generated a fluorescence signal only slightly lower than that of control *E. coli* cells.[113] The introduction of a physical barrier of organic molecules, *e.g.*, glycerol,[114] alginate,[115] amphiphilic phospholipids,[104,116] between the cells and a formed silica layer was also reported to protect the cell membranes and lead to increased cell viability up to 80–100% over nearly a month.

Biomimetic silicification is based on the formation of a thin silica shell around the cell when silica-forming proteins are utilized and can be performed under conditions that sustain cell viability.[108] Bacteria cells of genetically transformed *E. coli* able to express silicatein-α, a protein that catalyzes polymerization/polycondensation of the silicon alkoxide tetraethoxidesilane to poly(silicate), produced a silica capsule around themselves in the presence of silicic acid after induction of silicatein expression by isopropyl β-ᴅ-thiogalactopyranoside.[108] The formed silica shell did not affect the bacteria viability or growth behavior.[108]

Biomimetic silica formation of ~50 nm performed directly on yeast (*S. cerevisiae*) and bacteria (*E. coli*) cells and *Bacillus atrophaeus* spores showed a good cytocompatibility.[103] The cell modification included (a) LbL deposition of (PDDA/PSS)$_{10.5}$ layers to introduce positively charged quaternary amine groups from PDDA that served as catalytic sites for silicic acid derivatives (50 mM), and (b) silicification of the LbL coating at room temperature for 30 min.

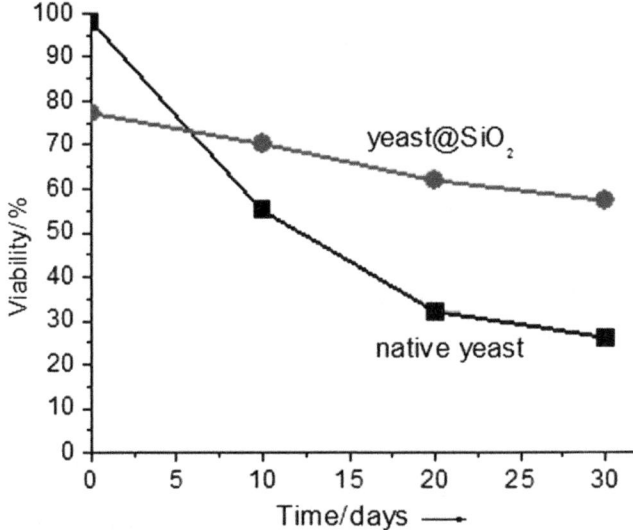

Figure 6.4 Viability of native yeast and yeast@SiO$_2$. Membrane integrity and metabolic activity of yeast cells were determined by fluorescent staining. Adapted with permission from S. H. Yang, K.-B. Lee, B. Kong, J.-H. Kim, H.-S. Kim, I. S. Choi, *Angew. Chem. Int. Ed.*, 2009, **48**, 9160. Copyright 2009, Wiley-VCH.

The viability of native and silica-encased yeast cells was profoundly different for cells stored in water at 48 °C for 30 days (Figure 6.4). The two-step treatment decreased the initial viability of silica-encased cells to about 77% that can be explained by the use of cationic PDDA and physical stress from the treatment. The observed viability decrease can also be partially due to free silicate diffusion toward the cell space.[102] The long-term cell viability, within 30 days, was maintained at the level of 56% for silica-coated cells, while it decreased to 24% for nonmodified cells. Thus, the biomimetic silicification involving LbL coating followed by silicification catalyzed by the cationic polyelectrolyte layers was only moderately toxic. Overall processes even allowed for the increase of the long-term viability of individual yeast cells, by stabilizing cellular membranes under physico-chemical pressure and by protecting the cellular structure from dehydration.[103]

The process of biomineralization on yeast cell surfaces was also utilized to produce cell-surrounding mineral shells of calcium phosphate[105] and calcium carbonate.[106] In the former case, the 4-bilayer LbL precoating of PDDA/poly(acrylic acid) (PDDA/PAA) or PAH/PSS was applied first to en-capsulate the cells and supply the carboxylate groups as active nucleation sites for calcium ions provided by 10 mM calcium chloride solution. The treatment of the LbL-coated yeast cells with 10 mM dibasic sodium phosphate at pH = 6.8 for 1 h led to formation of the mineral shell around cells that consisted of a 700-nm inner layer of amorphous calcium phosphate and a thin outer layer of crystallized octacalcium phosphate.[105] The cytocompatibility of this mineralization approach was evident from the cell viability data measured using fluorescent staining of live and dead cells with almost 100% of cells being alive on the first day after the treatment and with 85% cells surviving after a month of exposure of the encapsulated cells to distilled water.[105] Interestingly, when the demineralization of the coated cells was performed in a slightly acidic media at pH = 5.5 using a solution of hydrochloric acid, the cells were not damaged and started proliferating in the culture media. Although applicable to yeast cells, which can tolerate pH changes down to pH = 3,[117] such a demineralization approach, however, would not be possible for most mammalian cells leading to cell death.

The calcium carbonate micrometer-size shell was directly formed around yeast cells when the cells suspended in 0.33 M calcium chloride solution were exposed to 0.33 M sodium carbonate solution for 5 min under vigorous stirring.[106] In this case, the thick yeast cell wall directly provided nucleation sites for calcium ions and no LbL prelayers were required as the catalytic template. The viability of yeast cells after the treatment, though not quantitatively, was confirmed with the FDA essay. Notably, the high ionic stress due to high concentration of calcium ions was well tolerated by the treated cells and was not extremely cytotoxic. The demineralization was carried out in 0.1 M ethylenediaminetetraacetic acid disodium salt (EDTA) for 1 h and the cells liberated from the calcium carbonate shells could grow and demonstrated integrity of cellular membranes and enzymatic activity according to the FDA viability results.[106]

Formation of a thin titanium oxide shell around yeast (*S. cerevisiae*) and algae (*Chlorella vulgaris* FC-16 and *Chlorella sp.* C-141) cells using the biomineralization approach was shown to be relatively toxic to the cells.[110] The cells were coated with a 3-bilayer thin film assembled layer-by-layer from an arginine-lysine-rich peptide (1 mg mL^{-1}) and a small titania precursor molecule, titanium bis(ammonium lactate)dihydroxide (TIBALDH) (10 mM). At the sequential deposition of the peptide and TBALDH at the cell surfaces, the precursor was catalytically hydrolyzed, eventually resulting in the formation of a thin 35-nm TiO$_2$ shell. The chlorophyll autofluorescence and FDA essay tests revealed the viability of 69% after the treatment of *Chlorella vulgaris* cells. A similar procedure was extremely cytotoxic to yeast cells due to toxicity of TBALDH molecules to yeast.[110]

6.3.3.2 Formation of Nanoparticle Shells

Formation of metal nanoparticle shells around bacterial and fungal cells can be achieved through (a) metal reduction by bacterial extracellular structure[118], (b) direct deposition of bare or surface-coated nanoparticles on cell surfaces[119,120], and (c) LbL-mediated nanoparticle assembly.[27,30,62,65,66,] Very often, simple mixing of the cells and nanoparticles generates mixtures of cells and nanoparticles without the formation of a conformal coating. These approaches can also severely affect the cell viability due to toxicity of nanoparticles. For instance, when *Trichoderma asperellum* fungal spores or *S. cerevisiae* yeast cells were directly mixed with a suspension of silver nanoparticles, they were not able to proliferate, suggesting a highly toxic effect from the exposure.[66] Generally, nanoparticle shells around bacteria and yeast cells can be realized without much harm to the cells using an LbL approach during which a protective "cushion" of conformal film comprised of oppositely charged polyelectrolytes is formed to ensure a necessary distribution of surface charge and increase the number of binding sites required for strong inorganic nanoparticle attachment. A sequential deposition of several consecutive particle layers led to a rigid and functional particle-containing coating with a polyelectrolyte "glue" in between. As discussed above, a direct interaction of nanoparticles with cell surfaces might be toxic to cells depending on particle surface charge, particle coating, size and shape. Usually, a cationic polyelectrolyte is first deposited on the bacteria surface through ionic interactions between the polyelectrolyte and negatively charged surface of bacteria.[65] For example, bare gold or silver nanoparticles "sandwiched" between PAH layers were deposited on Gram-negative (*E. coli*) and Gram-positive (*Staphylococcus cohnii*) bacteria.[65] A considerable part of *Trichoderma asperellum* fungal spores and *S. cerevisiae* yeast cells remained vital after their surfaces were modified with sequential layers of PAH/PSS or bovine serum albumin/DNA from chicken erythrocytes with gold or silver nanoparticles between the polyelectrolyte layers and were capable of budding (yeast cells) or forming hyphae (*T. asperellum*).[66] Although the polyelectrolytes could serve as a protective barrier between the particles and a cell, leading to survival of most of the treated fungi, a partial cell death was still present and was probably due to some cytotoxicity from polycationic PAH. Similar effects were observed when bacterial spores of *B. subtilis* were coated with 72-nm silica particles deposited on top of the PLL/poly(glutamic acid)- (PLL/PGA) or (PDDA/PSS)-coated bacteria. The germination of the modified spores monitored by the percentage of the released dipicolinic acid was only 50% for (PLL/PGA)$_{4.5}$ and 70% for (PDDA/PSS)$_{4.5}$ while 90% for control (unmodified) bacterial spores. Studies on functionalization of yeast,[121] *Chlorella pyrenoidosa* algae cells,[30] bacteria[122] and nematodes[123] where PAH-coated magnetic nanoparticles (PAH@Fe$_2$O$_3$) were used instead of bare Fe$_2$O$_3$ nanoparticles demonstrated their nontoxicity to the cells. Compared to prokaryotic cells or yeast, animal cells are expected to be less robust due to the absence of a thick protective cell wall that prevents direct cytotoxic interactions of nanoparticles and a plasma membrane of animal cells. However, the assembly of nanoparticles directly onto HeLa cancer cell surfaces in one-step ionic attachment did not show extreme cytotoxicity. The modified cells showed viability of 83% close to 88% viability in control unmodified samples (Figure 6.5). Furthermore, the HeLa cells modified with PAH-coated Fe$_2$O$_3$ were able to grow and colonize suitable substrates.[67]

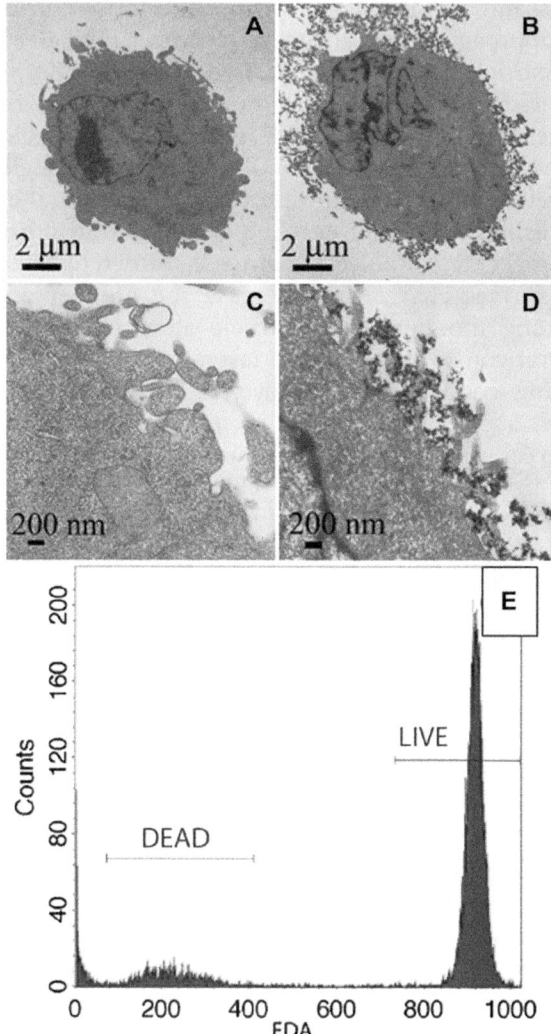

Figure 6.5   TEM images of (A and C) intact Hela cells and (B and D) Fe₂O₃@PAH-functionalized Hela cells. (E) Flow-cytometry data demonstrating the viability of FDA-stained functionalized HeLa cells.
Reprinted with permission from M. R. Dzamukova, A. I. Zamaleeva, D. G. Ishmuchametova, Y. N. Osin, A. P. Kiyasov, D. K. Nurgaliev, O. N. Ilinskaya, R. F. Fakhrullin, *Langmuir*, 2011, **27**, 14386. Copyright 2011, the American Chemical Society.

Carbon-based nanomaterials, such as multiwalled carbon nanotubes[62] or graphene oxide nanosheets[64] were used to functionalize yeast cells using the layer-by-layer approach. In both cases, the treatment resulted in a partial decrease in viability of the modified cells. Though not studied in detail, such toxic effects might be attributed to the properties of the nanomaterials used in cell surface modification and require more detailed insights.

6.4 Conclusion and Outlook

The impact of nanomaterials on biological structures is a topic of increasing interest and often requires a multidisciplinary approach for successful results. Although nanotechnology offers unique opportunities for cell surface engineering, it generates concerns about adverse effects of the nanocoatings on cellular viability and functions. One of the major worries is cytotoxicity of polymers and nanostructures used for cell modifications, as well as their metabolic products.

Upon modification, cells are exposed to a variety of chemical and physical stresses causing immediate or long-term toxic effects. The cell response to modification processes is evaluated by measuring cell death, the inhibition of cell proliferation, or cell metabolic activity. A variety of colorimetric assays are available for evaluating cell viability by measuring cell membrane integrity or probing cell activity. Eventually, several assays and approaches to evaluate toxicity of modification treatment should be utilized to prevent artifacts or ambiguous results. Structural and morphological differences between various types of cells lead to a different tolerance of the cells towards various cell surface modifications resulting in either cytocompatibility or cytotoxicity of the similar treatments. Microbial, yeast, or plant cells can better tolerate damaging osmotic pressure effects or deleterious impacts from cationic or toxic small molecule species due to their surrounding thick cell walls in contrast to fragile animal and human cells, in which these similar treatments can be extremely detrimental as targeted directly for the thin plasma membrane. Despite the challenges, the approaches to attenuate the toxicity of polycationic polymers are being developed. Thus, the toxicity of cationic polyelectrolytes dependent on the polymer charge density can be diminished by conjugating neutral molecules to the polyelectrolyte structures. Also, natural biopolymers have more similarities with the extracellular matrix, and are chemically versatile and have a good biocompatibility. The hydrogen-bonded multilayer assembly presents new opportunities for cytocompatible cell surface engineering as it allows for inclusion of polymers which carry no positive charge harmful to cells. Formation of thin solid films based on bioinspired mineralization can also decrease cytotoxic effects if used instead of a sol-gel mineralization approach. Modification of inorganic particles with molecules preventing a leak of cytotoxic ions or decreasing their solubility in culture media can be a way of minimizing nanoparticle cytotoxicity. In pancreatic islets modified with an ultrathin coating, islet necrosis can be avoided because of the rapid diffusion of nutrients and oxygen through the thin coating. Preservation of the islet integrity is a critical issue in modification of islet surfaces because islet cell functionality can be adversely affected when islets disintegrate or fuse in suspension culture. In this respect, the multilayer assembly may be beneficial in providing a conformal and cytocompatible coatings for individual islets. Though, despite the advantages, chemical modification of the cell membrane using covalent conjugation of molecules to the cell membrane proteins can result in disturbance of the cell physiology. In this respect, noncovalent modification methods can afford for cytocompatible coatings if nontoxic components are used.

Acknowledgement

This work was supported by the National Science Foundation under Award DMR-1306110.

References

1. E. Vega-Avila and M. K. Pugsley, *Proc. West. Pharmacol. Soc.*, 2011, **54**, 10.
2. S. R. Shin, H. Bae, J. M. Cha, J. Y. Mun, Y.-C. Chen, H. Tekin, H. Shin, S. Farshchi, M. R. Dokmeci, S. Tang and A. Khademhosseini, *ACS Nano*, 2012, **6**, 362.
3. J. P. Mosmann, *J. Immunol. Methods*, 1983, **65**, 55.
4. D. Gerlier and N. Thomasset, *J. Immunol. Methods*, 1986, **94**, 57.
5. E. Henriksson, E. Kjellén, P. Wahlberg, J. Wennerberg and J. H. Kjellström, *In Vitro Cell. Dev. Biol.*, 2006, **42**, 320.
6. W. Yan-Ping, L. Xiao-Yu, S. Chung-Qing and H. Zhi-Bi, *Acta Pharmacol. Sin.*, 2002, **23**, 263.
7. S. Al-Nasiry, N. Geusens, M. Hanssens, C. Luyten and R. Pijnenborg, *Hum. Reprod.*, 2007, **22**, 1304.
8. J. M. Clarke, M. R. Gillings, N. Altavilla and A. J. Beattie, *J. Microbiol. Methods*, 2001, **46**, 261.
9. M. Sato, Y. Murata, M. Mizusawa, H. Iwahashi and S. Oka, *Microbiol. Cult. Coll.*, 2004, **20**, 53.
10. R. D. Fields and M. V. Lancaster, *Am. Biotechnol. Lab.*, 1993, **11**, 48.
11. R. J. Gonzalez and J. B. Tarloff, *Toxicol. In Vitro*, 2001, **15**, 257.
12. S. A. Ahmed, R. M. Gogal, Jr. and J. E. Walsh, *J. Immunol. Methods*, 1994, **170**, 211.
13. H. X. Zhang, G. H. Du and J. T. Zhang, *Acta Pharmacol. Sin.*, 2004, **25**, 385.
14. J. O'Brien, I. Wilson, T. Orton and F. Pognan, *Eur. J. Biochem.*, 2000, **267**, 5421.
15. C. Baba, K. Yanagida, T. Kanzaki and M. Baba, *Antiviral Chem. Chemother.*, 2005, **16**, 33.
16. C. Legrand, J. M. Bour, C. Jacob, J. Capiaumont, A. Martial, A. Marc, M. Wudtke, G. Kretzmer, C. Demangel and D. Duval, *J. Biotechnol.*, 1992, **25**, 231.
17. A. J. Prija, S. P. Vijayalakshmi and A. M. Raichur, *J. Agric. Food Chem.*, 2011, **59**, 11838.
18. A. L. Hillberg and M. Tabrizian, *Biomacromolecules*, 2006, 7, 2742.
19. A. Diasporo, D. Silvano, S. Krol, O. Cavalleri and A. Gliozzi, *Langmuir*, 2002, **18**, 5047.

20. V. Kozlovskaya, S. Harbaugh, I. Drachuk, O. Shchepelina, N. Kelley-Loughnane, M. Stone and V. V. Tsukruk, *Soft Matter*, 2011, 7, 2364.
21. S. Krol, M. Nolte, A. Diasporo, D. Mazza, R. Mgrassi, A. Gliozzi and A. Fery, *Langmuir*, 2005, **21**, 705.
22. S. Harbaugh, N. Kelley-Loughnane, M. Davidson, L. Narayanan, S. Trott, Y. G. Chushak and M. O. Stone, *Biomacromolecules*, 2009, **10**, 1055.
23. L. E. Weigand, M. Sanchez, E.-B. Gunnesch, S. Zeiher, R. Schroeder and B. Suess, *RNA*, 2008, **14**, 89.
24. J. L. Carter, I. Drachuk, S. Harbaugh, N. Kelley-Loughnane, M. Stone and V. V. Tsukruk, *Macromol. Biosci.*, 2011, **11**, 1244.
25. B. Franz, S. S. Balkundi, C. Dahl, Y. M. Lvov and A. Prange, *Macromol. Biosci.*, 2010, **10**, 164.
26. S. Mansouri, Y. Merhi, F. M. Winnik and M. Tabrizian, *Biomacromolecules*, 2011, **12**, 585.
27. S. S. Balkundi, N. G. Veerabadran, D. M. Eby, G. R. Johnson and Y. M. Lvov, *Langmuir*, 2009, **25**, 14011.
28. M. Paidhungat, K. Ragkousi and P. Setlow, *J. Bacteriol.*, 2001, **183**, 4886.
29. D. Zhang, R. F. Fakhrullin, M. Özmen, H. Wang, J. Wang, V. N. Paunov, G. Li and W. E. Huang, *Microb. Biotechnol.*, 2011, **4**, 89.
30. R. F. Fakhrullin, L. V. Shlykova, A. I. Zamaleeva, D. K. Nurgaliev, Y. N. Osin, J. Garcia-Alonso and V. N. Paunov, *Macromol. Biosci.*, 2010, **10**, 1257.
31. S. Ho Yang, D. Hong, J. Lee, E. H. Ko and I. S. Choi, *Small*, 2013, **9**, 178.
32. R. F. Fakhrullin and Y. M. Lvov, *ACS Nano*, 2012, **6**, 4557.
33. H. Ai, M. Fang, S. A. Jones and Y. M. Lvov, *Biomacromolecules*, 2002, **3**, 560.
34. N. G. Veerabadran, P. L. Goli, S. S. Stewart-Clark, Y. M. Lvov and D. K. Mills, *Macromol. Biosci.*, 2007, **7**, 877.
35. J. T. Wilson, W. Cui, V. Kozlovskaya, E. Kharlampieva, D. Pan, Z. Qu, V. R. Krishnamurthy, J. Mets, V. Kumar, J. Wen, Y. Song, V. Tsukruk and E. L. Chaikof, *J. Am. Chem. Soc.*, 2011, **133**, 7054.
36. A. J. Swiston, C. Cheng, S. Ho Um, D. J. Irvine, R. E. Cohen and M. F. Rubner, *Nano Lett.*, 2008, **8**, 4446.
37. Y. Teramura and H. Iwata, *Soft Matter*, 2010, **6**, 1081.
38. H. Lodish, A. Berk, C. A. Kaiser, M. Krieger, A. Bretscher, H. Ploegh, A. Amon and M. P. Scott, *Molecular Cell Biology*, W. H. Freeman, New York, 2012, 7th edn.
39. B. Pham-Hoang, C. Romero-Guido, H. Phan-Thi and Y. Waché, *Appl. Microbiol. Biotechnol.*, 2013, **97**, 6635.
40. J. C. Kapteyn, H. Van Den Ende and F. M. Klis, *Biochim. Biophys Acta*, 1999, **1426**, 373.
41. O. Cabrera, D. M. Berman, N. S. Kenyon, C. Ricordi, P.-O. Berggren and A. Caicedo, *Proc. Nat. Acad. Sci. U.S.A.*, 2006, **103**, 2334.
42. M. Brissova, M. J. Fowler, W. E. Nicholson, A. Chu, B. Hirshberg, D. M. Harlan and A. C. Powers, *J. Histochem. Cytochem.*, 2005, **53**, 1087.

43. C. Brunot, L. Ponsonnet, C. Lagneau, P. Farge, C. Picart and B. Grosgogeat, *Biomaterials*, 2007, **28**, 632.
44. A. Beyerle, M. Irmler, J. Beckers, T. Kissel and T. Stoeger, *Mol. Pharmaceut.*, 2010, 7, 727.
45. G. Grandinetti, A. E. Smith and T. M. Reineke, *Mol. Pharmaceut.*, 2012, **9**, 523.
46. J. Cai, Y. Yue, D. Rui, Y. Zhang, S. Liu and C. Wu, *Macromolecules*, 2011, **44**, 2050.
47. M. Germain, P. Balaguer, J.-C. Nicolas, F. Lopez, J.-P. Esteve, G. B. Sukhorukov, M. Winterhalter, H. Richard-Foy and D. Fournier, *Biosens. Bioelectron.*, 2006, **21**, 1566.
48. Y. Teramura, Y. Kaneda, T. Totani and H. Iwata, *Biomaterials*, 2008, **29**, 1345.
49. J. T. Wilson, W. Cui and E. L. Chaikof, *Nano Lett.*, 2008, **8**, 1940.
50. D. Fischer, Y. Li, B. Ahlemeyer, J. Krieglstein and T. Kissel, *Biomatreials*, 2003, **24**, 1121.
51. C. M. Goodman, C. M. McCusker, T. Yilmatz and V. M. Rotello, *Bioconjugate Chem.*, 2004, **15**, 897.
52. M. Chanana, A. Gliozzi, A. Diasporo, I. Chodnevskaja, S. Huewel, V. Moskalenko, K. Ulrichs, H.-J. Galla and S. Krol, *Nano Lett.*, 2005, **5**, 2605.
53. S. Hong, P. R. Leroueil, E. K. Janus, J. L. Peters, M.-M. Kober, M. T. Islam, B. G. Orr, J. R. Baker, Jr. and M. M. Banaszak Holl, *Bioconjugate Chem.*, 2006, **17**, 728.
54. S. Hong, A. U. Bielinska, A. Mecke, B. Keszler, J. L. Beals, X. Shi, L. Balogh, B. G. Orr, J. R. Baker, Jr. and M. M. Banaszak Holl, *Bioconjugate Chem.*, 2004, **15**, 774.
55. A. Mecke, I. J. Majoros, A. K. Patri, J. R. Baker, Jr., M. M. Banaszak Holl and B. G. Orr, *Langmuir*, 2005, **21**, 10348.
56. S. M. Moghimi, P. Symonds, J. C. Murray, A. C. Hunter, G. Debska and A. Szewczyk, *Mol. Ther.*, 2005, **11**, 990.
57. P. Symonds, J. C. Murray, A. C. Hunter, G. Debska, A. Szewczyk and S. M. Moghimi, *FEBS Lett.*, 2005, **579**, 6191.
58. Z. Kadlecova, L. Baldi, D. Hacker, F. M. Wurm and H.-A. Klok, *Biomacromolecules*, 2012, **13**, 3127.
59. J. T Wilson, V. R. Krishnamurthy, W. Cui, Z. Qu and E. L. Chaikof, *J. Am. Chem. Soc.*, 2009, **131**, 18228.
60. R. B. Kolhatkar, K. M. Kitchens, P. W. Swaan and H. Ghandehari, *Bioconjugate Chem.*, 2007, **18**, 2054.
61. K. Fant, E. K. Esbjörner, A. Jenkins, M. C. Grossel, P. Lincoln and B. Norden, *Mol. Pharmaceut.*, 2010, 7, 1734.
62. A. I. Zamaleeva, O. R. Sharipova, A. V. Porfireva, G. A. Evtugyn and R. F. Fakhrullin, *Langmuir*, 2010, **26**, 2671.
63. S. R. Shin, H. Bae, J. M. Cha, J. Y. Mun, Y.-C. Chen, H. Tekin, H. Shin, S. Farshchi, M. R. Dokmeci, S. Tang and A. Khademhosseini, *ACS Nano*, 2012, **6**, 362.

64. S. H. Yang, T. Lee, E. Seo, E. H. Ko, I. S. Choi and B.-S. Kim, *Macromol. Biosci.*, 2012, **12**, 61.
65. M. Kahraman, A. I. Zamaleeva, R. F. Fakhrullin and M. Culha, *Anal. Bioanal. Chem.*, 2009, **395**, 2559.
66. R. F. Fakhrullin, A. I. Zamaleeva, M. V. Morozov, D. I. Tazetdinova, F. K. Alimova, A. K. Hilmutdinov, R. I. Zhdanov, M. Kahraman and M. Culha, *Langmuir*, 2009, **25**, 4628.
67. M. R. Dzamukova, A. I. Zamaleeva, D. G. Ishmuchametova, Y. N. Osin, A. P. Kiyasov, D. K. Nurgaliev, O. N. Ilinskaya and R. F. Fakhrullin, *Langmuir*, 2011, **27**, 14386.
68. J. Garcia-Alonso, R. F. Fakhrullin, V. N. Paunov, Z. Shen, J. D. Hardege, N. Pamme, S. J. Haswell and G. M. Greenway, *Anal. Bioanal. Chem.*, 2011, **400**, 1009.
69. P. H. Keen, N. K. Slater and A. F. Routh, *Langmuir*, 2012, **28**, 1169.
70. A. D. Lehmann, W. J. Parak, F. Zhang, Z. Ali, C. Roecker, G. U. Nienhaus, P. Gehr and B. Rothen-Rutishauser, *Small*, 2010, **6**, 753.
71. C. Kirchner, T. Liedl, S. Kudera, T. Pellegrino, A. Munoz Javier, H. E. Gaub, S. Stoelzle, N. Fertig and W. J. Parak, *Nano Lett.*, 2004, **5**, 331.
72. Y. Qiu, Y. Liu, L. Wang, L. Xu, R. Bai, Y. Ji, X. Wu, Y. Zhao, Y. Li and C. Chen, *Biomaterials*, 2010, **31**, 7606.
73. D. Hühn, K. Kantner, C. Geidel, S. Brandholt, I. De Cock, S. J. H. Soene, P. Rivera-Gil, J.-M. Montenegro, K. Braeckmans, K. Müllen, G. U. Nienhaus, M. Klapper and W. J. Parak, *ACS Nano*, 2013, **4**, 3253.
74. P. R. Leroueil, S. Hong, A. Mecke, J. R. Baker, B. G. Orr and M. M. Banaszak Holl, *Acc. Chem. Res.*, 2007, **40**, 335.
75. A. K. Suresh, D. A. Pelletier, W. Wang, J. L. Morrell-Falvey, B. Gu and M. J. Doktycz, *Langmuir*, 2012, **28**, 2727.
76. S. Kittler, C. Greulich, J. Diendorf, M. Koeller and M. Epple, *Chem. Mater.*, 2010, **22**, 4548.
77. T. Xia, Y. Zhao, T. Sager, S. George, S. Pokhrel, N. Li, D. Schoenfeld, H. Meng, S. Lin, X. Wang, M. Wang, Z. Ji, J. I. Zink, L. Maedler, V. Castranova, S. Lin and A. E. Nel, *ACS Nano*, 2011, **5**, 1223.
78. E. Navarro, F. Piccapietra, B. Wagner, F. Marconi, R. Kaegi, N. Odzak, L. Sigg and R. Behra, *Environ. Sci. Technol.*, 2008, **42**, 8959.
79. T. J. Brunner, P. Wick, P. Manser, P. Spohn, R. N. Grass, L. K. Limbach, A. Bruinink and W. J. Stark, *Environ. Sci. Technol.*, 2006, **40**, 4374.
80. A. Nel, T. Xia, L. Maedler and N. Li, *Science*, 2006, **311**, 622.
81. H.-F. Cui, S. K. Vashist, K. Al-Rubeaan, J. H. T. Luong and F.-S. Sheu, *Chem. Res. Toxicol.*, 2010, **23**, 1131.
82. V. E. Kagan, Y. Y. Tyurina, V. A. Tyurin, N. V. Konduru, A. I. Potapovich, A. N. Osipov, E. R. Kisin, D. Schwegler-Berry, R. Mercer, V. Castranova and A. A. Shvedova, *Toxicol. Lett.*, 2006, **165**, 88.
83. K. Pulskamp, S. Diabate and H. F. Krug, *Toxicol. Lett.*, 2007, **168**, 58.
84. X. Chen, U. C. Tam, J. L. Czlapinski, G. S. Lee, D. Rabuka, A. Zettl and C. R. Bertozzi, *J. Am. Chem. Soc.*, 2006, **128**, 6292.

85. D. Rabuka, M. B. Forstner, J. T. Groves and C. R. Bertozzi, *J. Am. Chem. Soc.*, 2008, **130**, 5947.
86. S. Krol, S. Guerra, M. Grupillo, A. Diasporo, A. Gliozzi and P. Marchetti, *Nano Lett.*, 2006, **6**, 1933.
87. *Multilayer Thin Films: Sequential Assembly of Nanocomposite Materials*, ed. G. Decher and J. B. Schlenoff, Wiley-VCH, Weinheim, 2012, 2nd edn.
88. S. De Koker, B. G. De Geest, C. Cuvelier, L. Ferdinande, W. Deckers, W. E. Hennink, S. De Smedt and N. Mertens, *Adv. Funct. Mater.*, 2007, **17**, 3754.
89. T. Bieber, W. Meissner, S. Kostin, A. Niemann and H. P. Elsasser, *J. Control. Release*, 2002, **82**, 441.
90. F. M. Menger, V. A. Seredyuk, M. V. Kitaeva, A. A. Yaroslavov and N. S. Melik-Nubarov, *J. Am. Chem. Soc.*, 2003, **125**, 2846.
91. J. Lee, S. H. Yang, S.-P. Hong, D. Hong, H. Lee, H.-Y. Lee, Y.-G. Kim and I. S. Choi, *Macromol. Rapid. Commun.*, 2013, DOI: 10.1002/macr.201300444.
92. Z. L. Zhi, B. Liu, P. M. Jones and J. C. Pickup, *Biomacromolecules*, 2010, **11**, 610.
93. V. Kozlovskaya, O. Zavgorodnya, Y. Chen, K. Ellis, H. M. Tse, W. Cui, J. A. Thompson and E. Kharlampieva, *Adv. Funct. Mater.*, 2012, **22**, 3389.
94. Y. Teramura, Y. Kaneda and H. Iwata, *Biomaterials*, 2007, **28**, 4818.
95. H. Iwata, T. Takagi, H. Amemiya, H. Shimizu, K. Yamashita, K. Kobayashi and T. Akutsu, *J. Biomed. Mater. Res.*, 1992, **26**, 967.
96. M. Amiji and K. Park, *J. Biomater. Sci., Polym. Ed.*, 1993, **4**, 217.
97. D. Xie, C. A. Smyth, C. Eckstein, G. Bilbao, J. Mays, D. E. Eckhoff and J. L. Contreras, *Biomaterials*, 2005, **26**, 403.
98. D. Y. Lee, K. Yang, S. Lee, S. Y. Chae, K.-W. Kim, M. K. Lee, D.-J. Han and Y. Byun, *J. Biomed. Mater. Res.*, 2002, **62**, 372.
99. S. Kizilel, A. Scavone, X. Liu, J.-N. Nothias, D. Ostrega, P. Witkowski and M. Millis, *Tissue Eng., Part A*, 2010, **16**, 2217.
100. L. Barani, E. Vasheghani-Farahani, H. A. Lazarjani, S. Hashemi-Najafabadi and F. Atyabi, *Biotechnol. Appl. Biochem.*, 2010, **57**, 25.
101. S. H. Yang, S. M. Kang, K.-B. Lee, T. D. Chung, H. Lee and I. S. Choi, *J. Am. Chem. Soc.*, 2011, **133**, 2795.
102. M. Perullini, M. Jobbagy, M. Bermudez Moretti, S. Correa Garcia and S. A. Bilmes, *Chem. Mater.*, 2008, **20**, 3015.
103. S. H. Yang, K.-B. Lee, B. Kong, J.-H. Kim, H.-S. Kim and I. S. Choi, *Angew. Chem. Int. Ed.*, 2009, **48**, 9160.
104. H. K. Baca, E. Carnes, S. Singh, C. Ashley, D. Lopez and C. J. Brinker, *Acc. Chem. Res.*, 2007, **40**, 836.
105. B. Wang, P. Liu, W. Jiang, H. Pan, X. Xu and R. Tang, *Angew. Chem. Int. Ed.*, 2008, **47**, 3560.
106. R. F. Fakhrullin and R. T. Minullina, *Langmuir*, 2009, **25**, 6617.
107. S. H. Yang, D. Hong, J. Lee, E. H. Ko and I. S. Choi, *Small*, 2013, **9**, 178.

108. W. E. G. Müller, S. Engel, X. Wang, S. E. Wolf, W. Tremel, N. L. Thakur, A. Krasko, M. Divekar and H. C. Schröder, *Biomaterials*, 2008, **29**, 771.
109. M. Perullini, M. Rivero, M. Jobbagy, A. Mentaberry and S. A. Bilmes, *J. Biotechnol.*, 2007, **127**, 542.
110. S. H. Yang, E. H. Ko and I. S. Choi, *Langmuir*, 2012, **28**, 2151.
111. C. F. Meunier, P. Dandoy and B.-L. Su, *J. Colloid Interface Sci.*, 2010, **342**, 211.
112. D. Anvir, T. Coradin, O. Lev and J. Livage, *J. Mater. Chem.*, 2006, **16**, 1013.
113. M. L. Ferrer, L. Yuste, F. Rojo and F. del Monte, *Chem. Mater.*, 2003, **15**, 3614.
114. M. L. Ferrer, Z. Y. Garcia-Carvajal, L. Yuste, F. Rojo and F. del Monte, *J. Mater. Chem.*, 2006, **18**, 1458.
115. P. Dandoy, C. F. Meunier, G. Leroux, V. Voisin, L. Giordano, N. Caron, C. Michiels and B.-L. Su, *PLoS ONE*, 2013, **8**, e54683.
116. E. C. Carnes, J. C. Harper, C. E. Ashley, D. M. Lopez, L. M. Brinker, J. Liu, S. Singh, S. M. Brozik and C. J. Brinker, *J. Am. Chem. Soc.*, 2009, **131**, 14255.
117. T. Imai and T. Ohno, *Appl. Environ. Microbiol.*, 1995, **61**, 3604.
118. W.-S. Kuo, C.-M. Wu, Z.-S. Yang, S.-Y. Chen, C.-Y. Chen, C.-C. Huang, W.-M. Li, C.-K. Sunc and C.-S. Yeh, *Chem. Commun.*, 2008, **37**, 4430.
119. A. Sugunan, P. Melin, J. Schnürer, J. G. Hilborn and J. Dutta, *Adv. Mater.*, 2007, **19**, 77.
120. Z. Li, S.-W. Chung, J.-M. Nam, D. S. Ginger and C. A. Mirkin, *Angew. Chem. Int. Ed.*, 2003, **42**, 2306.
121. R. W. Fakhrullin, J. Garcia-Alonso and V. N. Paunov, *Soft Matter*, 2010, **6**, 391.
122. D. Zhang, R. F. Fakhrullin, M. Özmen, H. Wang, J. Wang, V. N. Paunov, G. Li and W. E. Huang, *Microb. Biotechnol.*, 2011, **4**, 89.
123. R. T. Minullina, Y. N. Osin, D. G. Ishmuchametova and R. F. Fakhrullin, *Langmuir*, 2011, **27**, 7708.

Microelectronic Devices Based on Nanomaterial-Carrier Cells

VIVEK MAHESHWARI* AND SHEHAN SALGADO

Dept of Chemistry, The Nanotechnology Program; Waterloo Institute of Nanotechnology, University of Waterloo, 200 University Ave. West, Waterloo ON Canada N2L 3G1
*Email: vmaheshw@uwaterloo.ca

7.1 Introduction

Cells are structured materials with specific organization targeted to accomplish the task of sustaining life. The outer layer of cells is a plasma membrane or a cell wall, both of which act as barriers leading to a controlled environment inside the cell. This barrier composed of lipids, carbohydrates, and proteins is organic in composition. The natural environment of the cells dictates that ions and protons are the primary charged species that are the part of cellular processes. The evolution of ion and proton pumps and channels is a clear indication of their crucial role in sustaining life. Electrons, on the other hand, play a very limited role (mostly short range) due to the aqueous nature and organic composition of a cell's environment. This selection dictated by the natural environment of cells leads to significant challenges for development of electronic devices integrated with or based on cells. This is apparent in human environment, where most of our biological processes are biochemical and based on ions, but our civilization is driven in large part by electron-based processes including our basic source of energy – electricity. The integration of cells with an active microelectronic interface

RSC Smart Materials No. 9
Cell Surface Engineering: Fabrication of Functional Nanoshells
Edited by Rawil F Fakhrullin, Insung S Choi and Yuri Lvov
Published by the Royal Society of Chemistry, www.rsc.org

therefore requires the design of an interface for coupling an organic-ionic entity with an inorganic–electronic material.

The motivation for development of such cell–inorganic hybrids is dual purpose: 1) the cell's internal environment is dynamic and responds to its own needs and the stimuli received from the external factors. These dynamic changes result in both morphological adjustments (response from the cytoskeleton) in the cell and also variations in the chemical potential of the ions in the cell. To detect and monitor these cellular responses in real time an electronic interface is well suited, due to the progress made in micro-electronic circuits in terms of sensitivity and noise levels in these devices. Further, due to miniaturization, microelectronic devices on the size scale of cells are accessible. 2) A more ambitious outlook is using this electronic interface in reverse to dictate the cellular responses. This entails a more controlled and specific interfacing with the cell and precise translation of the electrical stimuli from the microelectronic device to a biorecognizable factor. Both these aims target to increase our ability to understand the basic unit of life, the cell. This also advances our ability to further build on this natural system and harness its complexity to use in man-made devices.

The complexity of developing a cell based microelectronic device lies first in the design of the inorganic material that will physically attach to the cell. Biocompatibility is the obvious issue to be addressed but equally crucial are the size of the basic unit that will attach to the cell and the translation mechanism by which the cells biosignal will be transformed to an electrical modulation. Due to size considerations, as cells are micrometer size in scale, nanomaterials are used for developing these interfaces.Three classes of nanomaterials: nanoparticles (0-dimensional, 0D), nanowires/nanorods (1-dimensional, 1D) and sheets (2-dimensional, 2D) are commonly used for this purpose.[1–3] The geometrical differences of the three materials lead to some significant distinctiveness in the possible devices and their functional characteristics. One apparent difference is for forming an electronic (or electrical) device the interface has to be electrically percolating, so a 0D interface will require a large no. of nanoparticles (NP) and hence will lead to a series of tunneling junctions (Figure. 7.1a).[4] 1D and 2D nanomaterial

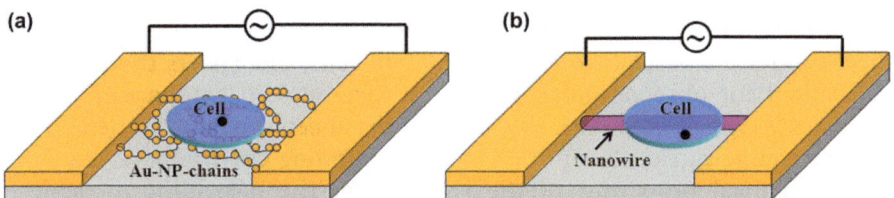

Figure 7.1   Schematic showing cell devices. (a) Chains of Au nanoparticles span the gap between two electrodes, forming a conductive pathway. The electrons flow by sequential tunneling across the nanoparticles. (b) Using a 1D nanomaterial like a nanowire the electron pathway is continuous. In both cases the cell is placed on top of the conductive pathway.

devices on the other hand can be made with a single continuous interface due to the high aspect ratios of these materials (Figure 7.1b).[1,5] This has several implications for the signal levels and sensitivity of the device in detecting the cellular signals. 0D nanomaterial-based devices will have a lower base signal due the sequential tunneling junctions along the electron percolating pathway of the nanoparticles. As shown in Figure 7.1a, the nanoparticles (self-assembled necklaces of Au nanoparticles) form a pathway for electron conduction and the cell is placed on top of them. The electron conduction results from the sequential tunneling of the electrons along the nanoparticles across the applied potential difference.[6] Though having a lower signal, the nanoparticles can act a sensitive detector for the dynamic behavior of the cells. The local tunneling barrier will be modulated due to the cells dynamic behavior and this should alter the current (electron) flow across the nanoparticles.

In the case of interface made with 1D and 2D nanomaterials the cell acts more as a biological gate that modulates the current due to its dynamic behavior.[1,7] The gating effect from the cell occurs due to modulations such as change in cell potential (as in the case of neurons) that directly alters the charge carrier dynamics in the interface.[3,8,9] Gating has also been observed due to changes in the cell's chemical potential and its mechanical characteristics. The ability to observe these cell dynamics in real time is a significant benefit provided by the electrical interface in comparison to chemical-fluorescent methods. There are advantages offered by both the 1D and 2D systems. 1D systems, typically materials such as nanowires and nanorods, have the advantage that due to their high aspect ratio a single element can lead to a percolating path for conduction of electrons. Further, due to their nanometer scale cross-section spatially local probing of the cell dynamics is possible. 1D systems have been used to generate extra cellular coupling with the cell membrane, probing, *e.g.* the transient dynamics and propagation of neural impulses in neurons.[3] Due to their small cross section 1D systems can also easily penetrate the cell membrane and lead to coupling with the cells interior. Nanowire (and nanotube) devices have demonstrated the ability to monitor a cell's internal chemical potential.[10–13]

2D nanomaterials, primarily graphene sheets, offer a premade electrically conducting large-area interface with nanometer-scale thickness. Graphene sheets micrometer scale in size are easily synthesized and hence a single sheet can have a surface area comparable to cells. The sheets therefore are well suited for observation of whole-cell dynamics in comparison to 1D elements.[1,5] The atomic-scale thickness of the sheets leads to the sensitivity of their charge-carrier dynamics to the surrounding environment. Further, the mechanical properties of the sheets offer the possibility to achieve electromechanical coupling with the cells. The bending modulus of material scales inversely to the power cube of thickness. Hence, graphene sheets with a thickness of ~1 nm easily bend and deform by stresses on the scale of kPa. In comparison, following the scaling with thickness 10 nm diameter wires will require MPa of stress to achieve similar scale of deformation.

This leads to the possibility of observing mechanical modulations in the cell using graphene sheets.[1] The other advantage offered by the large area and mechanical flexibility of the sheets is that cells can be wrapped in them forming a bio–inorganic hybrid.[14]

7.2 Device Configurations

7.2.1 IN and OUT Devices

The difference between an interface that couples to the exterior of the cells such as its plasma membrane from one that couples to the cell's interior are directly linked to cell dynamics. The exterior of the cell plays a dynamic role in its interaction with the surrounding. The cell's response is expressed in variables such as cell-membrane potential, dynamic modulation of its ion channels and its shape and size. The response of the cell as expressed on its external surface features provides a diverse array of signals that can be used to monitor the cell dynamics. The internal coupling in principle will have access to an array of internal cellular processes and hence an even more complex set of cell dynamics. The internal chemical potential of the cell, changes associated with expression of internal ion channels and pumps and dynamics of genetic expression are some of the cell dynamics that can be possibly linked to internal coupling devices. The **Out** devices can interface with the whole exterior surface area of the cell and hence can use larger-scale nanomaterials such as graphene, transistors made of CNT networks and semiconducting organic polymers for making the device.[5,7–9,15,16] In contrast, typically the requirement for internal coupling in the **In** devices leads to the use of single 1D nanowires and CNTs, where the nanomaterial penetrates the cells membrane, accessing the cells interior with minimally invasive effects. Recent examples have demonstrated in concept the **In** devices with internal coupling to the cells and dynamic recording of the cells response to a controlled external stimuli.[10,17] Using such internal coupling to target specific cell organelles such as mitochondria or the nucleus offers significant possibilities for the application of these internal cell devices. To achieve this, the 1D structure used to probe the interior of the cell will have to be further reduced to the scales of cell organelles. Targeting agents that lead to coupling with only specific cell machinery will also have to be used to achieve this aim.

The common challenge associated with both the techniques is the ability to relate the observed electrical signal to a specific cellular process (or processes). This is especially challenging in the case of placing the cell in an uncontrolled environment where multiple stimuli act on the cell and its response is also multifaceted. Developing an electrical interface that targets signal from only a specific cellular response–stimuli combination can provide devices that play a specific role in study and detection of cell dynamics. The challenge here is to make the interfacing with the cell specific.

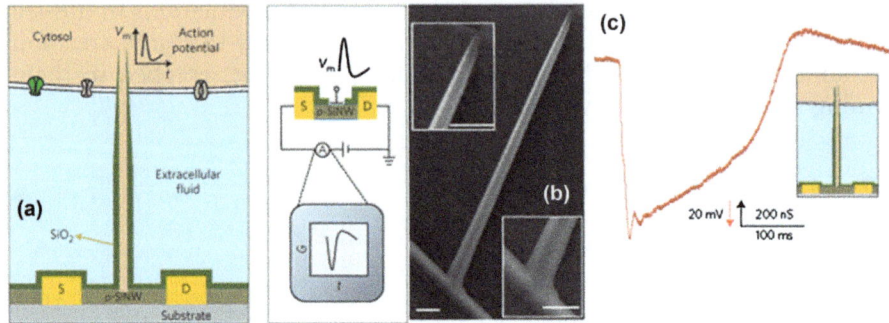

Figure 7.2   Schematic of an **In** cell device. (a) A Si nanotube brings the cells cytoplasmic fluid in contact with a p-type Si nanowire. (b) The inverted T junction of the nanowire and the connected nanotube. (c) A typical cell potential recorded from a cardiomyocyte cell. The change in the cell potential is associated with the rhythmic contractions of the cell. Reprinted with permission from ref. 10. Copyright (2011) Nature Publishing Group.

7.2.2   ON and UNDER Devices

On and **Under**: External interfaces with cell are more facile to accomplish. They can provide a multitude of information on cell dynamics that can lead to significant benefits in the study of cells. The external interface can be made by placing the nanomaterial on a substrate and then the cell is positioned on it forming the device (**Under** device, Figure 7.3).[3,7–10,16,18,19] The other approach is to make the interface on the surface of the cell (**On** device). In this case the nanomaterial is first interfaced with the cell surface and then the nanomaterial-cell hybrid is integrated with the microelectronic components to form the device (Figure 7.4).[1,2,4] The critical difference between the two is that in the **Under** device the interface has two interactions, one with the cell and the other with the substrate on which the device is assembled. This interaction can lead to constraints on the cell–interface interactions. For example, in the case of mechanical modulation due to change in cell size or shape, the nanomaterial interface in the **Under** approach will be constrained due to the substrate, while the nanomaterial in the **On** device will be able to respond to these mechanical changes. The advantage comes with the challenges, specifically the effect of the nanomaterial on the surface of the cell to the cell's interaction with its surroundings. The other challenge is that since the nanomaterial is first interfaced with the cell, it is necessary to gauge the ability of the nanomaterial to not only adhere to the cell surface but also its propensity to be internalized by the cell. The internalization of nanomaterials significantly impacts the cell's functioning and the effect needs to be critically gauged in such cases. The mitigation of these effects for the **On** device assembly needs particular strategies, some of which are discussed later in this chapter.

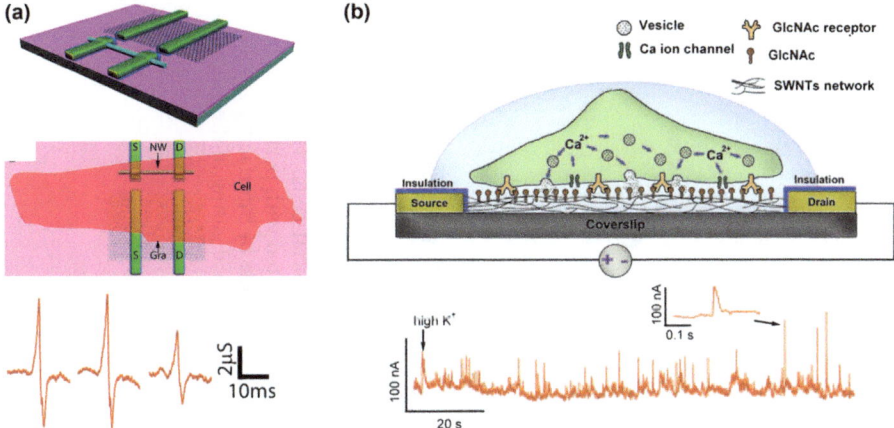

Figure 7.3 (a) **Under** cell devices made with graphene sheets and Si nanowire functioning as the interfacing material with the cell. The lower image shows the recording from cardiomyocyte cells placed on the graphene transistor shown in the top image. Reprinted with permission from ref. 5. Copyright (2010) the American Chemical Society. (b) An **Under** cell device made with a carbon nanotube network. PC12 cells are interfaced with the CNT networks forming the on cell device. The lower image shows that on stimulation by a high K^+ ion solution, the exocytosis from the cells results in gating and modulation of the current in the CNT network.
Reprinted with permission from ref. 8. Copyright (2009) Wiley-VCH Verlag GmbH & Co. KGaA, Weinheim.

The rest of this chapter will focus on devices built using cell nanomaterials, their principle of sensing and their future.

7.3 Principles of Cell-Nanomaterial Devices

Translating the cell dynamics to electrical signal modulations by the nanomaterial interface is governed by the characteristics of the material itself and the nature of the assembly. In particular, we will discuss two means of modulations, 1) The effect on tunneling barriers that may be present in the electron conduction pathway of the nanomaterial interface. 2) The gating of the electron conduction pathway due to cellular dynamics.

7.3.1 Tunneling Barrier Effect

As a simple model, if we consider tunneling between two species in the presence of an external electric field, the tunneling probability (P) is exponentially proportional to the height of the energy barrier between the two (Φ_c) and the width of the barrier (a). The probability also depends exponentially on the externally applied local electric field (E), the critical field for tunnelling (E_c) and on the dielectric constant of the media separating the

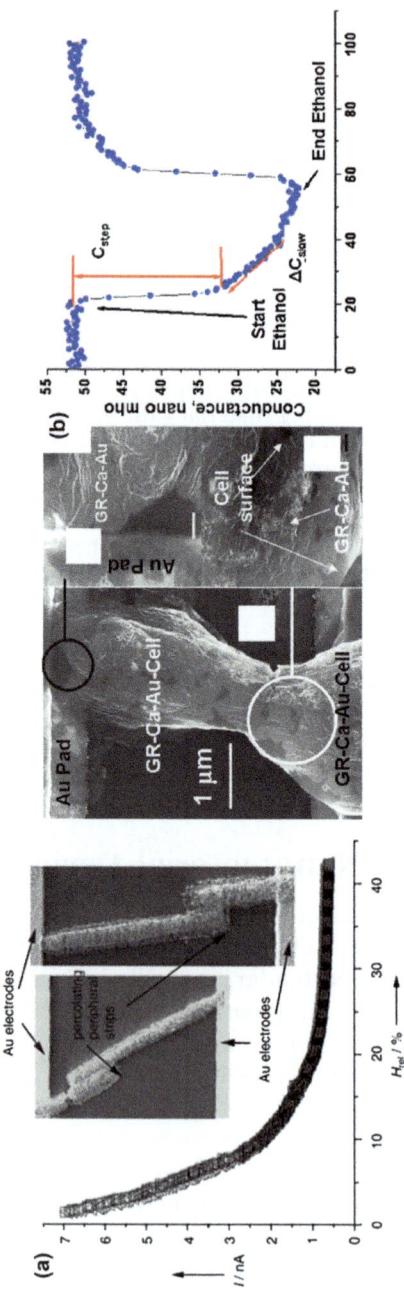

Figure 7.4 (a) **On** cell devices made by depositing Au nanoparticles in high density on the surface of bacterial cells. The change in humidity modulates the tunneling barrier and hence also the current as seen in the graph. Reprinted with permission from ref. 2. Copyright (2009) Wiley-VCH Verlag GmbH & Co. KGaA, Weinheim. (b) An **On** cell device made by interfacing graphene sheets with yeast cells. Osmotic stresses in the cell due to ethanol exposure lead to electromechanical gating of the graphene sheets. This modulates the current through them, shown on the right. Reprinted with permission from ref. 1. Copyright (2005) the American Chemical Society.

two species. With this basic formulation, if the external interface consists of a series of tunneling barriers, the current through it will be modulated due to any changes in the parameters governing the tunneling probability.[2,20,21]

$$P \propto \exp\left[-\Phi_c^{\frac{1}{2}} a\right] \approx \exp\left(-\frac{E_c}{E}\right)$$

Consider a device set-up with the interface consisting of tunneling junction in contact with the cell (Figure 7.1). An example of such a set-up is the use of Au nanoparticle chains to form a conductive pathway. The complete assembly is immersed in a solution required to maintain cell activity in a controlled environment. The cell is dynamic and this leads to modulation of the tunneling parameters in the interface. For example, the energy barrier can be altered due to cellular processes such as the flow of ions and small molecules. The ions carry charge and this can alter the magnitude of the energy barrier for tunneling. Further, the flow of ions will alter the local electrical double layer (EDL) which will be present on the interface as it is maintained at a potential (compared to the solution) for the flow of electrons. The EDL affects the local electric field of tunneling. Hence, the modulations in the current are reflective of the cell's behavior as we can directly correlate the change in the tunneling parameters, as noted above due to the cell's activity.

Further modulations in cells shape and size (*e.g.*, linked to cell's growth and cytoskeleton) can alter the physical dimensions of the tunneling gap. This again has an exponential effect on the tunneling probability. Such modulations in the physical size of the tunneling junction have been demonstrated by altering the humidity of the cell's environment.[2] This can be translated to devices where the cell is maintained in a controlled growth environment and its size is monitored by the modulations in current.

A challenge in the tunneling-based interface for cells is the signal levels, *i.e.* the magnitude of the current for voltages that can be applied under aqueous conditions without having faradic reactions in the system. In the case of using nanoparticles with micrometer-size cells this can lead to challenges. However, using nanorods or wires can significantly reduce the number of tunneling junctions in the device leading to a better signal-to-noise ratio. The trade-off is that a larger number of junctions also increases the sensitivity for detecting cellular modulations.

7.3.2 Gating Effect

A continuous interface made with nanomaterials such as nanowires or graphene sheets can detect cellular modulations due to corresponding changes in their charge-carrier dynamics. The interface offers a continuous path way for flow of electrons (or holes in the case of p-type nanomaterials). The nanomaterials in this case are usually a semiconductor such as doped Si nanowires and carbon nanotubes. 2D materials such as graphene where the type of charge carriers and their densities are easily modulated by external stimuli, also function on similar principles. The two basic modulations in such nonmetallic interfaces are the density of charge carriers or their mobility.[8,9,16,18,19]

Devices using doped Si nanowires have been demonstrated.[3,15,19] The devices in these cases were interfaced with neurons and myocyte cells and their modulation was recorded in response to the changes in the cell membrane potential. In the case of neurons this was due to the generation and propagation of neural impulses and for myocytes it corresponds to their rhythmic beating. In both cases, modulations of the cell membrane potential lead to gating of the nanowire and the dynamics of this modulation are recorded in the current across the nanowire. The interface–cell interaction is similar to a field effect transistor (FET). Similar configurations using semiconducting CNTs, other inorganic materials and organic semiconductors have been demonstrated.[7–9,16]

Recently, the FET type devices have been configured where the inside of the cell is probed using similar concept (Figure 7.2a). In this configuration, one end of a nanotube penetrates and is in contact with the cells interior, while its other end is linked to a nanowire (active wire) that translates between anode and cathode electrode. This configuration is similar to an inverted T-junction.[17] The nanotube in contact with the cell brings the cell's cytoplasmic fluid in contact with the active wire. The potential (chemical) of the cells interior relative to the outside is effectively translated by the gating of the active wire due to the contacting cellular fluid. The cells dynamics are reflected in the current modulation of the active wire. The FET-type devices rely on the gating effect due to the potential of the cell. Nanomaterials, due to their size constraints are able to act as effective transistors and also translate the cell dynamics into current modulations.[17,22]

Changes in a cell's shape, size and surface stresses are linked to its cytoskeleton. It is a dynamic system that also responds to a cell's interaction with its environment. Translation of changes in the mechanical state of the cell requires an electromechanically active nanomaterial interface. The nanomaterial should be able to respond to mechanical changes and its electrical properties have to be sensitive to such changes. In this respect, nanowire and nanoparticles have limitations as stresses in cellular systems are on the scale of kPa. The solid cross section of the nanowires and nanoparticles limits their deformation in compliance with the changes in a cell's mechanics. Further, unless the material has a high peizoresistive coefficient, the modulation in the current due to these stresses will be limited. Graphene sheets with nanometer-scale thickness are like paper and hence respond to inplane stresses by formation of wrinkles. Stresses in the range of kPa are sufficient to produce this response.[23,24] The thin cross section of the sheets also makes their charge-carrier dynamics sensitive to the applied stresses. The formation of wrinkles leads to local modulations in the band structure and also scattering of charge carriers and hence results in increased resistance with stress on the sheets. Devices with cells partially covered with the graphene sheets have been demonstrated where a decrease in current across the graphene sheets is observed as they wrinkle in response to a cell's surface stresses.[1]

7.3.3 Observing Ion Flow

Cell-based devices will operate in an aqueous environment as required to maintain a cell's biological activity. An omnipresent and routine cell process is the exchange of ions between the cell and its surroundings. A device that can detect this process can significantly expand our ability to monitor cell dynamics in real time. This will also expand the cell types that can be studied using such an electrical interface and also the range of stimuli that can be deciphered from the recorded signals. The detection of such an ion flow is challenging as the cell is already in an environment that has significant ionic strength (*e.g.* the nutrient broths used to grow the cells). Below, we discuss a proposed mechanism of sensing this ion flow on a graphene-covered cell.

The detection is based on a three-electrode system, a source and drain across the graphene-covered FET cell and a reference (*e.g.* Ag/AgCl) dipped in the electrolyte as the gate electrode.[25–27] A schematic of the system and the corresponding circuit elements is presented in Figure 7.5. Considering the graphene sheets as a planar surface, the capacitance across them, in relation to the reference electrode, consists of the sheets quantum capacitance (C_q) and the electrical double layer capacitance (C_d) (Figure 7.5) in series. For graphene sheets $C_q \sim 2~\mu F/cm^2$, and $C_d \sim 500~nF/cm^2$ (electrical double layer (EDL) of 1 nm thickness for physiological conditions).[25,26,28] Being in series, the smaller capacitance C_d, dominates the total capacitance. Therefore the gating characteristics of the sheets by the reference electrode will be dependent on C_d. The ions flowing across the cell boundary will alter the local ion concentration in the EDL (being just 1 nm thick), resulting in modulation of the electrical properties of the sheets on top due to two primary effects.

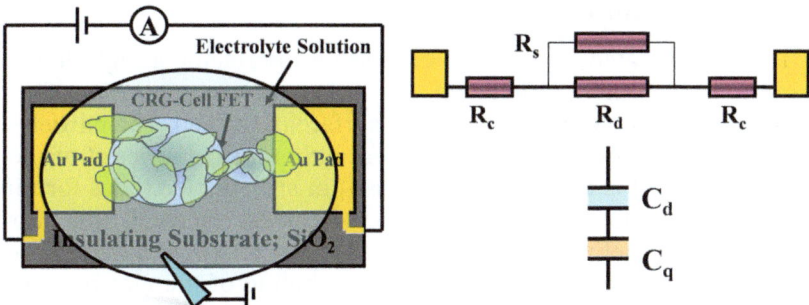

Figure 7.5   Schematic of a graphene-covered cell forming an **ON** device similar to Figure 7.4b. The schematic on the right shows a simplified electrical circuit for the device.

1. The double-layer capacitance (C_d) of the CRG sheets will be modulated resulting in a change of their surface potential and doping character- istics, altering the flow of current. C_d can be expressed as:[26]

   $1/C_d = x_o/(\varepsilon_o\varepsilon) + \lambda_D/[\varepsilon_o\varepsilon \cosh(ze\phi_o/2KT)]^*$, capacitance is per unit surface area here.

   x_o depends on the hydration radius of the ions, hence this term will be affected by the type of ions flowing across the surface (*e.g.* Na$^+$ or Ca^{2+}). λ_D, ϕ_o are both affected by the ion concentration. Hence C_d will be modulated by the flow of ions across the cell surface (therefore also CRG surface).

2. The graphene sheets that are interfaced with cells have a surface po- tential (or chemical potential). This surface potential is also dependent on the interaction between the sheets and the ions in the solution. Considering primarily cations K$^+$, Na$^+$, Mg^{2+} and Ca^{2+}, all these have considerable interaction energy with graphene sheets and also the defects (such as carboxylic groups at the edge of sheets when graphene is made by the chemical oxidation route) present in the sheets.[26] The interaction dependent on the concentration of the ions species will alter the surface potential of the sheets modulating its electrical properties.

7.4   Demonstrated Examples of ON Cell Devices

Two **On** cell devices are discussed below; in both cases the cell has a cell wall encapsulating its plasma membrane. The presence of a cell wall mitigates the effect of binding a nanomaterial to the cell surface. Such bacterial and yeast cells due to the presence of the cell wall present a logical starting point for development of **On** cell devices. Transition to mammalian cells and higher cell lines should progress with the lessons learnt from use of these more robust cells.

7.4.1   Gold Nanoparticles on Cells

This device, illustrated in 2005[2] was based on interfacing a high density of 10–13-nm size Au nanoparticles on the surface of bacterial cells. The par- ticles form a conductive pathway on the surface of the cell. The particles are not connected as a continuous pathway but rather form a series of tunneling junctions. This cell is deposited between 2-μm size Au electrodes forming a device. On application of bias to the two electrodes, electrons flow over the cell surface by tunneling across the Au nanoparticles (inset Figure 7.4a).

The current in the device results from tunneling of the electrons across adjacent Au nanoparticles and hence is sensitive to the local tunneling barrier (energy barrier) and the separation between these particles, the dis- tance for tunneling. The device was sensitive to the humidity in the en- vironment due to the coupling of the cell's size to the spacing between the

Au nanoparticles. The cell wall of the bacterial cells responds to the humidity, lower humidity leads to the dehydration of the cell wall and reduces the spacing between the nanoparticles, this leads to an increase in the tunneling current (Figure 7.3d). The response is dependent on the cell wall and its interaction with the humidity, the biological process of the cell is not coupled to the device. Hence even dead cells were able to provide the observed humidity sensitivity of the cell–Au–Np device. The theoretical calculation showed that due to the exponential sensitivity of the tunneling current to the separation distances, modulation on the scale of nanometers (and even less) in the cell wall coat were easily observed with this device.[2]

7.4.2 Chemically Reduced Graphene on Cells

Eukaryotic yeast cells (*Saccharomyces cerevisiae*) also have a cell wall and are relatively robust in comparison to mammalian cells. In this device, these cells were interfaced with chemically reduced graphene sheets (CRG) to form an **On** cell device. The interfacing is based on the negative zeta potential of the *S. Cerevisiae* cells, which allows for the binding with divalent cation (Ca^{2+}, Mg^{2+}) functionalized CRG sheets.[1,14] The sheets were on the surface of the cells and partially cover the cell, forming a conductive pathway. This basic set-up as illustrated in Figure 7.4b is the basis of the device.

The device was illustrated for electromechanical coupling between the CRG sheets and the cell.[1] The basis of the response presented was the osmotic stress to the cell causing a change to its shape and size. The result is surface stresses on the cell. This also causes stresses on the CRG sheets and result in their wrinkling. The wrinkles are scattering centers for the charge carriers and decrease the current level in the device under constant bias (Figure 7.4b). The basis of the response lies in the mechanical properties of the cell and the coupled graphene sheets.

The reported modulus of the CRG sheets is in the range of 0.3–0.6 TPa. Stresses in the range of 100 kPa easily strain the sheets resulting in the formation of wrinkles due to their nanometer-scale thickness. Since the sheets are deposited on the surface of yeast cells that have a cell wall modulus of ~150 MPa, volumetric strains of 1% are sufficient to wrinkle the CRG sheets. The charge carriers in CRG sheets are sensitive to scattering by inhomogeneities due to both the chemical and physical state of the sheets. Wrinkles result in distortions of the local energy structure of the sheets leading to increased scattering of the charge carriers. This decreases the conductivity of the CRG sheets due to wrinkles. The equivalent circuit of the CRG-Cell FET is shown in Figure 7.5; there are two primary sources of resistance, the contact resistance (R_c) and the resistance of the CRG sheets (R_d) on the surface of the cells. The solution resistance R_s is much greater than R_c and R_d combined, if the measurements are done in an aqueous environment. The applied bias to the device is limited to 200 mV to prevent any faradic currents in the solution. In aqueous solutions that can support the cell, the dielectric properties of the medium will remain unchanged. Therefore, the source of modulations in R_c will be due to the change in the contact area between the CRG sheets and the electrodes. The contact area (A_c) can alter due to changes in cell volume, considering a 10–15% reduction in cell volume this translates to a maximum change of 6–10% in surface area (isotropic change in cell volume). Using the simple approximation that $R_c \propto 1/A_c$, this leads to a maximum increase in R_c of 6–10%, while a 1% volumetric strain can lead to more than a 30–40% increase in R_d.[1] Further, the net resistance of the device, $R_t = R_c + R_d$ is dominated by R_d as the length of current transport over the cell surface is much greater than the contact length of the CRG sheets with the electrodes. Also, as the electrodes are made of noble metals such as Au, forming a metal–CGR contact, R_c is small.

Under such circumstances the mechanical modulations of the cell can translate to changes in current driven across the CRG sheets. It also offers the advantage that the material in this case is conformal to the cell.

7.5   Challenges and the Future

7.5.1   Challenges

Besides the challenges of interfacing the cell with the nanomaterial that acts as the device there are some important considerations for the future progress of such devices.

1. Nanomaterials are sensitive to surface conditions, this leads to both advantages and challenges. Signals can result from multiple stimuli in the case of cells being studied in uncontrolled environments. Most of the devices that have been demonstrated show the recording of cell signals on being exposed to a single controlled stimulus. To study cells in a more natural environment, they are exposed to multiple stimuli simultaneously and hence their response is a superposition of multiple signals. A device will need to both record these signals and analysis will be needed to decipher the superimposed pattern.

2. Long-term effect on the cells is a key point for interfacing of nanomaterials with cells. This needs to be carefully monitored for both **On** and **Under** devices. For the **On** cell devices this assumes significance as they cover at least a portion of the cell surface that would otherwise be in direct contact with the surroundings. Nanomaterials such as graphene, nanowires and nanoparticles are also diffusion barriers. Thus, just the transport barrier due to these materials can affect the cell. The biological implication of such an interface on, for example, functioning of the cell membrane and more downstream processes such as protein expression requires a more detailed study for the future progress of this

field. Preliminary reports on the devices have shown promise that strategies to mitigate adverse effect on cells are possible. It is important to note that the cell-toxicity challenge for the nanocell device field is very different from applications of drug delivery where the nano-materials become a part of the cell's internal environment. Strategies such as partial coverage of the cell surface show that cell propa-gation[1,14] and its metabolic functions are maintained, though more detailed studies are required on the long-term effects and downstream process. New strategies with multistructured nanomaterials may in future provide ways where the direct interfacing may be further modified to near field interfacing with the cells. For example, a mesh of nanowires may be placed close to the surface of the cells to record the cell dynamics without a physical contact with the cell surface, or gra-phene sheets with engineered holes can be placed on the cell surface to mitigate the effects of the diffusional barrier.

3. The type of nanomaterial used for the interfacing will present in each case a distinct set of possible cellular responses that can be recorded. The pattern of signal generated for a given cellular response will also be distinct with each type of nanomaterial. This would require a specific set of analysis tools for each type of nanocell device, complicating analysis and comparison between the devices.

7.5.2 The Future

Besides the development of devices, equally important is the need to develop methods to analyze the data that can map the observed modulation to a particular cellular response (or responses). This is crucial to make these devices as functional applications rather than an academic study. Appli-cation for such devices exist in multiple fields such as environmental monitoring, the mapping of the cells response can lead to tools for identification of pollutants in water and also deciding if their concen-trations pose a biological threat. More ambitious application will be in the field of screening the effect of drug molecules on cells. The ultimate aim is to integrate man-made circuits with biological machinery of the cell, which is the realms of real cyborg cells. There have been some advances in the field, such as a 3D network of nanowires that has been used in tissue engineering for a detailed study of the signals in this complex biological system (Figure 7.6a).[29] Besides the use of gating and tunneling devices, electrochemical devices are also being made using nanomaterials that allow the dynamic recording of cells electrochemical potential and currents.[11–13]

Cells used as templates with assembled inorganic nanomaterials have been used to demonstrate battery electrodes, memory devices and photosensors (Figure 7.6b).[30–32] The future will depend on the trans-lation of these small steps into an integrated approach to make a real cyborg cell.

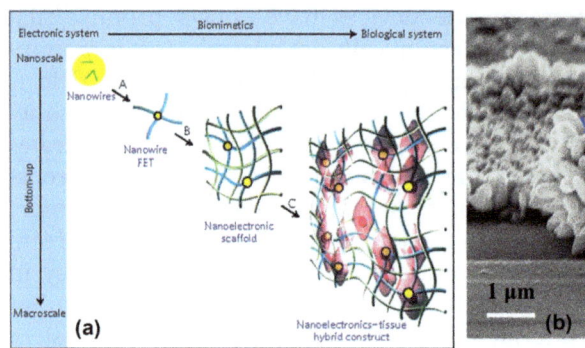
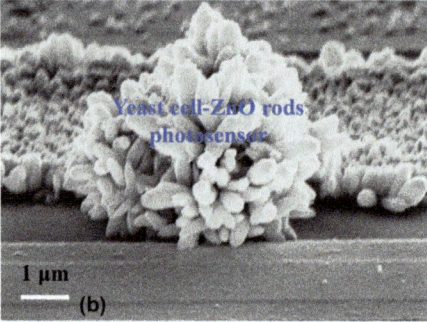

Figure 7.6 (a) Schematic of 3D network of nanowires that are used to cultivate cells in 3D forming a tissue. The nanowires are addressable and can be used to form a 3D map of the response from cells such as neurons and cardiomyocyte. Reprinted with permission from ref. 29. Copyright (2012) the Nature Publishing Group. (b) A single-cell photosensor made by performing electrochemical synthesis of ZnO rods on graphene covered cells. Reprinted with permission from ref. 30. Copyright (2013) the Royal Society of Chemistry.

References

1. R. Kempaiah, A. Chung and V. Maheshwari, *ACS Nano*, 2011, **5**, 6025.
2. V. Berry and R. F. Saraf, *Angew. Chem. Int. Ed. Engl.*, 2005, **44**, 6668.
3. F. Patolsky, B. P. Timko, G. Yu, Y. Fang, A. B. Greytak, G. Zheng and C. M. Lieber, *Science*, 2006, **313**, 1100.
4. V. Berry, S. Rangaswamy and R. F. Saraf, *Nano Lett.*, 2004, **4**, 939.
5. T. Cohen-Karni, Q. Qing, Q. Li, Y. Fang and C. M. Lieber, *Nano Lett.*, 2010, **10**, 1098.
6. I. S. Beloborodov, A. V. Lopatin, V. M. Vinokur and K. B. Efetov, *Rev. Mod Phys.*, 2007, **79**, 469.
7. T. S. Pui, H. G. Sudibya, X. Luan, Q. Zhang, F. Ye, Y. Huang and P. Chen, *Adv. Mater.*, 2010, **22**, 3199.
8. H. G. Sudibya, J. Ma, X. Dong, S. Ng, L. J. Li, X. W. Liu and P. Chen, *Angew. Chem. Int. Ed. Engl.*, 2009, **48**, 2723.
9. T. S. Pui, A. Agarwal, F. Ye, N. Balasubramanian and P. Chen, *Small*, 2009, **5**, 208.
10. X. Duan, R. Gao, P. Xie, T. Cohen-Karni, Q. Qing, H. S. Choe, B. Tian, X. Jiang and C. M. Lieber, *Nature Nanotechnol.*, 2012, 7, 174.
11. J. T. Robinson, M. Jorgolli, A. K. Shalek, M. H. Yoon, R. S. Gertner and H. Park, *Nature Nanotechnol.*, 2012, 7, 180.
12. C. Xie, Z. Lin, L. Hanson, Y. Cui and B. Cui, *Nature Nanotechnol.*, 2012, 7, 185.
13. F. J. Rawson, C. L. Yeung, S. K. Jackson and P. M. Mendes, *Nano Lett.*, 2013, **13**, 1.

14. R. Kempaiah, S. Salgado, W. L. Chung and V. Maheshwari, *Chem. Commun.(Camb.)*, 2011, **47**, 11480.

15. B. P. Timko, T. Cohen-Karni, G. Yu, Q. Qing, B. Tian and C. M. Lieber, *Nano Lett.*, 2009, **9**, 914.

16. V. Benfenati, S. Toffanin, S. Bonetti, G. Turatti, A. Pistone, M. Chiappalone, A. Sagnella, A. Stefani, G. Generali, G. Ruani, D. Saguatti, R. Zamboni and M. Muccini, *Nature Mater.*, 2013, **12**, 672.

17. R. Gao, S. Strehle, B. Tian, T. Cohen-Karni, P. Xie, X. Duan, Q. Qing and C. M. Lieber, *Nano Lett.*, 2012, **12**, 3329.

18. S. Ingebrandt, C. K. Yeung, M. Krause and A. Offenhausser, *Biosens. Bioelectron.*, 2001, **16**, 565.

19. T. Cohen-Karni, B. P. Timko, L. E. Weiss and C. M. Lieber, *Proc. Natl. Acad. Sci. USA*, 2009, **106**, 7309.

20. J. Kane, M. Inan and R. F. Saraf, *ACS Nano*, 2010, **4**, 317.

21. V. Maheshwari, D. E. Fomenko, G. Singh and R. F. Saraf, *Langmuir*, 2010, **26**, 371.

22. B. Tian, T. Cohen-Karni, Q. Qing, X. Duan, P. Xie and C. M. Lieber, *Science*, 2010, **329**, 830.

23. A. L. V. de Parga, F. Calleja, B. Borca, M. C. G. Passeggi, J. J. Hinarejos, F. Guinea and R. Miranda, *Phys. Rev. Lett.*, 2008, 100.

24. W. Z. Bao, F. Miao, Z. Chen, H. Zhang, W. Y. Jang, C. Dames and C. N. Lau, *Nature Nanotechnol.*, 2009, **4**, 562.

25. Y. Ohno, K. Maehashi, Y. Yamashiro and K. Matsumoto, *Nano Lett.*, 2009, **9**, 3318.

26. I. Heller, S. Chatoor, J. Mannik, M. A. G. Zevenbergen, C. Dekker and S. G. Lemay, *J. Am. Chem. Soc.*, 2010, **132**, 17149.

27. F. Chen, Q. Qing, J. L. Xia, J. H. Li and N. J. Tao, *J. Am. Chem. Soc.*, 2009, **131**, 9908.

28. J. L. Xia, F. Chen, J. H. Li and N. J. Tao, *Nature Nanotechnol.*, 2009, **4**, 505.

29. B. Tian, J. Liu, T. Dvir, L. Jin, J. H. Tsui, Q. Qing, Z. Suo, R. Langer, D. S. Kohane and C. M. Lieber, *Nature Mater.*, 2012, **11**, 986.

30. J. Tam, S. Salgado, M. Miltenburg and V. Maheshwari, *Chem. Commun.(Camb.)*, 2013, **49**, 8641.

31. Y. J. Lee, H. Yi, W. J. Kim, K. Kang, D. S. Yun, M. S. Strano, G. Ceder and A. M. Belcher, *Science*, 2009, **324**, 1051.

32. R. J. Tseng, C. L. Tsai, L. P. Ma and J. Y. Ouyang, *Nature Nanotechnol.*, 2006, **1**, 72.

CHAPTER 8

Artificial Spores

DAEWHA HONG, EUN HYEA KO AND INSUNG S. CHOI*

Center for Cell-Encapsulation Research, Department of Chemistry, KAIST, Daejeon 305-701, Korea
*Email: ischoi@kaist.ac.kr

8.1 Introduction: Cryptobiosis and Bacterial Endospores

All organisms in Nature face unavoidable environmental changes that require and testify to their adaptability.[1] During the long history of the Earth, some organisms have failed to adapt to changes in their environments and declined, whereas those that adopted intelligent strategies survived. Certain organisms have chosen an extreme strategy, called "cryptobiosis", in which most metabolic processes involved in their proliferation, reproduction, and development are shut down in response to adverse environmental conditions, such as nutrient deficiency, desiccation, freezing, and oxygen deficiency.[2]

In the cryptobiotic state, organisms remain dormant and protect their cellular components from external stresses. When a favorable environment is restored, the organisms return to the metabolically active state present before cryptobiosis. Several types of cryptobiosis exist: anhydrobiosis (induced by desiccation), cryobiosis (induced by freezing temperatures), osmobiosis (induced by high osmotic pressure), and anoxybiosis (induced by a lack of oxygen).

Commonly known organisms that undergo cryptobiosis include tardigrades, brine shrimp, nematodes, and rotifers.[3] Of these, the tardigrade (sometimes called the "water bear") is a representative model of

RSC Smart Materials No. 9
Cell Surface Engineering: Fabrication of Functional Nanoshells
Edited by Rawil F Fakhrullin, Insung S Choi and Yuri Lvov
© The Royal Society of Chemistry 2014
Published by the Royal Society of Chemistry, www.rsc.org

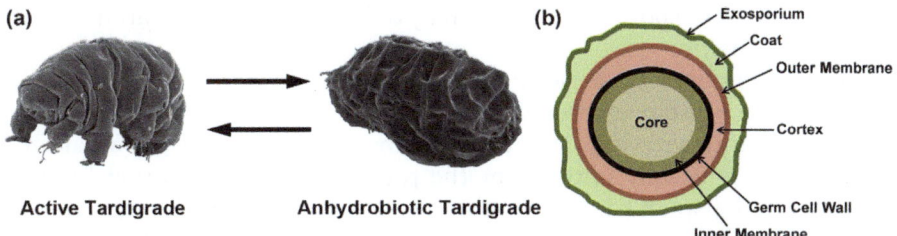

(a) Active Tardigrade   Anhydrobiotic Tardigrade

(b) Exosporium, Coat, Outer Membrane, Core, Cortex, Germ Cell Wall, Inner Membrane

Figure 8.1 (a) Tardigrades in active state and in the anhydrobiotic state. (b) The structure of bacterial endospore.

cryptobiosis.[4] This invertebrate animal, ranging in size from 0.1 to 1.0 mm, is found globally and lives in a variety of environments, from deep oceans to high mountains, attesting to its outstanding adaptability. Under dehydrated condition, tardigrades enter an anhydrobiotic state as a survival strategy (Figure 8.1a). During this process, they contract into a structure resembling a small tun. During the tun formation, the infolding of the intersegmental cuticle occurs, which reduces the body size. In addition, wax extrusion covers the surface to minimize the water loss by evaporation. Anhydrobiotic tardigrades are capable of surviving temperatures ranging from -273 °C to 151 °C, vacuum of 5×10^{-5} Pa, pressures up to 6000 atm, and exposure to chemicals such as alcohols.[5]

Some bacterial cells also manifest cryptobiotic behaviors, when the vegetatively growing cells change into (endo)spores in response to starvation. This process is called "sporulation", which includes the conversion of the metabolic state to a dormant state and the formation of shells that protect the cells from foreign aggression.[6] The protective shell is generally composed of cortex, coat, and exosporium.

During the dormant state, spores exhibit excellent resistance to various stressors, including dehydration, heat, ultraviolet (UV) and gamma radiation, and toxic chemicals.[7] The protective events of sporulation are accompanied by several phenomena.[8] First, a low water content and high levels of minerals are maintained in the spore core. Secondly, a repair system is activated at the molecular level that restores DNA or proteins damaged by external stressors. Thirdly, the reduced permeability of the spore presents a barrier to fatal chemicals and enzymes, thus reducing the possibility of threats to crucial molecules in the spore core, such as enzymes or DNA.

Although multiple factors are involved in the protective mechanisms of spores, the formation of the spore shell is probably the most dramatic event of sporulation (Figure 8.1b). The peptidoglycan layers of the cortex play a crucial role in preventing dehydration of the core. Over the cortex is the coat, which adds resistance to chemical and enzymatic threats. Without the coat, the cortex is vulnerable to peptidoglycan-degrading lysozymes. The exosporium is the outermost layer of the shell, which comes into contact with the environment and contains several enzymes involved in germination.

Thus, the shells of spores have a hierarchy of layers, and each stratum has an orthogonal function.

Dormant endospores sense the condition of the external environments, and when circumstances become favorable for growth, the spores are converted back into growing cells through the process called germination.[9] Germination is usually initiated by the presence of nutrients that bind to receptors in the spore's inner membrane. Although the precise mechanism has not been known so far, the following events take place: (1) release of small molecules including ions (H^+, Zn^{2+}, and Ca^{2+}) and dipicolinic acid from the spore core; (2) hydrolysis of the peptidoglycan cortex; (3) water uptake and expansion of the shrunk spore core.

Sporulation represents an intelligent strategy that protects cells temporarily from nutritionally unfavorable local conditions using dormancy. Spores can also relocate themselves using various hosts, wind, and water, which can coincidentally move them to fruitful environments for their germination and the resumption of vegetative growth.[10]

Similar protective shells are observed in the resting cyst of the ciliate *Maryna umbrellata* whose hierarchical shells composed of glass (*i.e.* silica) granules protect the cell from environmental stress, and also play a significant role in initiating the excystment process for their reproduction.[11]

On the other hand, the Nature-inspired, cytocompatible formation of ultrathin (< 100 nm), tough artificial shells on the surfaces of nonspore-forming cells shows the spore-like behavior that enhances the protection of the encapsulated cells from stresses as well as chemically controlling their metabolic activities.[12] These synthetic "cell-in-nanoshell" structures, designated "artificial spores", emulate the essential features of natural endospores, as follows: (1) the growth or division of living cells is controlled to some extent by tuning the thickness or physicochemical properties of the artificial shells; (2) resistance to external stressors is enhanced, with increased tolerance against malnutrition, osmotic pressure, enzymatic attack, chemical stressors, or UV, which is lethal to native, uncoated cells. In this chapter, the current status of artificial spores is described, and their applications are discussed.

8.2   Criteria and Approaches for Fabrication of Artificial Spores

Artificial spores are usually constructed by introducing robust and permeable nanometric shells onto the surfaces of individual living cells. The minimum requirements for the artificial shells are as follows. (1) Mechanical durability: an artificial shell requires some degree of mechanical durability to withstand the force of cell division; to resist external physical stresses, such as osmotic pressure and dehydration; and to maintain the original cell structure under these stresses. (2) Selective permeability: permselectivity is desirable to block foreign aggressors (*e.g.*, lytic enzymes and macrophages), to ensure that the cell is protected while allowing the passage of small molecules, such as nutrients and gases, and to ensure the viability of the cell.

It is, however, difficult in practical terms to construct robust shells, because the chemical treatment required usually involves harsh reaction conditions, which potentially threaten living cells. Therefore, the development of bio- and cytocompatible methods that do not jeopardize the original viability of the cells during the formation of robust shells is particularly important.

Several cytocompatible methods have been proposed for constructing robust shells (Table 8.1). For example, layer-by-layer (LbL) assembly has been used to coat individual hepatocyte carcinoma cells with biopolymer pairs,

Table 8.1 Approaches to the fabrication of artificial spores.

Approach	Materials	Cell Type	Properties	Refs
Layer-by-Layer	Fibronectin/ gelatin Collagen/ laminin	*Hepatocyte carcinoma*	Resistance (shear stress)	13
	Halloysite nanotubes (above 100 nm)	*Saccharomyces cerevisiae*	Division control	16
	TA/PVPON	*Saccharomyces cerevisiae*	Division control	19
	PEI-C/HA-C	*Saccharomyces cerevisiae*	Division control	20
	PMA-co-NH$_2$ (crosslinked)	*Saccharomyces cerevisiae*	Division control	27
	PAH/PSS	*Saccharomyces cerevisiae*	Resistance (lysosomal digestion)	31
Mussel-inspired polymerization	Polydopamine	*Saccharomyces cerevisiae*	Division control, Resistance (lytic enzyme)	14
Bioinspired mineralization	Silica	*Saccharomyces cerevisiae*	Division control, Resistance (malnutrition)	15
	Calcium phosphate (microshells)	*Saccharomyces cerevisiae*	Division control, Resistance (osmotic pressure, lytic enzyme)	25
	Silica/titania	*Chlorella sp.*	Resistance (heat)	34
Electrostatic deposition	Ca^{2+}-functionalized reduced graphene oxide	*Saccharomyces cerevisiae*	Division control, Resistance (osmotic pressure)	17
Crystallization	Silica	*Saccharomyces cerevisiae*	Resistance (heat)	33
	Lanthanide phosphate (about 500 nm)	Zebrafish	Resistance (UV)	37
Hydrogel	CeO$_2$-loaded silica matrix	*Chlorella vulgaris*	Resistance (UV)	38

such as fibronectin/gelatin or collagen (type IV)/laminin.[13] In this example, the use of biological recognition between the members of the biopolymer pairs satisfies both the rigidity requirement and the cytocompatibility requirement. The coated cells showed greater resistance to physical stress (*i.e.* centrifugation) than uncoated cells, indicating that the LbL methods is potentially useful for the construction of artificial shells in the realization of artificial spores. Mussel-inspired polymerization[14] and bioinspired mineralization[15] have also produced reasonably rigid shells for the possible construction of artificial spores. Halloysite clay nanotubes,[16] graphene,[17] and titania[18] are also candidate substances for the construction of robust shells.

8.3   Control of Cell Division

Natural endospores maintain their dormant state while external conditions are unfavorable. The nanoencapsulation of individual living cells within artificial shells have shown that cell division can be controlled, because the shells act as a physicochemical barrier that retards cell division to some extent. This control can be classified as passive control or active control. Passive control refers to systems in which the proliferation of the cell is simply delayed by increasing the thickness of the shell or reinforcing the shell. Active control is the stimulus-responsive control of cell-division timing by using an external signal that modulates the status of the shell. Natural endospores germinate in response to their external environments, so the development of an active system is much more desirable than the development of a passive system, although more challenging.

8.3.1   Passive Control of Cell Division

Perhaps the most striking, yet classic demonstration of the control of cell division exerted by nanoshell formation is the delay of cell division caused by increased shell thickness. LbL assembly would be a simple approach to the control of cell division, because the thickness of artificial shells is easily tuned by varying the number of LbL layers. For example, the lag phase of yeast cells was gradually extended by increasing the number of tannic acid (TA)/poly(*N*-vinylpyrrolidone) (PVPON) bilayers (Figure 8.2a).[19]

The orthogonal control of cell division has also been demonstrated in the LbL approach, in which the lag phase of yeast cells was extended not only by varying the number of shell layers but also by crosslinking the layers (Figure 8.2b). Catechol-grafted polyethyleneimine (PEI-C) and catechol-grafted hyaluronic acid (HA-C) were used for LbL coating of yeast, and the formed polyelectrolyte multilayers (PEMs) were subject to oxidative crosslinking under slightly basic conditions (pH 8.5).[20] The crosslinking enhanced the rigidity of the artificial shells. The lag phase of yeast cells was first investigated by varying the number of LbL bilayers (5, 10, or 15 bilayers). As expected, the increased number of the bilayers delayed the division

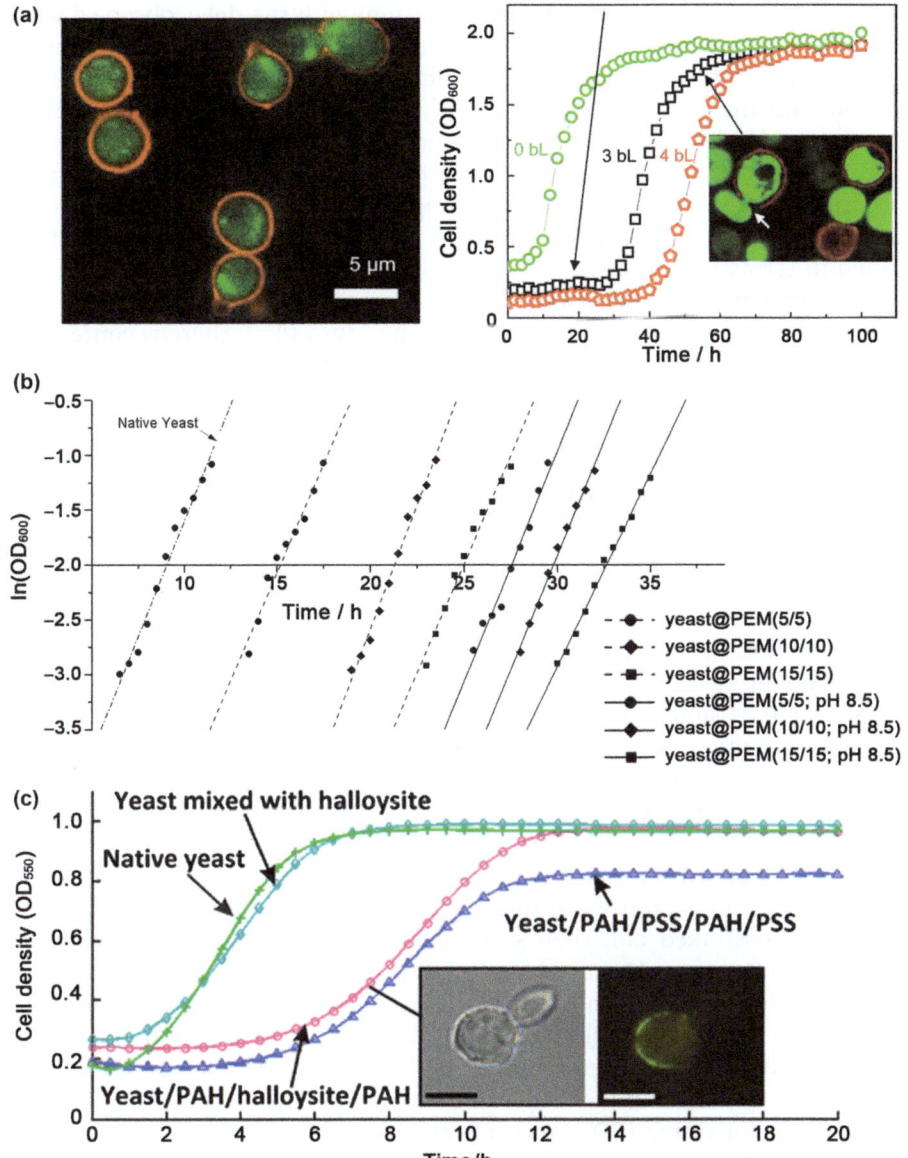

Figure 8.2 Passive control of cell division *via* LbL approach. (a) The number of tannic acid/poly(*N*-vinylpyrrolidone) (PVPON) LbL layers controlled the cell division. (b) The crosslinking of the LbL layers composed of catechol-grafted polyethyleneimine (PEI-C) and catechol-grafted hyaluronic acid (HA-C), as well as the number of the layers, controlled the cell division. (c) Halloysite was incorporated into the LbL shells for the yeast encapsulation, which retarded the cell division compared with the native yeast.

Reproduced with permission from the Royal Society of Chemistry (Copyright 2011, 2013)[16,19] and Wiley (Copyright 2013).[20]

timing of the cells more, which was consistent with the delay observed with the PEM composed of TA/PVPON. The crosslinking dramatically prolonged the "dormant" state compared with that of uncrosslinked cells. Yeast cells coated with the crosslinked 5 bilayers (yeast@PEM(5/5; pH 8.5)) showed a more extended lag phase than the cells coated with the 15 bilayers that were not crosslinked (yeast@PEM(15/15)). This result showed that the cross-linking of the LbL layers increased the rigidity of the shell, which repressed cell division more effectively. Moreover, the control of cell division is fine tuned in terms of both the thickness and the robustness of the shell.

Another aspect of LbL assembly is its property of ubiquity: virtually any type of charged materials can be incorporated into PEM, thereby conferring unprecedented properties on the artificial shells that originate from the selected material itself. Halloysite nanotube (HNT),[16] magnetic nanoparticle (MNP),[21] and graphene[22] have been integrated into PEM during the course of LbL assembly. Other interactions, such as hydrogen bonding, can also be utilized for the LbL construction of artificial shells, exemplified by that interaction between TA and PVPON.

HNTs was introduced into LbL assembly by utilizing the anionic properties of HNTs.[16] The negatively charged surfaces of yeast cells (ζ-potential: -31.2 mV) were coated with cationic poly(allylamine hydrochloride) (PAH), which altered the surface potential to $+29$ mV. The positively charged surface made it possible to deposit the negatively charged HNTs. Another bilayer of PAH and anionic poly(sodium styrene sulfonate) (PSS) was deposited onto the yeast surfaces to secure the nanotubes to the cell walls and maintain their assembled structures on the cells. The yeast cells coated with PAH/HNT/PAH/PSS showed the delayed cell division compared with native cells (Figure 8.2c).

Although the LbL approach is a simple and feasible method to control cell division, a repetitive process is unavoidable and time consuming. Moreover, even the crosslinked LbL shell still lacks the rigidity required for the sufficient retardation and confinement of cell growth.

Covalently bonded organic nanoshells were fabricated on individual yeast cells based on the mussel-inspired polymerization of dopamine.[14] Dopamine, a minimal biomimetic building block inspired by mussel adhesive proteins, polymerizes under slightly basic aqueous conditions, and the resulting polydopamine (PD) can coat most surfaces. This one-shot method with universal applicability was used on yeast cell surfaces to facilitate the formation of tough and rigid shells. When dopamine was polymerized on the surfaces of yeast cells for 3 h, single PD-coated yeast cells (yeast@PD$_1$) were formed. The same procedure was then repeated with yeast@PD$_1$ cells to form double-coated yeast cells (yeast@PD$_2$), and the growth curves of yeast, yeast@PD$_1$, and yeast@PD$_2$ were evaluated (Figure 8.3a). The lag phases of yeast@PD$_1$ and yeast@PD$_2$ lasted for 36 h and 84 h, respectively, which indirectly supported that the mussel-inspired coating method produced fairly rigid organic nanoshells compared with those produced with conventional LbL methods.

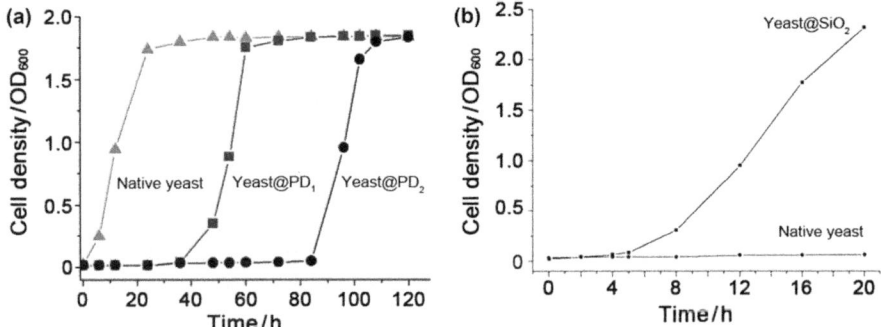

Figure 8.3 Passive control of cell division *via* a non-LbL approach. (a) Individual yeast cells were nanoencapsulated with polydopamine shells, and their division was controlled by the shell thickness. (b) Silica shells were formed on the yeast cell surfaces by bioinspired silicification, which prolonged the lag phase to some extent compared with the native yeast. Reproduced with permission from the American Chemical Society (Copyright 2011)[14] and Wiley (Copyright 2009).[15]

An alternative method for delaying cell division is to use inorganic materials in the encapsulation process. The formation of hierarchically organized inorganic structures occurs in Nature, where catalytic peptides/proteins harmonize with inorganic precursors under physiological conditions. The bioinspired approach, which utilizes the mechanistic chemical principles of biomineralization processes, is more attractive than conventional chemical methods for preserving the original viability of living cells, because bioinspired mineralization occurs under physiologically mild conditions.

The siliceous cell walls of diatoms offer one chemical approach for the construction of artificial spores that is cytocompatible and generates the robust shells that can control the cell division.[23] Individual living yeast cells were encapsulated within silica shells, inspired by the silicification of diatoms.[15] The encapsulation process was achieved in two sequential biocompatible steps: LbL self-assembly and bioinspired silicification. Because quaternary amine groups are reported to be a catalytic template for bioinspired silicification,[24] poly(diallyldimethylammonium chloride) was chosen as one of the PEM components. After bioinspired silicification, 50-nm thick silica shells were formed on the yeast-cell surfaces. The resultant cell-in-nanoshell structures persisted in their "dormant" state and did not divide for 20 h (Figure 8.3b).

Although the examples cited emulate the basic features of natural endospores, their processes are predominantly passive, simply delaying or retarding cell division relative to that of native cells. However, natural endospores always sense their surrounding environments and degrade their shells when conditions are appropriate for their reproduction. Therefore, the advanced features of artificial spores exert temporal control over cell division in response to external signals.

8.3.2   Active Control of Cell Division

Even when dormant, endospores actively interact with their environment to monitor the possibility of germination. Germination is usually initiated when nutrients bind to the receptors of the spores. This is followed by sequential processes that ultimately stimulate cell growth and division. Among these processes, shell degradation is a major event, converting cellular metabolism from the dormant state to the active state. However, the current status of artificial spores makes it difficult to design degradable shells, because cytocompatible processes are required at all steps of the encapsulation process and during shell degradation.

Only a few examples of the stimulus-responsive degradation of artificial shells have been developed. Ionic calcium phosphate microshells were degraded with HCl (Figure 8.4a),[25] and calcium carbonate (CaCO$_3$) microshells with ethylenediaminetetraacetic acid (EDTA).[26] EDTA was used as an artificial "germinant" that could cytocompatibly dissolve the CaCO$_3$ shells. pH-Responsive organic shells were also proposed that would actively control cell

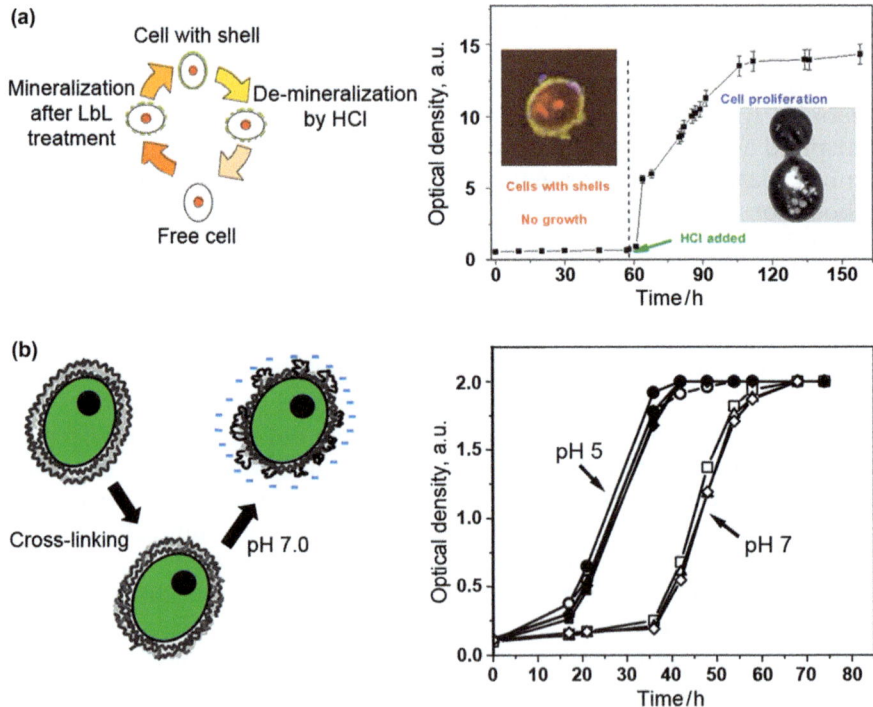

Figure 8.4   Active control of cell division. (a) HCl dissolved the calcium phosphate microshell that encased individual yeast. (b) pH-Responsive shells controlled the cell division.
Reproduced with permission from Wiley (Copyright 2008)[25] and the American Chemical Society (Copyright 2012).[27]

division (Figure 8.4b):[27] the pH-responsive polymer shrunk and swelled according to the pH values, and its introduction onto cell surfaces was used to actively control their division. In the crosslinked PEM shell of amine-bearing poly(methacrylic acid) (PMA-*co*-NH$_2$) derivatives, the deprotonation of the free carboxylic acid group above pH 5.0 increased the net negative charge on the shells, increasing the shell thickness and thus reducing the transport of nutrients, ultimately delaying cell growth.

The issues in the active control of cellular metabolism (*i.e.* cell division) are to design and synthesize programmable shells that sense specific molecules, thus allowing their degradation on demand. Clues and solutions will be found in the field of drug-delivery systems, in which the stimuli-driven release of cargo drugs is particularly important. For example, macrocapsules were enzymatically degraded for cargo release. The strategy was based on the introduction of dopamine-modified poly(L-glutamic acid) (PGA) conjugates (PGA$_{PDA}$), which was cleaved by proteases (Figure 8.5a).[28] The capsules of thiol/disulfide-bearing PMA (PMA$_{SH(disulfide)}$) showed the controlled release of their drug entities when treated with glutathione, which selectively reduced and broke down disulfide linkages to thiol groups (Figure 8.5b).[29] Using hydrogels that were photochemically degraded with nitrobenzyl groups is another possible way to develop artificial shells that allow the active, stimuli-driven control of cell division (Figure 8.5c).[30] The most important criterion that must be satisfied by these candidate systems is that the vitality of the living cells is ensured during the whole processes of encapsulation and degradation.

8.4 Resistance to Stress

Perhaps the most noteworthy characteristic of natural endospores is their remarkable protection of cells against otherwise fatal stressors. Artificial spores are also equipped with unprecedented resistance to stressors (osmotic pressure, malnutrition, toxic enzymes, high temperatures, or UV radiation), which would otherwise cause fatal damage to native cells. This resistance depends on the appropriate manufacture and composition of the enveloping shells.

8.4.1 Resistance to Malnutrition and Osmotic Pressure

In Nature, natural endospores are usually generated when the cells sense a deficiency of nutrients, and initiate sporulation to avoid wasting energy on nonessential metabolism. Therefore, maintaining a cell's dormant state while preserving its vitality under malnutritive conditions is a valuable trait that artificial spores could emulate.

The survival rates of the silica-encapsulated yeast cells (yeast@SiO$_2$) and native yeast cells were compared (Figure 8.6a) in a nutrient-deficient environment (pure water at 4 °C).[15] After incubation for 30 days, yeast@SiO$_2$ showed a three-fold increase in viability compared with the native yeast.

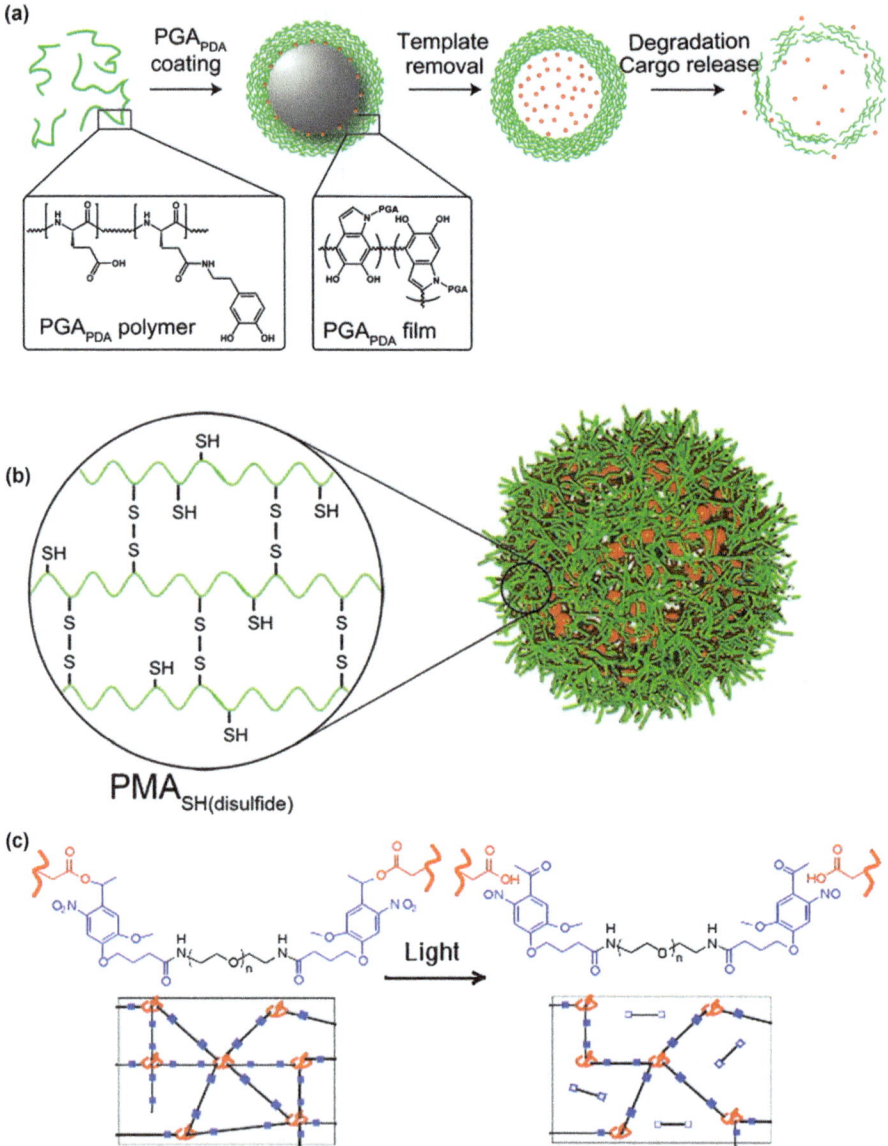

Figure 8.5 Potential materials for stimulus-responsive shells for cell division. (a) PGA$_{PDA}$ was cleaved by protease. (b) PMA$_{SH(disulfide)}$ was cleaved with glutathione for the controlled drug release. (c) Hydrogels bearing nitrobenzyl groups were degraded photochemically.
Reproduced with permission from Wiley (Copyright 2011),[28] the American Chemical Society (Copyright 2011),[29] and AAAS (Copyright 2009).[30]

The silica nanoshells probably improved the survival rate of the cells by retaining an intact cytoplasmic environment by stabilizing the cell membrane under physicochemical pressures.

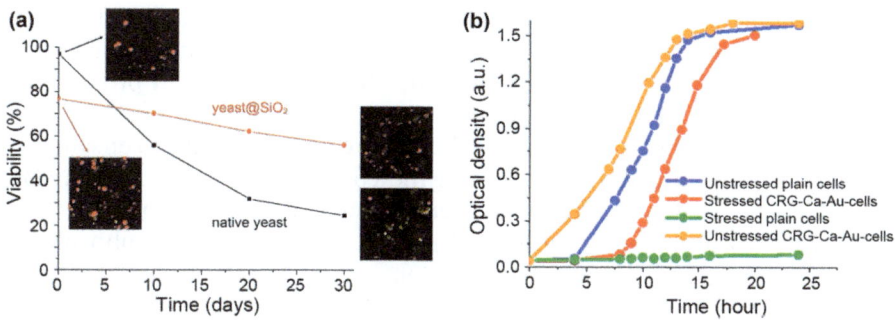

Figure 8.6 Resistance to malnutrition and osmotic pressure. (a) Enhanced survival rate of yeast@SiO$_2$ was observed under nutrient-free conditions. (b) Yeast cells encapsulated with CRG-Ca-Au showed the protection capability against osmotic pressure.
Reproduced with permission from Wiley (Copyright 2009)[15] and the Royal Society of Chemistry (Copyright 2011).[17]

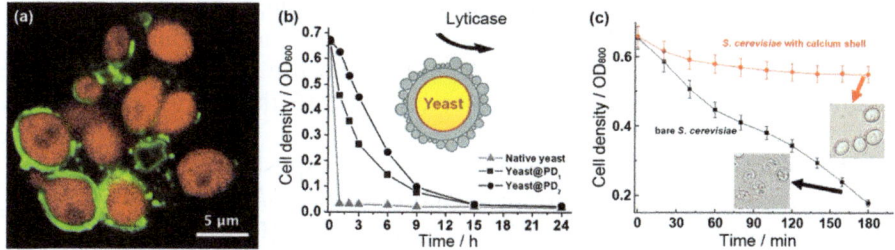

Figure 8.7 Resistance to lytic enzymes. (a) LbL-Coated yeast cells with PAH and PSS showed the protection capability against lysosomal enzymes. Yeast cells encapsulated within (b) polydopamine and (c) calcium phosphate shells showed the enhanced resistance to lyticase.
Reproduced with permission from Wiley (Copyright 2003, 2008)[25,31] and the American Chemical Society (Copyright 2011).[14]

The protection of yeast cells against osmotic pressure was achieved by encapsulating the cells in chemically reduced graphene oxide sheets functionalized with Ca^{2+}-modified gold nanoparticles (CRG-Ca-Au). The CRG-Ca-Au yeast cell was able to grow after suspension in pure water (seven days at 10 °C), whereas the native yeast failed to grow (Figure 8.6b).[17]

8.4.2 Resistance to Lytic Enzymes

Another advantage of artificial spores is their capacity to survive other biological factors that threaten the survival of the encapsulated cells. The precise control of shell porosity, which prevents the access of hazardous enzymes but allows the passage of small molecules, is necessary for their protective efficacy, and is therefore an important issue.

LbL-coated yeast cells were incubated with the protozoan *Paramecium primaurelia* whose lysosomal enzymes digest yeast cells (Figure 8.7a).

The protection of yeast cells from digestion was achieved by controlling the number of LbL layers as well as salt concentrations.[31] This study suggested the possibility that tighter control over the shell porosity could protect the encapsulated cells from lytic enzymes.

Yeast cells encapsulated within PD shells were fairly resistant to enzymatic attack by lyticase, a cell-wall-lysing enzyme complex: ~70% of yeast@PD$_1$ and ~90% of yeast@PD$_2$ survived after 1-h incubation, whereas >90% of the native yeast was lysed (Figure 8.7b).[14] Micrometric calcium phosphate shells showed a similar effect (Figure 8.7c).[25]

8.4.3 Resistance to Heat

Fungi, especially in their dormant states as spores, conidia, or sclerotia, are particularly resistant to heat, because trehalose, a nonreducing disaccharide, rapidly accumulates in response to heat shock and increases the thermoresistance of the spore.[32] The capacity of natural spores to counteract heat was also emulated in artificial spores. For example, silica nanoparticle-coated yeast cells showed moderately enhanced thermotolerance compared with native yeast cells (Figure 8.8a). After 2-h exposure to 52 °C, the viability of the coated yeast was 68%, whereas that of the native yeast was about 43%.[33]

The synergistic effects of the silica–titania composite shells in thermotolerance have been demonstrated (Figure 8.8b). *Chlorella* cells were encapsulated individually within SiO$_2$–TiO$_2$ shells with (RKK)$_4$D$_8$ (R: arginine; K: lysine, D: aspartic acid) as a biomimetic catalyst for both titanium bis(ammonium lactato)dihydroxide (TiBALDH) (TiO$_2$ precursor) and silicic acid (SiO$_2$ precursor).[34] The resulting ultrathin SiO$_2$–TiO$_2$ shells enhanced the thermotolerance of *Chlorella*: with the fact that the normal growing

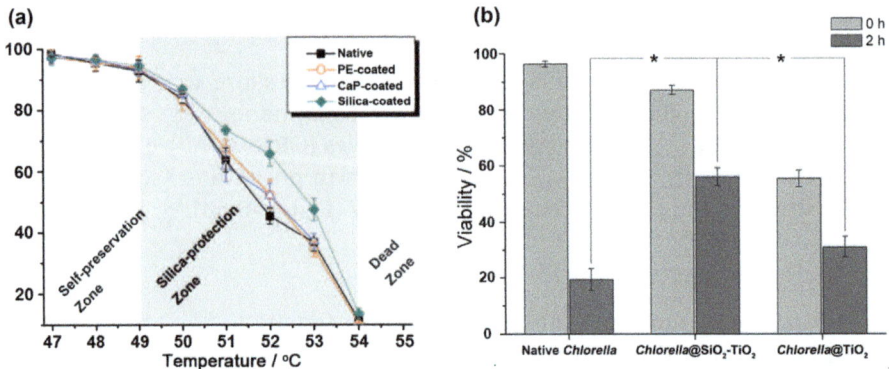

Figure 8.8   Resistance to heat. (a) Silica-nanoparticle-decorated yeast cells showed enhanced thermotolerance compared with native ones. (b) *Chlorella* cells encapsulated within silica–titania composite shells showed enhanced thermotolerance.
Reproduced with permission from Wiley (Copyright 2010, 2013).[33,34]

temperature for *Chlorella* is 23 °C (cf. 30 °C for yeast cells), the cells were heated to 45 °C, which is a fairly stressful condition for *Chlorella* cells. After 2 h, the majority of native *Chlorella* cells were dead ($\sim$81%), but more than a half of the SiO_2–TiO_2-encapsulated *Chlorella* (*Chlorella*@SiO_2-TiO_2) survived the elevated temperature (56%). The survival ratio for *Chlorella*@SiO_2-TiO_2 was calculated to be 0.64 ($= 56.0\%/87.2\%$), whereas that for native *Chlorella* was 0.20 ($= 19.3\%/96.3\%$), indicating that the SiO_2–TiO_2 shells caused a three-fold increase in thermoprotection. By comparison, the survival ratio for *Chlorella*@TiO_2 was 0.56 ($= 31.1\%/55.4\%$), indicating the synergistic effects of the SiO_2–TiO_2 composite.

8.4.4 Resistance to UV

Natural endospores showed the enhanced resistance to UV radiation. For example, the spores of *Bacillus subtilis* are reported to be 5–50 times more resistant to UV radiation than the growing cells.[35] Their protection against UV originates mainly from the DNA repair systems that restore the mutated photoproducts of DNA rather than from the shell itself, which protects the inner cell from other stressors. The accumulation of dipicolinic acid is also involved in the repair system.[36]

Metabolically engineering nonspore-forming cells to confer UV resistance is extremely challenging, and UV resistance is, therefore, usually established by designing shells whose materials filter and/or block UV light, which is lethal to living cells.

Inorganic shells were formed to protect zebrafish embryos from UV radiation during the course of their development.[37] Fertilized zebrafish eggs were individually encapsulated within thick ($\sim$500 nm) ionic lanthanide phosphate ($LnPO_4$) shells, and their embryonic development under long-term UV radiation was compared with that of native embryos (Figure 8.9a). The percentage hatchability of the embryos within the shells was higher than that of the native embryos, and the rate of malformation was also lower for the encased embryos.

Other materials have also been used to protect living algae by filtering out harmful UV, while allowing the passage of visible light for their photosynthetic processes. Living *Chlorella vulgaris* cells were entrapped within a silica hydrogel matrix with embedded CeO_2 nanoparticles.[38] The shielding effect against harmful UV originated from the nanometer-sized ($<$ 10 nm) CeO_2 particles, which also minimized the unwanted scattering of visible light. When irradiated with UV-B light, the green intensity of the algae within the CeO_2-loaded silica hydrogel remained virtually unchanged, whereas a dramatic reduction in intensity was observed with the CeO_2-free silica hydrogels, indicating the photo-oxidative death of the algae (Figure 8.9b). The CeO_2-loaded silica hydrogel also showed an efficient radical scavenging capacity, attributed to the presence of the Ce^{3+}–Ce^{4+} redox pair. The survival of the encapsulated algae in the presence of H_2O_2 was greater than that of the control. Although these examples deviate slightly from the original

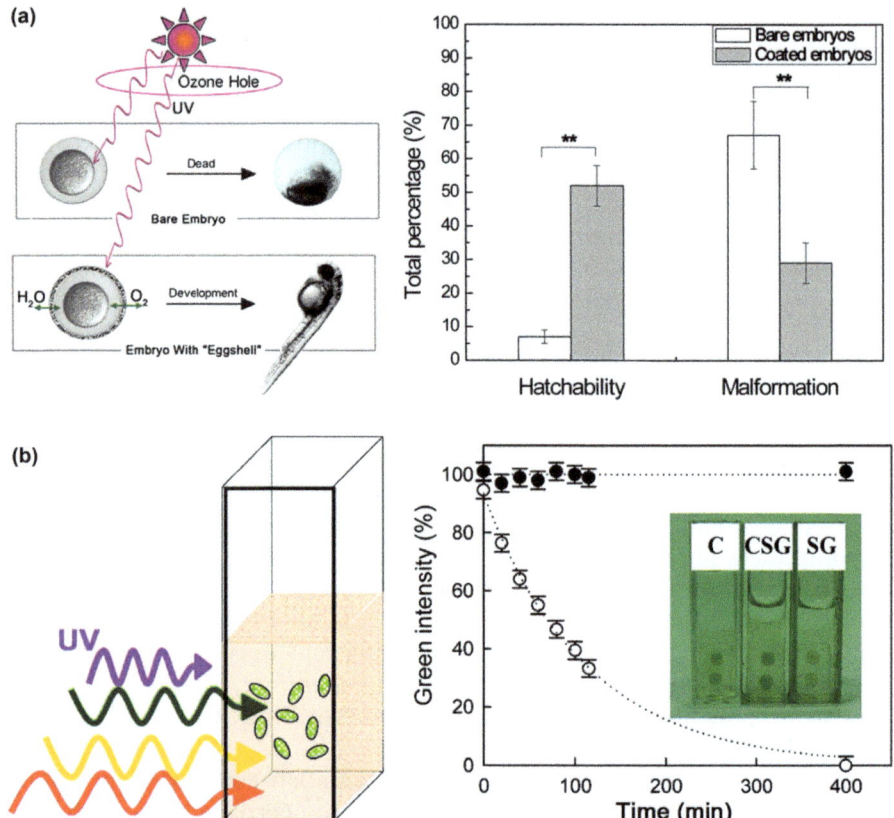

Figure 8.9   Resistance to UV. (a) Fertilized zebrafish embryos encapsulated within lanthanide phosphate (LnPO₄) shells showed enhanced hatchability and reduced malformation under UV radiation. (b) Green intensity of *Chlorella vulgaris* entrapped inside silica hydrogel (SG, ○) and CeO₂-loaded silica hydrogel (CSG, ●) as a function of UV exposure time. Inset: Quartz cells containing *Chlorella vulgaris* cultures inside SG and CSG after a UV-irradiation period of 6.5 h. A nonirradiated SG control (C) is also presented.
Reproduced with permission from PLOS (Copyright 2010)[37] and the American Chemical Society (Copyright 2011).[38]

concept of artificial spores (*i.e. single* cells encapsulated within robust and ultrathin shells), their materials and approaches provide possible tools for designing artificial shells with UV resistance.

8.5   Applications

The features of artificial spores, in principle, need not be limited to the properties of natural endospores, but encasing individual living cells within artificial materials will endow the encapsulated cells with additional, exogenous properties, which expands the applications to many diverse biotechnological fields. For example, functional shells can be utilized for cell-based sensors, high-throughput screening, and cell therapy.[39] The encapsulation of individual living cells also provides a useful model system for studying the biology of single cells.

8.5.1 Functional Artificial Shells for Biotechnological Applications

Forming functionalizable/functional shells on living cells allows the cells to be incorporated into materials or onto surfaces, which is a crucial step in the development of cell-based sensors, high-throughput screening, and cell therapy. It is particularly important for single-cell-based sensors, where the encapsulated cells are immobilized at designated sites on the surfaces. One way to achieve this is to endow the shell itself with specific functionality analogs to that of exosporium in natural endospores.

A major obstacle to conferring functionalities onto inorganic shells is their chemical inertness and/or the harsh reaction conditions for functionalizations. For example, the well-known silane chemistry for silica shells is lethal to living cells, and the ionic crystals are not amenable to functionalization.

A bioinspired silicification process involving changes in the silica precursors suggests that functionalizable groups can be incorporated under biocompatible conditions. Thiol groups (SH) was incorporated into biomimetic silica shells (SiO_2^{SH}) by using (3-mercaptopropyl)trimethoxysilane (MPTMS) as the silica precursor.[40] MPTMS was selected as the model additive, because it was reported to be polycondensed simultaneously with silicic acid under physiologically mild conditions. The utility of individual yeast cells within SiO_2^{SH} shells (yeast@SiO_2^{SH}) was thus extended, because thiol groups reacted with a variety of maleimide derivatives under cytocompatible conditions. For example, the fluorescent dye fluorescein was introduced into the SiO_2^{SH} shell with a fluorescein-linked maleimide. Biotin-linked maleimide was also introduced into the SiO_2^{SH} shell, allowing the biospecific interaction with streptavidin. The biospecific interaction led to the site-specific immobilization of living cells onto a surface: the biotin-functionalized yeast@SiO_2^{SH} was immobilized on defined areas of poly(polyethyleneglycol methacrylate) (poly[PEGMA]) films that were patterned with streptavidin (Figure 8.10a).

Another approach to the functionalization of inorganic shells is to identify and utilize the inherent reactivity of materials, as in the case of photocatalytic titania (TiO_2). Individual *Chlorella* cells were encased in TiO_2 shells: the resulting TiO_2 shells were further functionalized using the specific interaction between TiO_2 and the 1,2-hydroxybenzyl group (Figure 8.10b).[18]

Organic shells can also give a possibility of chemical functionalizability. For example, yeast@PD reacted with the amine or thiol groups *via* 1,4-conjugate addition under slightly basic conditions.[14]

In the practical application of cell-based sensors, the replacement and removal of immobilized cells is desirable for the fabrication of reusable

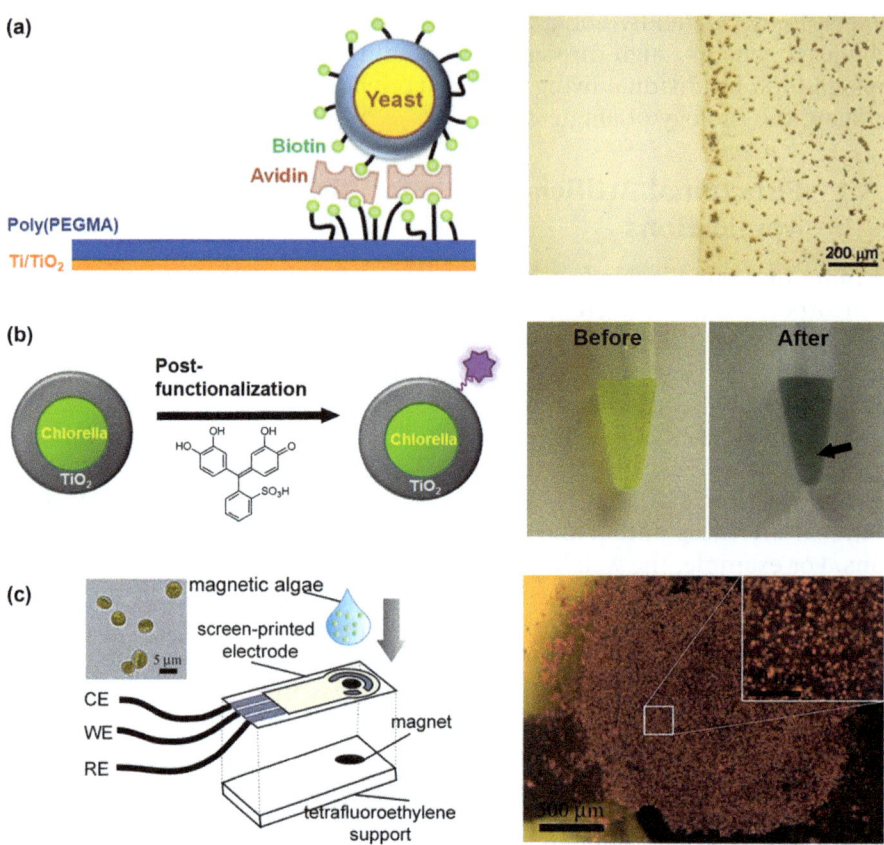

Figure 8.10 Functional artificial shells. (a) Silica shells bearing thiol groups on yeast were functionalized with maleimide-biotin groups. Postfunctionalization with streptavidin enabled the site-selective immobilization of yeast onto the biotin-presenting surfaces. (b) The titanium oxide shell on yeast was functionalized with pyrocatechol violet. (c) *Chlorella* cells coated with magnetic nanoparticles showed reversible immobilization on screen-printed electrodes (CE: counter electrode, WE: working electrode, RE: reference electrode).
Reproduced with permission from Wiley (Copyright 2011),[40] the American Chemical Society (Copyright 2012),[18] and the Royal Society of Chemistry (Copyright 2011).[41a]

biosensors. Therefore, it is necessary to develop a reversible system in which encapsulated cells associate with a surface platform. Various microbial and mammalian cells were magnetically functionalized with PAH-stabilized MNPs.[41] In a demonstration of cell-based sensors, *Chlorella pyrenoidosa* microalgal cells were coated with PAH-stabilized MNPs for their reversible immobilization on screen-printed electrodes, where they detected herbicides *via* an electrical signal (Figure 8.10c).[41a] The cells were disengaged when the underlying magnet was removed.

8.5.2 Biological Platform for Studying the Biology of Single Cells

The concept of artificial spores involves the nanoencapsulation of individual living cells, whose artificial shells confer enhanced protection against external stressors, such as heat, radiation, lytic enzymes, or osmotic pressures. The compartmentalization of a single cell within an artificial shell also provides a useful model system for investigating cellular behaviors at the single-cell level.

The biological studies of single cells and their communications would be facilitated further by the development of artificial shells that have sufficient aqueous spaces inside to allow the three-dimensional movement of single cells as well as possessing selective permeability to control the entry of signaling molecules.

Isolating individual cells from other cells can provide important information on cell-to-cell communication. For example, quorum sensing (QS), in which the collective behavior of individuals benefits the group, is an interesting phenomenon, and is considered one of the survival mechanisms of micro-organisms.[42] Multiple factors, including cell density, cell dimensions, and the mass transport of autoinducers, are related to QS. However, most of the studies have been performed with large numbers of cells, thereby obscuring whether mass transport and/or cellular dimensions themselves can initiate QS pathways.

The cell-directed assembly of lipid–silica nanostructures allowed the confinement of individual *Staphylococcus aureus* bacterial cells in a cytocompatible manner. Individual cells within silica hemispheres on a solid substrate activated their QS pathways, suggesting that other factors, such as cellular dimensions and diffusional characteristics, influenced QS at the single-cell level.[43] The precise control of cell density, shell volume, mass-transport properties, and the spatial distribution of cells will allow further investigation of QS.

8.6 Conclusions

The cytocompatible nanoencapsulation of nonspore-forming cells within ultrathin (<100 nm), robust, and rigid shells has rudimentarily mimicked the cryptobiotic features of endospores, which survive under harsh environmental conditions, including chemicals, heat, UV radiation, desiccation, and malnutrition. The value of artificial spores is not limited to their spore-like behavior; they also offer a technology platform for single-cell-based applications and a basic model for investigating the biology of single cells. The purpose of this emerging field does not only seek to mimic the behavior of natural spores, but also to understand cellular responses to the environmental changes at the single-cell level. In addition, the techniques developed for cytocompatible and cytoprotective nanoshell formation would offer basic tools for multiple applications in biotechnology.

References

1. K. Donohue, R. R. de Casas, L. Burghardt, K. Kovach and C. G. Willis, *Annu. Rev. Ecol. Syst.*, 2010, **41**, 293.
2. J. S. Clegg, *Comp. Biochem. Physiol., B*, 2001, **128**, 613.
3. (a) D. R. Nelson, *Integr. Comp. Biol.*, 2002, **42**, 652; (b) J. E. Morris and B. A. Afzelius, *J. Ultrastruct. Res.*, 1967, **20**, 244; (c) A. M. Treonis and D. H. Wall, *Integr. Comp. Biol.*, 2005, **45**, 741; (d) M. Caprioli, A. K. Katholm, G. Melone, H. Ramlbv, C. Ricci and N. Santo, *Comp. Biochem. Physiol., A*, 2004, **139**, 527.
4. L. Sømme, *Eur. J. Entomol.*, 1996, **93**, 349.
5. R. O. Schill, *The Epoch Times*, 2009.
6. P. T. McKenney, A. Driks and P. Eichenberger, *Nat. Rev. Microbiol.*, 2013, **11**, 33.
7. A. Driks, *Trends Microbiol.*, 2002, **10**, 251.
8. P. Setlow, *J. Bacteriol.*, 1992, **174**, 2737.
9. P. Setlow, *Curr. Opin. Microbiol.*, 2003, **6**, 550.
10. W. L. Nicholson, N. Munakata, G. Horneck, H. J. Melosh and P. Setlow, *Microbiol. Mol. Biol. Rev.*, 2000, **64**, 548.
11. (a) W. Foissner, *Acta Protozool.*, 2009, **48**, 223; (b) W. Foissner, B. Weissenbacher, W. D. Krautgartner and U. Lütz-Meindl, *J. Eukaryot. Microbiol.*, 2009, **56**, 519.
12. (a) S. H. Yang, D. Hong, J. Lee, E. H. Ko and I. S. Choi, *Small*, 2013, **9**, 178; (b) D. Hong, M. Park, S. H. Yang, J. Lee, Y.-G. Kim and I. S. Choi, *Trends Biotechnol.*, 2003, **21**, 338.
13. A. Matsuzawa, M. Matsusaki and M. Akashi, *Langmuir*, 2013, **29**, 7362.
14. S. H. Yang, S. M. Kang, K.-B. Lee, T. D. Chung, H. Lee and I. S. Choi, *J. Am. Chem. Soc.*, 2011, **133**, 2795.
15. S. H. Yang, K.-B. Lee, B. Kong, J.-H. Kim, H.-S. Kim and I. S. Choi, *Angew. Chem. Int. Ed.*, 2009, **48**, 9160.
16. S. A. Konnova, I. R. Sharipova, T. A. Demina, Y. N. Osin, D. R. Yarullina, O. N. Ilinskaya, Y. M. Lvov and R. F. Fakhrullin, *Chem. Commun.*, 2013, **49**, 4208.
17. R. Kempaiah, S. Salgado, W. L. Chunga and V. Maheshwari, *Chem. Commun.*, 2011, **47**, 11480.
18. S. H. Yang, E. H. Ko and I. S. Choi, *Langmuir*, 2012, **28**, 2151.
19. V. Kozlovskaya, S. Harbaugh, I. Drachuk, O. Shchepelina, N. Kelley-Loughnane, M. Stone and V. V. Tsukruk, *Soft Matter*, 2011, 7, 2364.
20. J. Lee, S. H. Yang, S.-P. Hong, D. Hong, H. Lee, H.-Y. Lee, Y.-G. Kim and I. S. Choi, *Macromol. Rapid Commun.*, 2013, **34**, 1351.
21. R. F. Fakhrullin, J. García-Alonso and V. N. Paunov, *Soft Matter*, 2010, **6**, 391.
22. S. H. Yang, T. Lee, E. Seo, E. H. Ko, I. S. Choi and B.-S. Kim, *Macromol. Biosci.*, 2012, **12**, 61.
23. M. Sumper, *Science*, 2002, **295**, 2430.

24. (a) S. H. Yang, J. H. Park, W. K. Cho, H.-S. Lee and I. S. Choi, *Small*, 2009, **5**, 1947; (b) J. J. Yuan, O. O. Mykhaylyk, A. J. Ryan and S. P. Armes, *J. Am. Chem. Soc.*, 2007, **129**, 1717.

25. B. Wang, P. Liu, W. Jiang, H. Pan, X. Xu and R. Tang, *Angew. Chem. Int. Ed.*, 2008, **47**, 3560.

26. R. F. Fakhrullin and R. T. Minullina, *Langmuir*, 2009, **25**, 6617.

27. I. Drachuk, O. Shchepelina, M. Lisunova, S. Harbaugh, N. Kelley-Loughnane, M. Stone and V. V. Tsukruk, *ACS Nano*, 2012, **6**, 4266.

28. C. J. Ochs, T. Hong, G. K. Such, J. Cui, A. Postma and F. Caruso, *Chem. Mater.*, 2011, **23**, 3141.

29. Y. Yan, Y. Wang, J. K. Heath, E. C. Nice and F. Caruso, *Adv. Mater.*, 2011, **23**, 3916.

30. A. M. Kloxin, A. M. Kasko, C. N. Salinas and K. S. Anseth, *Science*, 2009, **324**, 59.

31. R. Magrassi, P. Ramoino, P. Bianchini and A. Diaspro, *Microscopy*, 2010, **73**, 931.

32. C. De Virgilio, U. Simmen, T. Hottiger and A. Wiemken, *FEBS Lett.*, 1990, **273**, 107.

33. G. Wang, L. Wang, P. Liu, Y. Yan, X. Xu and R. Tang, *ChemBioChem*, 2010, **11**, 2368.

34. E. H. Ko, Y. Yoon, J. H. Park, S. H. Yang, D. Hong, K.-B. Lee, H. K. Shon, T. G. Lee and I. S. Choi, *Angew. Chem. Int. Ed.*, 2013, **52**, 12279.

35. P. Setlow, *Environ. Mol. Mutagen.*, 2001, **38**, 97.

36. T. Douki, B. Setlow and P. Setlow, *Photochem. Photobiol. Sci.*, 2005, **4**, 591.

37. B. Wang, P. Liu, Y. Tang, H. Pan, X. Xu and R. Tang, *PLoS ONE*, 2010, **5**, e9963.

38. C. Sicard, M. Perullini, C. Spedalieri, T. Coradin, R. Brayner, J. Livage, M. Jobbágy and S. A. Bilmes, *Chem. Mater.*, 2011, **23**, 1374.

39. (a) P. Banerjee and A. K. Bhunia, *Trends Biotechnol.*, 2009, **27**, 179; (b) D. W. Green, G. Li, B. Milthorpe and B. Ben-Nissan, *Mater. Today*, 2012, **15**, 60.

40. S. H. Yang, E. H. Ko, Y. Y. Jung and I. S. Choi, *Angew. Chem. Int. Ed.*, 2011, **50**, 6115.

41. (a) A. I. Zamaleeva, I. R. Sharipova, R. V. Shamagsumova, A. N. Ivanov, G. A. Evtugyn, D. G. Ishmuchametova and R. F. Fakhrullin, *Anal. Methods*, 2011, **3**, 509; (b) M. R. Dzamukova, A. I. Zamaleeva, D. G. Ishmuchametova, Y. N. Osin, A. P. Kiyasov, D. K. Nurgaliev, O. N. Ilinskaya and R. F. Fakhrullin, *Langmuir*, 2011, **27**, 14386.

42. C. M. Waters and B. L. Bassler, *Annu. Rev. Cell Dev. Biol.*, 2005, **21**, 319.

43. E. C. Carnes, D. M. Lopez, N. P. Donegan, A. Cheung, H. Gresham, B. S. Timmins and C. J. Brinker, *Nat. Chem. Biol.*, 2010, **6**, 41.

CHAPTER 9

Artificial Multicellular Assemblies from Cells Interfaced with Polymers and Nanomaterials

ANUPAM A. K. DAS,[a] RAWIL F. FAKHRULLIN[b] AND
VESSELIN N. PAUNOV*[a]

[a] Department of Chemistry, University of Hull, Hull, HU6 7RX, UK;
[b] Department of Microbiology, Kazan Federal University, Kazan, Republic
of Tatarstan, Russian Federation, 420008
*Email: V.N.Paunov@hull.ac.uk

9.1 Introduction

There are many examples in Nature where in biological system cells assemble in higher structures following different patterns or as a result of specific cell adhesion. For example, lichens have evolved as multicellular assemblies involving layers of fungal cells and photosynthetic algae in a symbiotic relationship that enables them to survive in harsh environmental conditions. Similar cell assembly occurs in higher organisms driven by specific bioadhesion between the cell membranes. Currently available technologies are still far from being able to control the cell interactions using the same bioligands, which does not allow direct control in reproducing the natural processes of cell assembly in many tissue engineering applications. However, recent advances in bionanotechnology have allowed

RSC Smart Materials No. 9
Cell Surface Engineering: Fabrication of Functional Nanoshells
Edited by Rawil F Fakhrullin, Insung S Choi and Yuri Lvov
Published by the Royal Society of Chemistry, www.rsc.org

mimicking of these cell-assembling processes existing in Nature by using cells whose membranes have been modified using a variety of novel materials, including polymers and nanoparticles. The interfacing of individual cells with nanomaterials without impacting their viability is a huge challenge but it also allows additional control over the way they interact with their environment and brings multiple technological opportunities for assembling of such modified cells into a variety of multicellular structures with many possible applications that go well beyond tissue engineering. This chapter aims to give a comprehensive overview of the emerging area of the fabrication of multicellular assemblies mediated by the cell interactions with polymers and nanomaterials. Here, we will focus on the recent progress in this rapidly growing research field where cells functionalized with various kinds of nanomaterials have been explored to make artificial cell structures that can be used in tissue engineering,[1] biosensors,[2,3] toxicity microscreening devices,[4] microelectronics[5] and in bioanalytical chemistry.[6] The fabrication of artificial multicellular assemblies using surface-modified cells and nanomaterials is a truly interdisciplinary area that combines chemistry, biology, physics and material sciences. Recent research highlights in this field include microorganisms being used as templates for deposition of polyelectrolyte multilayers,[7] polyelectrolyte-functionalized particles[8] and nanoparticles[9] that result in an enhancement of their functions in areas like high-resolution biomedical imaging, gene sequencing for molecular diagnostics and whole-cell biosensor-based electronic devices. The methods used for deposition of nanoparticles and polymers on the cell surface depend on the specifics of the application of the functionalized cells, which also determine whether the cell viability needs to be preserved. For example, in tissue-engineering applications, the preservation of the viability of the functionalized cells is of paramount importance for the tissue/organ regeneration. Cell viability has also been an issue in the case of cells interfaced with magnetic nanoparticles used as biosensing elements in microfluidic devices.[10] However, there are cases where the viability of the produced "cyborg" cells is not important, like in the case of cells used as conductive microbridges.[11] In this chapter we will cover briefly the main categories of cyborg cells as building blocks for cell assembly in higher structures. We will consider some of the template-driven methods for the fabrication of multicellular assemblies involving interactions with nanoparticles and polymer coatings as well as external fields. We will also outline some of the recent developments in 3D-tissue bioprinting with prefabricated cell spheroids as well as multicellular structures formed by self-folding of 3D cell-laden microstructures by cell traction force (origami folding).[12] We will give an overview of several reports of gel and polyelectrolyte-mediated multicellular structures. Finally, we will discuss some of the applications of the cell structures formed by nanomaterials and polyelectrolyte-mediated interactions.

9.2   Cyborg Cells as Building Blocks for Cell Assembly

There is a variety of approaches used in the process of making three-dimensional structures of cyborg cells that usually starts with modifications of the cell-media interface. These can involve direct deposition of nanomaterials on the cell surface, polymer-mediated functionalization of cells and formation of inorganic mineral shells on cell surfaces turn the cells into suitable building blocks for making artificial multicellular structures with applications for bioelectronic, biosensors, cell-based therapies and tissue engineering. Cyborg cells have been fabricated by depositing nanoparticles, nanorods, nanotubes, nanosheets and various other colloidal particles on the cell surface.[13] The cell surface can be modified beforehand with other materials in order for the efficient absorption of the particles on the surface. Microbial cell walls are mostly negatively charged at neutral pH due to partial dissociation of carboxylic groups from carbohydrates and membrane proteins on the cell surface. This allows for surface-charge-mediated absorption of various oppositely charged particles and polymers on the cell surface and establishes a simple route for altering the cell/media interface. For example, Berry *et al.* reported the deposition of cationic poly(lysine)-coated nanoparticles on the surface of a negatively charged bacteria.[9] They also used cationic surfactants like cetyl trimethyl ammonium bromide (CTAB) to modify the surface charge of gold nanoparticles that promoted their adhesion to the surface of the bacterial cell walls by imparting positive charge onto the nanoparticles within a wide range of pH values.[9] An alternative method involves the deposition of polyelectrolytes and nanoparticles[14] on the bacterial cell wall. Current techniques for polyelectrolyte deposition allows 2–10 layers of polyelectrolytes with known concentrations to be deposited on the cells. Cyborg cells have been produced by coating with various cationic and anionic polyelectrolytes, calcium phosphate and silica nanoparticles, and mineral shells like calcium carbonate. The layer-by-layer (LbL) assembly of polyelectrolytes on the cell surface is achieved *via* sequential absorption of oppositely charged polyelectrolytes that are bound to each other through attractive electrostatic interactions and can later be used for fabrication of cellular assemblies or additional functionalization with other nanomaterials.[15] There is a variety of cells that have already been used after surface modification with polyelectrolytes and nanoparticles that includes yeast,[10] bacteria,[16] fungi,[8] human cell lines,[17] stem cells and even round worms like *C. elegans*.[18]

The deposition of nanoparticles in combination with various polyelectrolytes has various advantages over direct deposition of nanoparticles on the cell surface.[2,8,14] The polyelectrolytes facilitate the adhesion of the nanoparticles to the cell surface, hence providing the structure with stability and also preventing the internalization of nanoparticles through the cell wall into the cytoplasm through endocytosis. Figure 9.1 shows various kinds of combination of polyelectrolytes and nanoparticles used to interface different types of living cells. For example, Suo *et al.*[20] reported human blood platelets

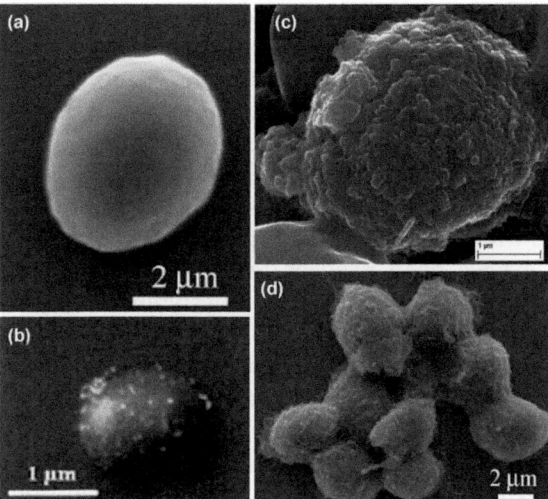

Figure 9.1 Deposition of nanoparticles and polymers on living cells. (a) Uncoated yeast cell.[19] (b) *E. coli* cells coated with PAH-AuNP-PAH layers.[6] (c) Yeast cell coated with PAH-stabilized magnetic nanoparticles.[14] (d) Yeast cells coated with PAH/PSS doped with carbon nanotubes.
Reprinted with permission from Ref. 19 Copyright (2010) American Chemical Society. Reprinted from Ref. 6 with kind permission from Springer Science and Business Media, Copyright (2009) Springer. Reproduced with permission from Ref. 14 Copyright (2010) The Royal Society of Chemistry.

coated with polyelectrolytes and silica nanoparticles (polyelectrolyte mediated silication). Zamaleeva *et al.*[19] successfully produced viable cells that were coated with polyelectrolytes and doped with nanotubes. Silica shells thicker than 50 nm have been deposited on yeast.[21] Yeast cells have also been coated with gold and magnetic nanoparticles and stabilized using PAH without significant impact on the cell viability.[6,14] The functionalization of cells with polymers and/or nanoparticles not only creates additional mechanisms for directed deposition of the modified individual cells onto patterns and templates but also gives better control over the interaction forces between the cells that lead to the formation of artificial multicellular assemblies. These strategies have recently enabled researchers to assemble artificial cell structures of various geometries like free-standing biofilms,[22] cellosomes,[23–25] linear multicellular structures[26] and 3D multicellular composites.[27]

9.3 Template-Mediated Multicellular Assemblies

Surface-functionalized cells have been used successfully in preparation of 2D and 3D cellular assemblies and multicellular clusters, both template-mediated and "on surface" assembled free standing structures.[2,23,25,28,29] The 2D cellular assemblies have been produced by microcontact printing using polyelectrolyte-coated yeast cells.[30] Cell–polyelectrolyte assemblies were first used to make microcapsules with multilayer polyelectrolyte films.[31–33] The template is a sacrificial core particle that can be dissolved once the cells have been deposited and bound together that produces a hollow-shell structure of spherical, spheroid or other geometries with a wide range of applications. Typically, the following types of particles have been used as templates for preparation of microcapsules: melamine formaldehyde[34–36], polystyrene,[37] silica particles,[38] organic[39] and inorganic microcrystals.[40] By analogy with colloidosomes,[41] which consist of a hollow membrane of colloid particles, cell assemblies of hollow-shell architecture, termed "cellosomes"[23,42] have been recently reported. These represent cellular assemblies made of a monolayer of living cells held together by colloid interactions. They are also formed using a sacrificial core to obtain hollow-shell cellosomes membrane followed by dissolving the core, while maintaining the viability of the cells. Cellosomes can be considered as simple models of multicellular organisms.[23] Two different types of templates have been used to fabricate cellosomes. Microbubbles were templated with yeast cells coated with cationic polyelectrolyte that were additionally bond together by anionic polyelectrolyte followed by the dissolution of the microbubbles.[42] This approach uses the fact that the microbubbles carry a net negative surface charge that attracts the polycation-coated cells and forms a shell of adsorbed cells on the bubbles surface.

In the air-bubbles templating method[24] the cells were functionalized using a combination of polyelectrolytes (PAH/PSS/PAH) in order to charge reverse the cell's surface potential and promote their spontaneous self-assembly on the surface of the air micro bubbles as shown in Figures 9.2a and b that form upon vortexing of the cell suspension. The coated air microbubbles were then sealed by further addition of polyanion (PSS). Figures 9.2c and d show these structures after being filtered out which following further rehydration and dissolution of the air bubble cores form yeastosomes (Figure 9.2e). The obtained cellosome structure strongly resemble primitive colonial organisms like *Volvox*.[24] Interestingly, they also look very similar to recently discovered fossilized multicellular clusters of nonmarine eukaryotes (Figure 9.2f), obtained after dissolution of ancient sediment rock formations.[43] The viability of the yeastosomes has been demonstrated by exposing them to fluorescein diacetate (FDA) that causes the cell to emit green fluorescence due to accumulation of fluorescein in the cell cytoplasm due to intracell hydrolysis of the FDA by nonspecific esterases.[24]

Recently, the fabrication of cellosome assemblies of yeast on calcium carbonate microcrystals doped with polyelectrolytes and magnetic nanoparticles has been reported.[25] In this approach, calcium carbonate microcrystals of rod-like and rhombohedra-like morphology were coated with magnetic nanoparticles. Yeast cells were deposited onto these magnetically

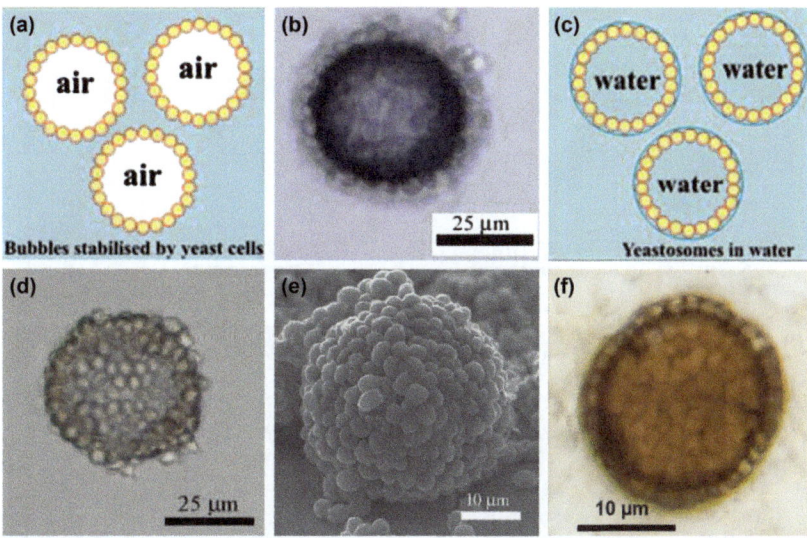

Figure 9.2 Controlled organization of functionalized cells into multicellular clusters. (a) Schematic of spherical cellosomes with air bubble as template. (b) Optical image of an air-bubble template coated with yeast cells treated with PAH/PSS/PAH. (c) Schematic of spherical cellosomes in water. (d) 3D reconstructed image of a cellosome infused with water. (e) SEM image of spherical multicellular cluster assembled on air microbubbles.[42] (f) Fossilized structures of primitive nonmarine eukaryotes obtained from dissolved ancient sediment rocks, showing remarkable resemblance between artificially assembled cellosomes and naturally evolved primitive multicellular aggregates.[43]
Reproduced with permission from Ref. 42 Copyright 2010 The Royal Society of Chemistry. Image (f) reprinted with permission from Ref. 43 Copyright 2011 Macmillan Publishers Ltd.

responsive templates using polyelectrolyte-mediated binding by the LbL method. Oppositely charged polyelectrolyte coats were selected as terminal layers of the cells and the microcrystals to promote electrostatic adhesion of the cells on the template. The incubation of the microcrystals in a suspension of an excess of oppositely charged cells resulted in the formation of multicellular assemblies as shown in Figure 9.3. The template, Figures 9.3e and f, was then dissolved using ethylene diamine tetra acetic acid (EDTA) solution that resulted in a structure that retained its initial shape of the microcrystals template, as shown in Figures 9.3g and h. Needle-like cellosomes from yeast cells (yeastosomes) were obtained by templating aragonite microcrystals with polyelectrolyte-coated yeast, while a similar procedure yielded rhombohedra-shaped yeastosomes by using calcite crystals as templates.[23,25] These structures were magnetically modified by per-coating the microcrystals with magnetite nanoparticles that made them responsive to external magnetic field and allowed their extraction from the cell suspension with an external magnet.

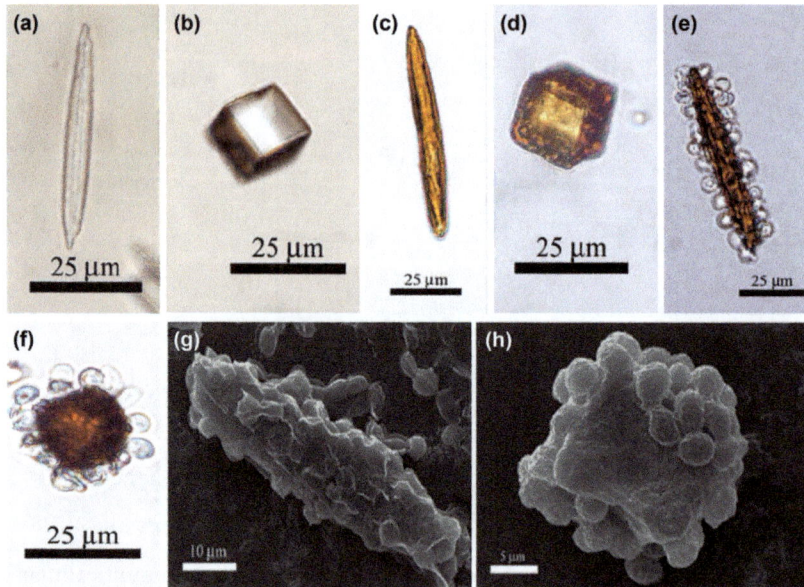

Figure 9.3   Optical microscopic images of (a) bare aragonite and (b) calcite microcrystals. (c) Rod-like aragonite and (d) rhombohedra-like calcite templates coated with polyelectrolytes and magnetic nanoparticles
PAH/MNP/PAH/PSS/PAH. (e) Needle-like and (f) rhombohedra-like
core–shell capsules coated with yeast cells. SEM image of (g) Rod-like
and (h) rhombohedra-like cellosomes.[23]
Reproduced from Ref. 23 with permission from the PCCP Owner
Societies Copyright 2010 The Royal Society of Chemistry.

9.4   Multicellular Assembly Driven by External Fields

Living cells have been functionalized with magnetic particles using various
processes in order to control specific spatial organization of the living cells
using an external magnetic field. However, retaining the viability of the
coated cells during the deposition of the magnetic particles is a major issue.
The current approaches to manipulate cells require to either attaching to the
magnetic particles to the surface of the living cells or their internalization by
the cells.[28,44–47] In the works by Ino *et al.*[44,48] a novel technology was developed based on a physical method, where magnetic force was used to
fabricate cell patterns on a nonfunctionalized surface with high resolution.
Magnetite cationic liposomes (MCLs) were used in the process of magnetic
force-based tissue engineering (Mag-TE) where various kinds of cells were
magnetically labeled using MCLs and a magnetic force was applied to make
3D multilayered cell sheet-like structures[48–51] and various tubular structures.[28] The magnetically labeled fibroblast with MCLs and was seeded into
a tissue culture dish with a steel plate and a magnet. The fibroblast cells
were lined up in the shape of a line and show high-resolution cell micro-
patterning. Several other patterns have been produced using this process of

Mag-TE including cells in a curved line or other desired shapes.[44] The Mag-TE process is not limited to a specific cell culture surface but any different culture surface can be used, which includes biological gels and non-absorbent surfaces. There are various other methods designed to produce fully viable magnetized cells. Fakhrullin *et al.*[14] successfully produced magnetized yeast cells using polyelectrolyte-coated nanoparticles of magnetite stabilized by tetramethyl ammonium hydroxide with which to make a stable coating on the cell surface and efficiently retain the cell's viability. This method has the potential beyond just the stabilization of yeast cells with magnetic nanoparticles as it can be used in the transportation and positioning of the cells using an external magnetic field, which in turn will help in the formation of viable desired shaped multicellular structures.[10] A magnet was shown to be used in order to extract the magnetic cells from a suspension, as shown in Figure 9.4e.

Tanase *et al.* used ferromagnetic nanowires to control the cell localization, single-cell trapping and ordered multicellular assemblies with the help of micromagnet arrays.[52] A microscopic slide was patterned with micro-magnetic arrays from the bottom of the parallel-plate flow chamber and with the introduction of uniform external magnetic fields and the magnetic nanowires the cells were lined up as shown in Figures 9.4a and b. This process allows the cells to be patterned accordingly just by controlling the external magnetic fields and the fluid flow. The other advantage is the control of the cellular alignment by reversing the external field direction to invert the sign of the wire micromagnet direction and in turn change the alignment of the cells without any selective modification or chemical functionalization of the substrate or the cell surface.[52] The above method requires cell surface modification or the absorption of magnetic nanoparticles into the cell, which might cause long-term cytotoxicity effects, depending on the cell type.

Krebs *et al.*[53] developed a method for spatial organization of cells using freely suspended inert and cytocompatible magnetic nanoparticles. They used the process of negative magnetophoresis to form linear, oriented cell structures under a uniform magnetic field (Figures 9.4h–j). The cells were suspended in an aqueous ferrofluid that upon application of magnetic field aligns its ferrofluid particles that arrange the suspended cells in the form of chains due to their induced magnetic dipoles. These authors show that after the ferrofluid removal, the cells adhered to the substrate surface survive the alignment process and grow as shown in Figure 9.4i. Figure 9.4j shows chains of fluorescently labeled human umbilical vein endothelial cells (HUVECs) that grow in length after the ferrofluid removal. This process can be efficiently used in the tissue-engineering field like in the process of vasculogenesis.[53]

There are various conventional methods for assembling living cells in a suspension by methods such as sedimentation,[54] free-volume restriction,[55] capillary forces[56] and convective evaporation.[57] These processes are indiscriminate, inefficient and difficult to control for preparation of multicellular

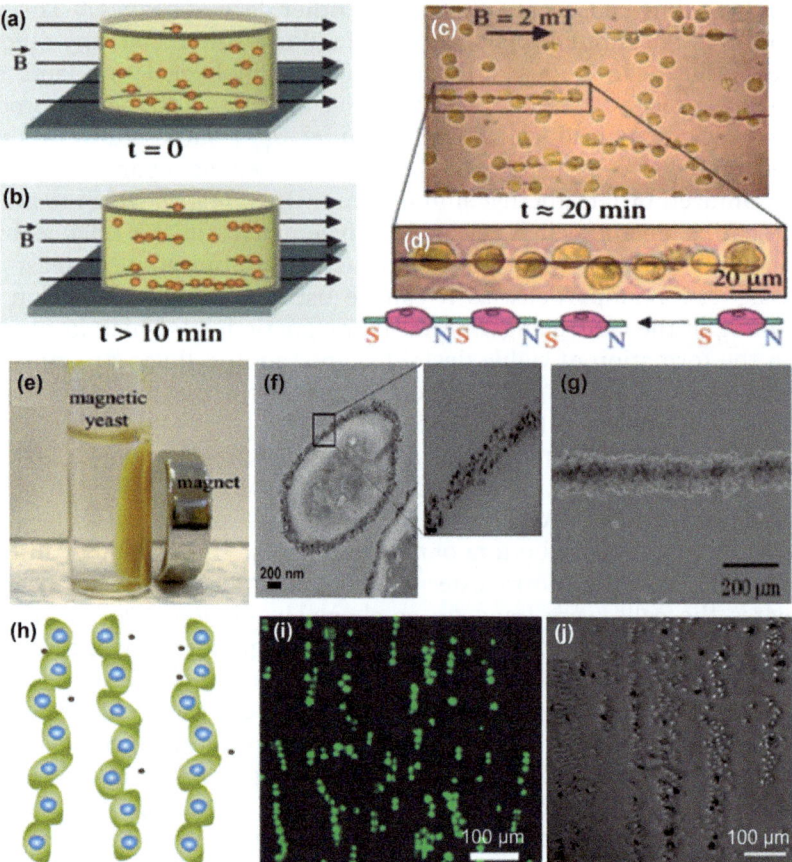

Figure 9.4   Magnetic cell chaining:[52] (a,b) Schematics of nanowires bound to suspended cells and aligned in a magnetic field B; (c) Cell-chain formation process due to magnetic dipole-dipole interactions; (d) Optical image of the cell chains formed at the bottom of the culture dish. Magnetically functionalized living cells:[14] (e) the assembly of the magnetized yeast cells after application of a permanent magnet. (f) Yeast cells coated with PAH-stabilized MNPs. (g) Linear cell pattern using steel plates and magnets.[44] Magnetophoresis-driven cell assembly:[53] (h) Scheme, (i) fluorescence and (j) optical microscopy image demonstrating linear arrangement of magnetically functionalized human endothelial cells assembled in external magnetic field.
Reproduced with permission from Refs. 44, 52, 53 Copyright 2010 The Royal Society of Chemistry; Copyright © 2007 Wiley Periodicals, Inc.; Copyright (2009) the American Chemical Society.

assemblies. The process of magnetophoresis and the use of magnetic nanoparticles deposition on the living cells requires careful design of the magnetic-particle coatings and process of the functionalization of the cells, which might cause cell-viability issues.[14,17] The alternative to this process is the use of an alternating current (AC) electric field to arrange living cells into desired structures. An AC field can be used to achieve a quick controlled and scalable multicellular assembly. The term dielectrophoresis(DEP) means the interaction between the assembled particles due to an induced AC electric field.[58,59] Shang *et al.*, showed how DEP can be used in combination with electrical detection methods to temporarily direct individual bacterial cells across micrometer-sized electrode junctions. However, as the cells can only be arranged temporarily by the AC field, a new cell-assembly method was developed as the combination of physical manipulation (DEP) and biomolecular recognition using biotin and avidin.[60] Avidin has multiple binding sites for biotin and an extremely high binding constant. This allowed the cells or the gold nanorods functionalized with biotin to be assembled with the application of DEP between the electrodes functionalized with avidin and when the AC filed was removed they remained in position because of the adhesion produced by the biotin and avidin interaction.[60] An obvious drawback of this process is that it is not easy to coat any living cells with biotin. Small *et al.*[26] used DEP and a combination of hydrogel trapping and polyelectrolyte-mediated binding to permanently "arrest" the position of living yeast cells in long linear aggregates after the removal of the electric field. The yeast cells were pretreated using PAH and carboxymethyl cellulose (CMC), cationic and anionic polyelectrolytes respectively, by using the LbL method. The gelation of the agarose solution trapped the cell assemblies that were further treated with oppositely charged polyelectrolyte (PAH) introduced electrophoretcially in the gel to permanently bind the cells in the linear structure. The agarose gel matrix was then dissolved, which in turn produced free-standing yeast cell strings that were locked in linear structures (Figure 9.5g) when using any templates or scaffold structures. The viability of the cells was confirmed using fluorescein diacetate solution (Figure 9.5h).

Gupta *el al.*[61,62] demonstrated the permanent assembly of cells by using DEP and functionalized synthetic microparticles binding to the cells through biospecific interactions. The chips were designed with a two co-planar electrode system with the application of the electric field in one direction resulted in the formation of live yeast cell chains, as shown in Figure 9.5c. They illustrate that at low frequencies smaller particles are captured in between the cell junctions by positive dielectrophoresis. By using this effect to capture surface-functionalized colloidal particles they demonstrated how permanent cell structures can be assembled on a chip. On a chip with a four-electrode system the application of electric field in two perpendicular directions resulted in the formation of a 2D ordered array of cells, as illustrated in Figure 9.5d.

9.5 Assembly of Tissue Spheroids

Cellular spheroids are 3D cell-culture models that are being actively explored for the reconstruction of organs and tissues by bioprinting methods. The tissue spheroids can be produced using various processes in which conditions are engineered so that cells tend to aggregate together, rather than

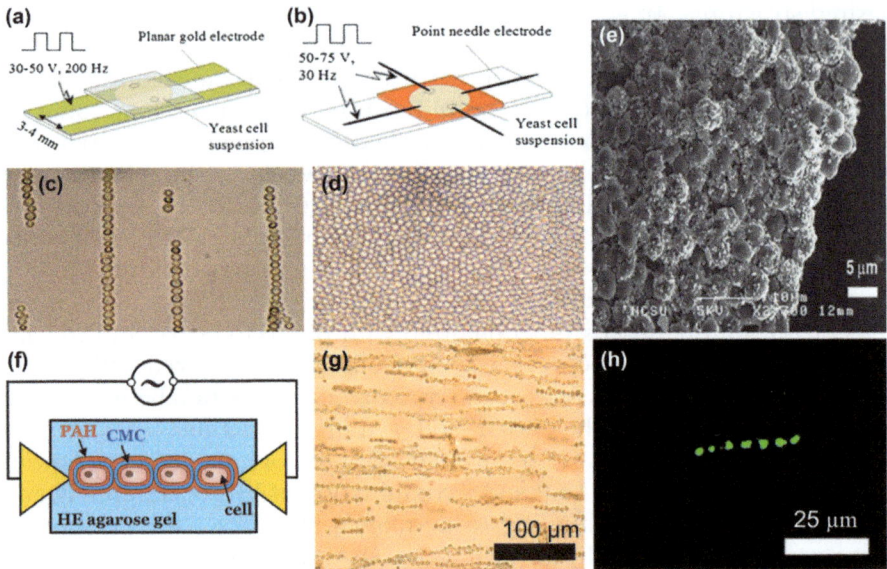

Figure 9.5 Schematics of the (a) two-electrode configuration and (b) four-electrode setup, showing (c) chains from live yeast cells assembled with a two-electrode system.[61] (d) 2D ordered arrays from the live yeast cells formed by using the four-electrode setup. (e) SEM image of a closely packed fixed yeast membrane. (f) Schematics of the preparation of cell strings, using DEP assembly combined with a hydrogel trapping method and polyelectrolyte mediated binding.[26] (g) Optical microscope image of a yeast cells undergoing DEP assembly. (h) Fluorescence microscopy images of the obtained free-standing yeast cell strings after treatment with FDA.
Reproduced with permission from Refs. 26, 61 Copyright 2013 The Royal Society of Chemistry.

adhere to the culture substrate. Cellular spheroids can be produced by various simple methods like spinner culture, rotary culture, nonadhesive surfaces, *etc.*[63] However, it is difficult to control the size of the cell spheroids with these simple methods, hence, to produce uniform cellular spheroids process like the hanging-drop culture[64,65] and culturing in microwells[66] are used, as shown in Figures 9.6a–h. The other methods of uniform spheroid formation involves the microfabrication process combined with the external forces that results in the technique of microrotational flow[67] and the magneto-Archimedes effect.[68] The hanging-drop technique was originally designed to cultivate stem-cell embryoid bodies and involves deposition of cell-suspension drops onto the underside of the lid of a culture dish. The drop volume is typically around 15–30 µL containing around 300–3000 cells, as shown in Figure 9.6a. The lid is then inverted and the drops hold in its position by surface tension. Gravity makes the cells concentrate at a point inside the drop resulting in the formation of spheroids at the liquid/air interface, as shown in Figure 9.6b. This technique has been used to produce

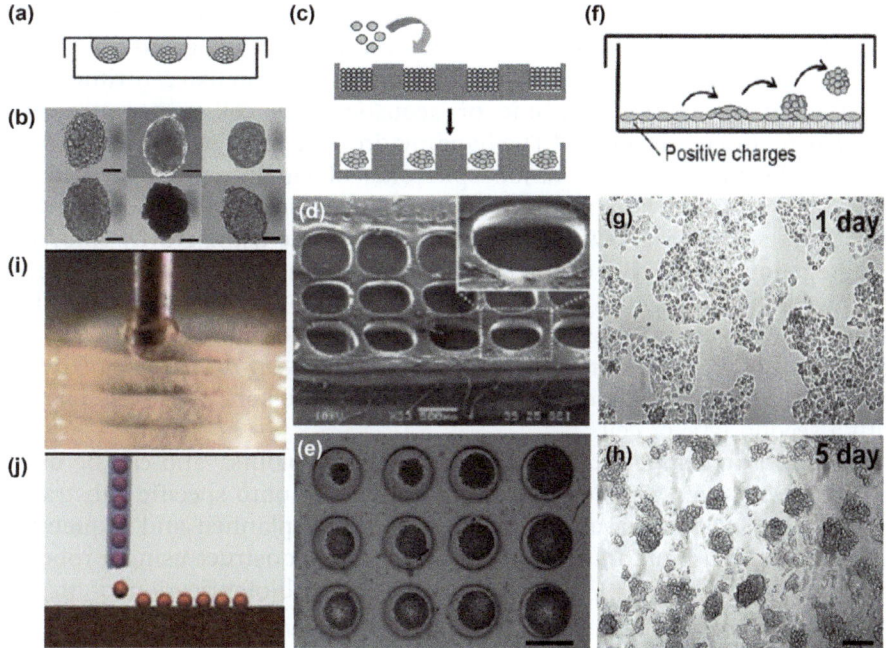

Figure 9.6 Methods for multiple spheroid generation.[74] (a) Hanging-drop culture. (b) Morphology of hanging-drop spheroids formed in the hanging-drop culture. From left to right, upper row: SK-hep1, Hep3B, Huh7; Lower row HepG2, primary rat hapatocytes, PLC/PRF/5. Scale bar is 500 µm. (c) Micromolding techniques. (d) A scanning electron microscope image of PEG microstructures fabricated by photolithography. (e) Spheroid formation in the PEG microwells. (f) Hepatocyte self-assembly in Primaria dishes. (g) and (h) Self-organization of rat hepatocytes into spheroids on Primaria dishes on days 1 and 5, respectively. Bar, 150 µm. Bioprinter deposition of tissue spheroids:[76] (i) general view; (j) Tissue spheroids digital dispensing in air.
Reprinted with permission from Refs. 74, 76 Copyright © 2008 Wiley-VCH Verlag GmbH & Co. KGaA, Weinheim; Copyright 2009 Elsevier.

multicellular tissue spheroids of defined sizes, cell numbers and compositions[69–71] and demonstrate the interaction between two different cell types in a tissue spheroid.[72,73] In other techniques like micromolding, the cells are suspended in a fabricated device made of microwells of defined size and aspect ratio, molded using nonadhesive inert materials like agarose or polyethylene glycol (PEG), as shown in Figures 9.6c and d. After loading in the device, the cells are redistributed using gravity and hydrodynamic forces resulting in the formation of aggregates according to the microwell geometry, as illustrated in Figure 9.6e.[74] Another method for multiple spheroid generation involves the use of various functionalized surfaces. The Primaria dishes that have a positive surface are used for culturing primary hepatocytes. The cells adhere to the surface, as shown in Figure 9.6f, to form a monolayer with spreading cell morphology. The hepatocyte cells undergo contraction, migration and transition to 2D cells sheets and then into 3D aggregates after 5 days of incubation as shown in Figures 9.6g and h.[75] There are other methods of multiple tissue spheroid generation like centrifugation pellet culture, electric, magnetic or acoustic force cell aggregation enhancement and many more.[74] The hanging-drop technology has been widely used in the production of tissue spheroids of single culture as well as multiculture tissue spheroids.[71] The ultrastructure analysis of HepG2-derived microtissues shown in Figure 9.7k was made using gravity enforced reaggregation in hanging drops that showed the integration of single cells into a compact liver-like structure.[71] Kelm and Fussenegger[71] have also shown how to generate tissue spheroids composed of multiple cell types by the process of gravity-enforced reaggregation in hanging drops, as illustrated in Figures 9.7i and j, where premanufactured tissue spheroids were coated with a second type of cell.

These multiple spheroids were represented as "bioink" and can be used for precise, automated release or digital punching onto specific substrates like sprayed hydrogel layers ("biopaper") and the planned and sequential bioassembly and biofabrication into living organ construct using a robotic dispenser ("bioprinters").[76] The method for the biofabrication of tissue spheroids on a large scale for tissue engineering or organ printing depends on various factors, for example, the technology should be scalable, the spheroids must be size standardized for a free-flowing dispensing through the bioprinter nozzle. In addition, the spheroids shouldn't be damaged in the bioprinting process and the tissue spheroids shouldn't lose the capacity to fuse among each other in order to grow into a viable organ. Hence, facing the fact that most of the processes are not scalable[74] with the present technology, the development of a process for large-scale biofabrication of tissue spheroids will become a rapidly growing field in scaffold-free tissue engineering.[76] Recent advances in fabrication of cellosomes[23,24] can obviously lead to possible upscaling of the processes of formation of tissue spheroids.

The tissue fusion process is one of the important steps in the generation of a full organ. It has been said before that the tissue fusion process is fluidic in nature.[76] Fusion has been defined before as "melting together" and that shows the fluidic nature of the tissue spheroids. The fusion is a process that occurs during the embryonic development and can be done *in vitro*.[77] The kinetics in the fusion of two droplets has been showed to be similar to the kinetics of the fusion of two rounded embryonic heart cushion tissue explants using the hanging-drop technique of multiple spheroid generation.[78] The measurement of viscosity and surface tension of tissue spheroids proved that they are fluidic-like structure.[78] The tissue fusion process is driven by surface tension forces as explained by Steinberg's "differential adhesion hypothesis" and proves that it is a phenomenon of fluid mechanics.[79] The other factors responsible in the tissue fusion process are number, motility, viability, redistribution, activation of the cell-adhesion receptors and formation of the extracellular matrix that can impede cell motility.[80] The exact parameters behind efficient tissue spheroid formation are still to be elucidated.[76] Figures 9.7a–h show various-shaped vascular tissue constructs that can be used in the biofabrication of vascularized histotypical tissue spheroids and in turn the formation of part or segment of the kidney intraorgan vascular tree.It can be concluded that the microtissue-based approach can be used as the future for organ transplant generation and shows a lot of promise, but in order for it to be possible there are some serious technological challenges that need to be systematically addressed.[76] From this point of view, there is a lot of room for future developments where cell

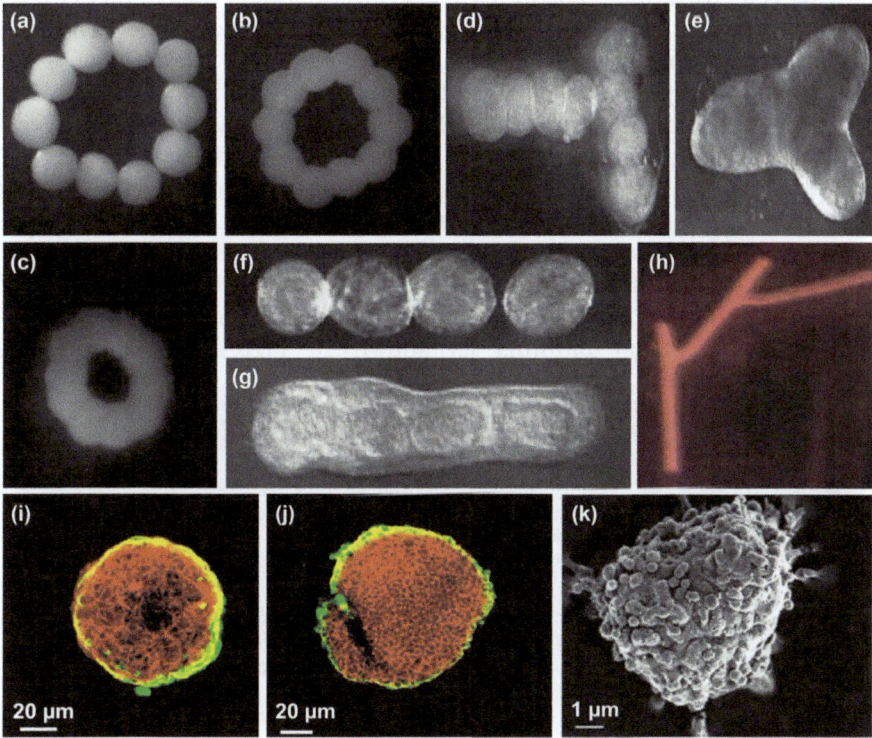

Figure 9.7 Bioassembled vascular tissue construct of tissue spheroids.[76] (a–c) Sequential steps in the formation of ring like vascular tissue construct during the tissue-fusion process. (d) Fabrication of branched vascular segments from unilumenalvascular tissue spheroids in collagen type 1 hydrogel. (e) The vascular tissue spheroids after the tissue-fusion process. (f) Vascular tissue spheroids arranged in a straight line structure placed in a collagen type 1 hydrogel. (g) The vascular tissue spheroids after the tissue fusion process. (h) Bioprinted segment of vascular tree of a kidney intraorgan.[76] Premanufactured feeder spheroids coated with a second cell type by cultivation in hanging drops:[71] (i) human dermal fibroblast coated with keratinocytes (HaCaT), (j) HepG2 coated with HUVECs. Feeder tissues were stained with F-actin (red) HUVECs (green) and HaCaT (green). (k) SEM image of a reaggregated HepG2 spheroid.
Reproduced with permission from Refs. 71, 76 Copyright 2004, 2009 Elsevier.

cultures modified with polyelectrolytes and nanoparticles[13] can play an important role for microfabrication of tissue spheroids,[81] design of novel fusion mechanisms[25] and improvement in the deposition of the prefabricated spheroids in the bioprinting process.[76]

9.6   Hydrogel and Polyelectrolyte-Mediated Multicellular Assemblies

Growing cells in various biocompatible scaffolding materials allows control of the cells spatial configuration and their interactions with each other and the extracellular matrix, and has many applications in the field of tissue engineering, energy production and many others.[12,82] The selection of scaffolding biomaterials determines the mechanical, physical and chemical properties of the extracellular matrix and also the viability of the cells immobilized in the system. Hydrogels are the most commonly used for efficient cell immobilization and have found applications in cell assembly in tissue engineering,[1,83] hydrogen production[82] and others. Polyelectrolytes have also been used for efficient arrangement of cells in multilayered biofilms[84] as well as cell monolayers[83,85] immobilized on various substrates or as free-standing artificial biofilms as shown in Figure 9.8.[22,29] Various

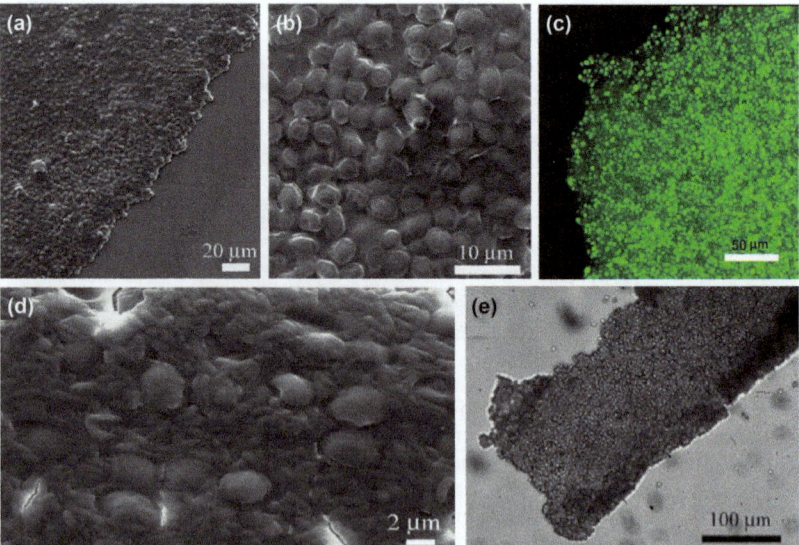

Figure 9.8   SEM image of artificial free-standing yeast biofilms at increasing magnification (a and b) PAH/DNA/PAH/DNA/PAH/Yeast/PAH/DNA/PAH/DNA. (c) CLSM image of FDA treated artificial free-standing yeast biofilm. (d) SEM image of an artificial free standing symbiotic biofilm made of yeast (large oval cells) and *E.coli* (rod-shaped cells). (e) Optical microscope image of ribbon like artificial free-standing yeast biofilm.[29] Reproduced with permission from Ref. 29 Copyright2011 Elsevier.

combinations of cationic and anionic polyelectrolytes were used to make a sandwich of the yeast cells on a sacrificial calcium carbonate core in order to make free standing artificial biofilms, as shown in Figures 9.8a and b. The biofilm retained its viability which was shown by using FDA as a staining agent in Figure 9.8c. The same process was used to make free-standing symbiotic artificial yeast and *E. coli* biofilm, as shown in Figure 9.8d.[29] These processes have huge application in the field of energy production or various other product extractions.In the natural environment, micro-organisms normally exist embedded in a complex extracellular matrix based on glyco-proteins and carbohydrates that forms a biofilm. This matrix protects the cells from adverse conditions, supports and provides them with nutrients for efficient growth and survival. These conditions can be mimicked in artificial biofilms by assembling the cells in the matrix of a biocompatible material that allows one to collect the products of the cell activity more efficiently. Recently, microalgae have been explored as an alternative source of biofuels including the ability of specific strains to produce hydrogen. It has been shown that assembling microalgae in the form of biofilms using sodium alginate matrix significantly enhances their hydrogen production yield. *Clamydomonas reinhadrtii* cells have been successfully immobilized in al-ginate hydrogel layers on a substrate that showed significantly higher hydrogen output compared to suspension culture.[82] The microalgae cell retained its viability in the process.

Extracellular matrix as a hydrogel laden with cells has been widely used in tissue engineering to construct artificial organs in regenerative medi-cine.[86-88] Matsunaga *et al.*[88] successfully assembled millimeter-thick macroscopic tissues with complex microstructures by arranging a large number of collagen gel-based microtissue units as shown in Figures 9.9d and f. They prepared monodisperse collagen beads using the process of axisymmetric flow-focusing device (AFFD) (Figure 9.9a). Cell-loaded beads were obtained by seeding the cells on these collagen beads making a monoculture or by encapsulating the cells and seeding a different kind of cell on the surface to obtain a coculture. These cell beads are then arranged in a specifically designed mold in order to obtain a 3D tissue structure and then it was released from the mold to form a free fabricated tissue.[30] An alternative method to produce 3D cell-laden microstructures is known as *cell origami*,[12] whose formation is driven by the cell-traction force. Cells are successfully patterned on microplates coated with 2-methacryloyloxyethyl phosphoryl choline (MPC) as this polymer promotes cell adhesion and prevents protein adsorption. The cells were coated on these microplates, as shown in Figure 9.9g, with a gelatine layer in between for their release from the microplate after the self-folding process. A flexible joint was introduced in the center in order to control the folding process. Figures 9.9h and i shows a regular dodecahedron-shaped cell assembly produced by the process of *cell origami* driven by the cell-traction force.[12]

The LbL method of sequential adsorption of oppositely charged poly-electrolytes on flat substrates has resulted in the wide range of applications

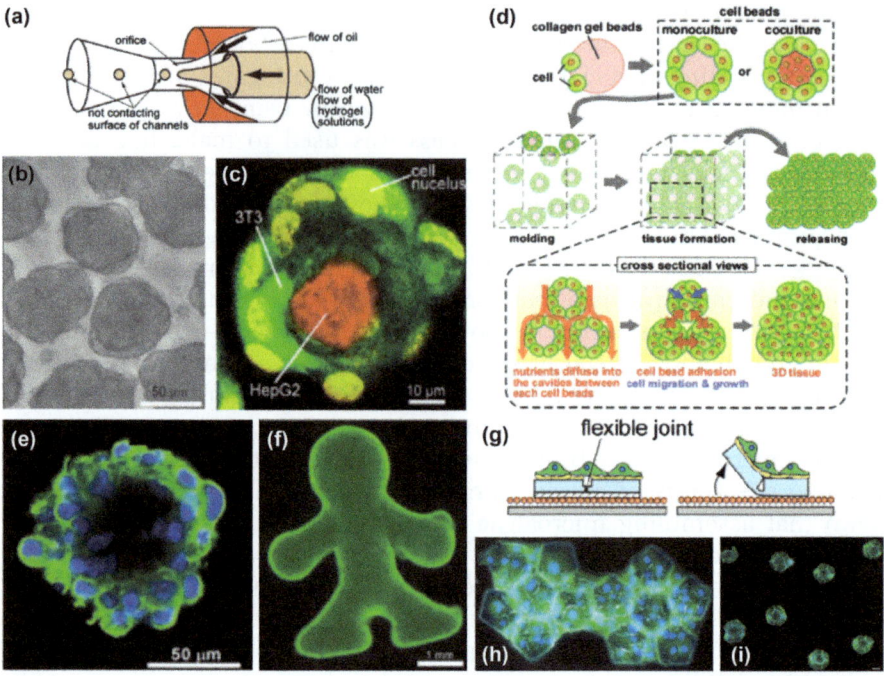

Figure 9.9   (a) Schematic diagram of the AFFD process[63] used to produce monodisperse droplets. (b) Bright-field image of monocultured cell beads
based on 3T3 cells on collagen beads. (c) Fluorescence confocal microscope image of cocultured cell beads, NIH 3T3 cells over the collagen
gel beads, encapsulating HepG2 cells. (d) Schematics of the formation
of monoculture and coculture cell beads made by culturing cells on the
surface of the monodisperse collagen gel beads or collagen gel beads
encapsulating another cell type. The cell beads are stacked into a
designed silicone mold in order to form 3D tissues. (e) Fluorescent
confocal microscope image of a monoculture cell beads. (f) Formation
of a millimeter-scale 3D tissue made by molding monodisperse
cell beads with the fluorescent image showing cell viability.[88] Cell-
origami techniques for programmed assembly of multicellular structures:[12] (g) Schematics of a folding microplate with a flexible joint. (h) A
regular dodecahedron. (i) A fluorescent image of a self-folded regular
dodecahedron.
Reproduced with permission from Refs. 12, 63, 88 Copyright 2011
Wiley-VCH Verlag GmbH & Co. KGaA, Weinheim; Copyright 2013 The
Royal Society of Chemistry; Copyright 2012 PLOS One.

in the field of biomaterials, tissue engineering, medical coatings, drug delivery, biotechnology, *etc.*[84] A similar strategy was recently used by Konnova
et al.[29] for fabrication of multilayered free-standing biofilms by LbL deposition of alternating layers of polyelectrolytes and yeast cells over a sacrificial layer of calcium carbonate nanocrystals on flat substrates, followed
by dissolution of the $CaCO_3$ underlay. This led to the formation of artificial
free-standing biofilm structures that show tendency to self-fold when the

outer polyelectrolyte coats on both sides of the biofilm had opposite charge. They also made symbiotic biofilms consisting of two different micro-organisms, yeast and *E.coli.* This approach allows one to make artificial co-lonial micro-organisms composed of several different unicellular species with many different applications. Rajagopalan *et al.* assembled nanoscale poly-electrolyte multilayer scaffold on the top of living mammalian cells.[85] This enabled them to successfully make 3D hepatocyte–polyelectrolyte–fibrioblast constructs, hepatocyte–polyelectrolyte–endothelial constructs and hepato-cyte-polyelectrolyte-hepatocyte constructs resulting in maintaining their morphology as well as enhancing their functional capability.[85] Krol *et al.* showed a different approach by encapsulating the cells with polyelectrolytes and then its subsequent deposition on a patterned surface while main-taining the viability of the cells.[30] They used the process of microcontact printing of polyelectrolytes to successfully pattern coated and uncoated cells into the desired cell assemblies. Matsusaki *et al.*, successfully made more than three multicellular layers by preparing fibronectin–gelatine nanofilms on the surface of each layer to mimic the natural extracellular matrix as fibronectin and gelatine are natural components of the extracellular matrix.[83]

9.7 Applications of Artificial Multicellular Assemblies

Functionalized multicellular assemblies have various applications. Here, we will focus on several different conceptual studies that show how various techniques of multicellular assemblies can be used to find potential appli-cations in tissue engineering, biosensing, alternative energy production, *etc.*

There are various cell-immobilization techniques that use chemical crosslinking, adsorption on a substrate, *etc.*, which can significantly com-promise the viability of the immobilized cells. However, LbL processes of cell deposition mediated by polyelectrolytes or hydrogel-mediated immobiliza-tion of cells, magnetic nanoparticles and field-driven assemblies have been demonstrated to solve this problem and produce efficient multicellular vi-able assemblies, as discussed in the previous sections. The poleyelectrolyte and magnetic nanoparticles-mediated cellular assemblies have been used to construct biosensors. Magnetically functionalized *Chlorella pyrenoidosa* microalgae[3] and yeast[10] were used as sensing elements in a whole-cell bio-sensor. The PAH-stabilized magnetic nanoparticles deposited on the cell surface enable the precise control of the cells in a device using a micro-magnet and their assembly in designated chambers in a microfluidic device[4] and on screen-printed electrodes.[3] The biosensors developed by the above process were successfully used in determining the level of genotoxicity and triazine herbicide screening.

The process of the directed assembly of tissue spheroids (Figure 9.10a) has huge scope of application in tissue engineering by using special robotic

(a)  **(b)**

Figure 9.10 Principles of the bioprinting technology. (a) Scheme of layer-by-layer tissue speroid deposition and tissue-fusion process in the bioassembly of tubular tissue construct using the method of bioprinting of the self-assembled tissue spheroids. (b) 3D dispensing laboratory bioprinter-"LBP".
Reproduced with permission from Ref. 76 Copyright 2009 Elsevier.

bioprinters (see Figure 9.10b). The LbL tissue spheroid deposition combined with the tissue fusion process (Figure 9.10a) will enable a solid scaffold-free organ printing process.[76]

Brandy *et al.* produced the multicellular structure of yeast cells template on air bubbles suspended in aqueous solution which remarkably resemble the fossilized multicellular structures are believed to be one of the first nonmarine multicellular organisms discovered in ancient river sediments.[42] Fakhrullin *et al.* and Brandy *et al.* envisaged that similar pathway of formation of clusters from primitive unicellular micro-organisms on mineral particles might have been responsible for the origin of multicellular life in Nature.[23–25]

Acknowledgements

Anupam Das appreciates financial support from the University of Hull for his PhD study. Vesselin Paunov acknowledges partial support from EU COST action CM1101 for the preparation of this manuscript. This work was partially (RFF) funded by the subsidy of the Russian Government to support the Program of Competitive Growth of Kazan Federal University among the World's Leading Academic Centres. Rawil F. Fakhrullin acknowledges the funding by RFBR 12-03-93939-G8, RFBR 12-04-33290_mol_ved_a and RFBR 14-04-01474_a grants.

References

1. N. N. Kachouie, Y. A. Du, H. Bae, M. Khabiry, A. F. Ahari, B. Zamanian, J. Fukuda and A. Khademhosseini, *Organogenesis*, 2010, **6**, 234–244.
2. R. F. Fakhrullin and Y. M. Lvov, *ACS Nano*, 2012, **6**, 4557–4564.
3. A. I. Zamaleeva, I. R. Sharipova, R. V. Shamagsumova, A. N. Ivanov, G. A. Evtugyn, D. G. Ishmuchametova and R. F. Fakhrullin, *Anal. Methods-Uk.*, 2011, **3**, 509–513.

4. J. Garcia-Alonso, R. F. Fakhrullin, V. N. Paunov, Z. Shen, J. D. Hardege, N. Pamme, S. J. Haswell and G. M. Greenway, *Anal. Bioanal. Chem.*, 2011, **400**, 1009–1013.
5. V. Berry and R. F. Saraf, *Angew. Chem. Int. Edit.*, 2005, **44**, 6668–6673.
6. M. Kahraman, A. I. Zamaleeva, R. F. Fakhrullin and M. Culha, *Anal. Bioanal. Chem.*, 2009, **395**, 2559–2567.
7. H. Ai, M. Fang, S. A. Jones and Y. M. Lvov, *Biomacromolecules*, 2002, **3**, 560–564.
8. R. F. Fakhrullin, A. I. Zamaleeva, M. V. Morozov, D. I. Tazetdinova, F. K. Alimova, A. K. Hilmutdinov, R. I. Zhdanov, M. Kahraman and M. Culha, *Langmuir*, 2009, **25**, 4628–4634.
9. V. Berry, A. Gole, S. Kundu, C. J. Murphy and R. F. Saraf, *J. Am. Chem. Soc.*, 2005, **127**, 17600–17601.
10. J. Garcia-Alonso, R. F. Fakhrullin and V. N. Paunov, *Biosens. Bioelectron.*, 2010, **25**, 1816–1819.
11. V. Berry, S. Rangaswamy and R. F. Saraf, *Nano Lett.*, 2004, **4**, 939–942.
12. K. Kuribayashi-Shigetomi, H. Onoe and S. Takeuchi, *Plos One*, 2012, 7.
13. R. F. Fakhrullin, A. I. Zamaleeva, R. T. Minullina, S. A. Konnova and V. N. Paunov, *Chem. Soc. Rev.*, 2012, **41**, 4189–4206.
14. R. F. Fakhrullin, J. Garcia-Alonso and V. N. Paunov, *Soft Matter*, 2010, **6**, 391–397.
15. A. I. Zamaleeva, I. R. Sharipova, A. V. Porfireva, G. A. Evtugyn and R. F. Fakhrullin, *Langmuir*, 2009, **26**, 2671–2679.
16. D. Y. Zhang, R. F. Fakhrullin, M. Ozmen, H. Wang, J. Wang, V. N. Paunov, G. H. Li and W. E. Huang, *Microb. Biotechnol.*, 2011, **4**, 89–97.
17. M. R. Dzamukova, A. I. Zamaleeva, D. G. Ishmuchametova, Y. N. Osin, A. P. Kiyasov, D. K. Nurgaliev, O. N. Ilinskaya and R. F. Fakhrullin, *Langmuir*, 2011, **27**, 14386–14393.
18. R. T. Minullina, Y. N. Osin, D. G. Ishmuchametova and R. F. Fakhrullin, *Langmuir*, 2011, **27**, 7708–7713.
19. A. I. Zamaleeva, I. R. Sharipova, A. V. Porfireva, G. A. Evtugyn and R. F. Fakhrullin, *Langmuir*, 2010, **26**, 2671–2679.
20. Z. Y. Suo, R. Avci, X. H. Yang and D. W. Pascual, *Langmuir*, 2008, **24**, 4161–4167.
21. S. H. Yang, K. B. Lee, B. Kong, J. H. Kim, H. S. Kim and I. S. Choi, *Angew. Chem. Int. Edit.*, 2009, **48**, 9160–9163.
22. S. A. Konnova, M. Kahraman, A. I. Zamaleeva, M. Culha, V. N. Paunov and R. F. Fakhrullin, *Colloid Surface B*, 2011, **88**, 656–663.
23. R. F. Fakhrullin, M. L. Brandy, O. J. Cayre, O. D. Velev and V. N. Paunov, *Phys. Chem. Chem. Phys.*, 2010, **12**, 11912–11922.
24. M. L. Brandy, O. J. Cayre, R. F. Fakhrullin, O. D. Velev and V. N. Paunov, *Soft Matter*, 2010, **6**, 3494–3498.
25. R. F. Fakhrullin and V. N. Paunov, *Chem. Commun.*, 2009, 2511–2513.
26. W. R. Small, S. D. Stoyanov and V. N. Paunov, *Biomater. Sci.*, 2013.
27. G. Frasca, F. Gazeau and C. Wilhelm, *Langmuir*, 2009, **25**, 2348–2354.

28. A. Ito, K. Ino, M. Hayashida, T. Kobayashi, H. Matsunuma, H. Kagami, M. Ueda and H. Honda, *Tissue Eng.*, 2005, **11**, 1553–1561.

29. S. A. Konnova, M. Kahraman, A. I. Zamaleeva, M. Culha, V. N. Paunov and R. F. Fakhrullin, *Colloids Surf. B: Biointerfaces*, 2011, **88**, 656–663.

30. S. Krol, M. Nolte, A. Diaspro, D. Mazza, R. Magrassi, A. Gliozzi and A. Fery, *Langmuir*, 2005, **21**, 705–709.

31. H. Mohwald, *Colloid Surface A*, 2000, **171**, 25–31.

32. G. B. Sukhorukov, E. Donath, H. Lichtenfeld, E. Knippel, M. Knippel, A. Budde and H. Mohwald, *Colloid Surface A*, 1998, **137**, 253–266.

33. E. Donath, G. B. Sukhorukov, F. Caruso, S. A. Davis and H. Mohwald, *Angew. Chem. Int. Edit.*, 1998, **37**, 2202–2205.

34. C. Y. Gao, E. Donath, H. Mohwald and J. C. Shen, *Angew. Chem. Int. Edit.*, 2002, **41**, 3789–3793.

35. K. Katagiri and F. Caruso, *Macromolecules*, 2004, **37**, 9947–9953.

36. A. S. Angelatos, B. Radt and F. Caruso, *J. Phys. Chem. B*, 2005, **109**, 3071–3076.

37. H. P. Yap, J. F. Quinn, S. M. Ng, J. Cho and F. Caruso, *Langmuir*, 2005, **21**, 4328–4333.

38. Y. J. Wang, A. M. Yu and F. Caruso, *Angew. Chem. Int. Edit.*, 2005, **44**, 2888–2892.

39. F. Caruso, D. Trau, H. Mohwald and R. Renneberg, *Langmuir*, 2000, **16**, 1485–1488.

40. D. V. Volodkin, A. I. Petrov, M. Prevot and G. B. Sukhorukov, *Langmuir*, 2004, **20**, 3398–3406.

41. O. J. Cayre, P. F. Noble and V. N. Paunov, *J. Mater. Chem.*, 2004, **14**, 3351–3355.

42. M.-L. Brandy, O. J. Cayre, R. F. Fakhrullin, O. D. Velev and V. N. Paunov, *Soft Matter*, 2010, **6**, 3494–3498.

43. P. K. Strother, L. Battison, M. D. Brasier and C. H. Wellman, *Nature*, 2011, **473**, 505–509.

44. K. Ino, A. Ito and H. Honda, *Biotechnol. Bioeng.*, 2007, **97**, 1309–1317.

45. Y. Sahoo, A. Goodarzi, M. T. Swihart, T. Y. Ohulchanskyy, N. Kaur, E. P. Furlani and P. N. Prasad, *J. Phys. Chem. B*, 2005, **109**, 3879–3885.

46. L. Hakho, L. Yong, E. Alsberg, D. E. Ingber, R. M. Westervelt and D. Ham, Solid-State Circuits Conference, 2005. Digest of Technical Papers. ISSCC. 2005 IEEE International, 2005.

47. Q. A. Pankhurst, J. Connolly, S. K. Jones and J. Dobson, *J. Phys. D Appl. Phys.*, 2003, **36**, R167–R181.

48. K. Ino, A. Ito, H. Kumazawa, H. Kagami, M. Ueda and H. Honda, *J. Chem. Eng. Jpn.*, 2007, **40**, 51–58.

49. A. Ito, M. Hayashida, H. Honda, K. I. Hata, H. Kagami, M. Ueda and T. Kobayashi, *Tissue Eng.*, 2004, **10**, 873–880.

50. A. Ito, E. Hibino, C. Kobayashi, H. Terasaki, H. Kagami, M. Ueda, T. Kobayashi and H. Honda, *Tissue Eng.*, 2005, **11**, 489–496.

51. A. Ito, K. Ino, T. Kobayashi and H. Honda, *Biomaterials*, 2005, **26**, 6185–6193.

52. M. Tanase, E. J. Felton, D. S. Gray, A. Hultgren, C. S. Chen and D. H. Reich, *Lab Chip*, 2005, **5**, 598–605.
53. M. D. Krebs, R. M. Erb, B. B. Yellen, B. Samanta, A. Bajaj, V. M. Rotello and E. Alsberg, *Nano Lett.*, 2009, **9**, 1812–1817.
54. J. X. Zhu, M. Li, R. Rogers, W. Meyer, R. H. Ottewill, W. B. Russell and P. M. Chaikin, *Nature*, 1997, **387**, 883–885.
55. P. N. Pusey and W. Vanmegen, *Nature*, 1986, **320**, 340–342.
56. P. A. Kralchevsky and N. D. Denkov, *Curr. Opin. Colloid In.*, 2001, **6**, 383–401.
57. B. G. Prevo and O. D. Velev, *Langmuir*, 2004, **20**, 2099–2107.
58. H. A. Pohl, *J. Appl. Phys.*, 1951, **22**, 869–871.
59. H. A. Pohl, *J. Appl. Phys.*, 1958, **29**, 1182–1188.
60. L. Shang, T. L. Clare, M. A. Eriksson, M. S. Marcus, K. M. Metz and R. J. Hamers, *Nanotechnology*, 2005, **16**, 2846–2851.
61. S. Gupta, R. G. Alargova, P. K. Kilpatrick and O. D. Velev, *Soft Matter*, 2008, **4**, 726–730.
62. S. Gupta, R. G. Alargova, P. K. Kilpatrick and O. D. Velev, *Langmuir*, 2010, **26**, 3441–3452.
63. Y. Morimoto and S. Takeuchi, *Biomater. Sci.*, 2013, **1**, 257–264.
64. Y. C. Tung, A. Y. Hsiao, S. G. Allen, Y. S. Torisawa, M. Ho and S. Takayama, *Analyst*, 2011, **136**, 473–478.
65. W. G. Lee, D. Ortmann, M. J. Hancock, H. Bae and A. Khademhosseini, *Tissue Eng. Part C-Me*, 2010, **16**, 249–259.
66. M. Kato-Negishi, Y. Tsuda, H. Onoe and S. Takeuchi, *Biomaterials*, 2010, **31**, 8939–8945.
67. H. Ota, R. Yamamoto, K. Deguchi, Y. Tanaka, Y. Kazoe, Y. Sato and N. Miki, *Sens. Actuators B-Chem.*, 2010, **147**, 359–365.
68. Y. Akiyama and K. Morishima, *Appl. Phys. Lett.*, 2011, 98.
69. R.-Z. Lin, L.-F. Chou, C.-C. Chien and H.-Y. Chang, *Cell Tissue Res.*, 2006, **324**, 411–422.
70. J. M. Kelm, N. E. Timmins, C. J. Brown, M. Fussenegger and L. K. Nielsen, *Biotechnol. Bioeng.*, 2003, **83**, 173–180.
71. J. M. Kelm and M. Fussenegger, *Trends Biotechnol.*, 2004, **22**, 195–202.
72. T. Nicholas, D. Stefanie and N. Lars, *Angiogenesis*, 2004, 7, 97–103.
73. J. M. Kelm, E. Ehler, L. K. Nielsen, S. Schlatter, J. C. Perriard and M. Fussenegger, *Tissue Eng.*, 2004, **10**, 201–214.
74. R.-Z. Lin and H.-Y. Chang, *Biotechnol. J.*, 2008, **3**, 1172–1184.
75. E. S. Tzanakakis, L. K. Hansen and W. S. Hu, *Cell Motil. Cytoskel.*, 2001, **48**, 175–189.
76. V. Mironov, R. P. Visconti, V. Kasyanov, G. Forgacs, C. J. Drake and R. R. Markwald, *Biomaterials*, 2009, **30**, 2164–2174.
77. J. M. Perez-Pomares and R. A. Foty, *Bioessays*, 2006, **28**, 809–821.
78. K. Jakab, B. Damon, F. Marga, O. Doaga, V. Mironov, I. Kosztin, R. Markwald and G. Forgacs, *Dev Dynam*, 2008, **237**, 2438–2449.
79. M. S. Steinberg, *Curr. Opin. Genet. Devel.*, 2007, **17**, 281–286.

80. A. Neagu, K. Jakab, R. Jamison and G. Forgacs, *Phys. Rev. Lett.*, 2005, 95.
81. Y.-W. Chang, P. He, S. M. Marquez and Z. Cheng, *Biomicrofluidics*, 2012, **6**, 24118–241189.
82. S. N. Kosourov and M. Seibert, *Biotechnol. Bioeng.*, 2009, **102**, 50–58.
83. M. Matsusaki, K. Kadowaki, Y. Nakahara and M. Akashi, *Angew. Chem. Int. Edit.*, 2007, **46**, 4689–4692.
84. C. J. Detzel, A. L. Larkin and P. Rajagopalan, *Tissue Eng. Part B-Re*, 2011, **17**, 101–113.
85. P. Rajagopalan, C. J. Shen, F. Berthiaume, A. W. Tilles, M. Toner and M. L. Yarmush, *Tissue Eng.*, 2006, **12**, 1553–1563.
86. A. P. McGuigan and M. V. Sefton, *Proc. Natl. Acad. Sci.*, 2006, **103**, 11461–11466.
87. D. A. Bruzewicz, A. P. McGuigan and G. M. Whitesides, *Lab Chip*, 2008, **8**, 663–671.
88. Y. T. Matsunaga, Y. Morimoto and S. Takeuchi, *Adv. Mater.*, 2011, **23**, H90–H94.

CHAPTER 10

Magnetic Decoration and Labeling of Prokaryotic and Eukaryotic Cells

IVO SAFARIK,[*a,b] ZDENKA MADEROVA,[a] KRISTYNA POSPISKOVA,[b] KATERINA HORSKA[a] AND MIRKA SAFARIKOVA[a]

[a] Department of Nanobiotechnology, Institute of Nanobiology and Structural Biology of GCRC, Na Sadkach 7, 370 05 Ceske Budejovice, Czech Republic; [b] Regional Centre of Advanced Technologies and Materials, Palacky University, Slechtitelu 11, 783 71 Olomouc, Czech Republic
*Email: ivosaf@yahoo.com

10.1 Introduction

The majority of prokaryotic and eukaryotic cells can interact with a wide range of nano- and microparticles and films. The modified cells usually maintain their viability, but the presence of foreign material on their surfaces, in protoplasm or in intracellular organelles can provide additional functionalities. Cells modified using different procedures can be employed as whole-cell biosensors, whole-cell biocatalysts, applied in toxicity microscreening devices and also as efficient adsorbents of different types of organic and inorganic xenobiotics.[1–3]

Various nanoparticles have been used to modify cells surfaces, such as gold, silver, palladium or silica ones, carbon nanotubes, etc.[1] However,

RSC Smart Materials No. 9
Cell Surface Engineering: Fabrication of Functional Nanoshells
Edited by Rawil F Fakhrullin, Insung S Choi and Yuri Lvov
Published by the Royal Society of Chemistry, www.rsc.org

modification of cells with magnetic nano- and microparticles is exceptionally important and magnetically modified cells have been used in many applications. This review chapter focuses on the description of various procedures and materials to prepare magnetically responsive prokaryotic and eukaryotic cells and shows the most important examples of their applications.

10.2   Magnetic Decoration and Labeling of Cells

The magnetization of originally diamagnetic cells can be usually performed by the attachment of magnetic nano- or microparticles on the cell surface; during phagocytosis magnetic particles can be internalized into protoplasm. Alternatively, magnetic molecular labels or paramagnetic ions can be used for a cell's magnetic modification.

Several expressions such as decoration and labeling are used to describe the preparation of magnetically modified cells. The term "decorated cells" usually describes the cells covered with large numbers of nanoparticles, in some cases also in several layers; the particles usually do not cross the cell membrane or cell wall. On the contrary, target cells can be magnetically labeled just by a single (or only a few) magnetically responsive particle(s) attached to the cell wall or membrane. However, magnetic labeling also represents the situation when magnetic nanoparticles cross the cell membrane/wall and enter the cytoplasm or a periplasmic space. The term "magnetic labeling" is broader and includes a specific type of magnetic labeling called "magnetic decoration" of cells. The equivalent expressions to "magnetically labeled cells" used throughout this review are "magnetically responsive cells" or "magnetically modified cells". The common characteristics of all magnetically modified cells are their specific interactions with external magnetic field.

In general, magnetic nano- and microparticles have found many interesting applications in various areas of biosciences, medicine, biotechnology and environmental technology. Different types of responses of such materials to an external magnetic field enable various applications, namely selective separation, targeting and localization of magnetically responsive nano- and microparticles and other relevant materials using an external magnetic field (*e.g.* using an appropriate magnetic separator, permanent magnet, or electromagnet), heat generation (which is caused by magnetic particles subjected to a high-frequency alternating magnetic field), increase of a negative T_2 contrast by magnetic iron oxides nanoparticles during magnetic resonance imaging or great increase of apparent viscosity of magnetorheological fluids when subjected to a magnetic field. In addition, magnetic iron oxide(s) nanoparticles can exhibit peroxidase-like activity.[2,4]

An absolute majority of cells exhibits diamagnetic behavior. However, there is an exceptional group of bacteria called magnetotactic bacteria having the ability to synthesize intracellular biogenic magnetic nanoparticles (based either on magnetite (Fe_3O_4) or greigite (Fe_3S_4)) that enables their magnetic separation and movement; it is generally assumed that magnetosomes (organelles containing individual magnetic nanoparticles covered with a lipid bilayer) are involved in magnetoreception.[5] Also magnetotactic algae of the genus *Anisonema* (Euglenophyceae) have been isolated from a coastal mangrove swamp in northeastern Brazil; the magnetic response of the cells was caused by many magnetite particles organized in chains.[6] Paramagnetic behavior of erythrocytes (red blood cells) is caused by the presence of methaemoglobin containing iron atoms in the cell haemoglobin in the ferric state.[7]

As already stated, an absolute majority of prokaryotic and eukaryotic cells is diamagnetic. In those cases, when an ability to respond to external magnetic field should be added, several basic procedures can be used for the modification of the cells surface, such as the nonspecific attachment of magnetic nanoparticles (*e.g.*, by the magnetic fluid treatment),[8] by binding of maghemite particles[9] or magnetite particles[10–14] on the cell surface, by covalent immobilization of magnetic particles on cell surface or vice versa,[15,16] by specific interactions with immunomagnetic nano- and microparticles,[17–19] magnetic quantum dots[20] or magnetoliposomes,[21] by the biologically driven precipitation of paramagnetic compounds on the cell surface,[22] by crosslinking of the cells or isolated cell walls with a bifunctional reagent in the presence of magnetic particles[23] or by entrapment (together with magnetic particles) into biocompatible polymers.[24,25] As can be seen, in most cases the magnetic properties of the modifiers are caused by the presence of nano- or microparticles of magnetite (Fe_3O_4) or maghemite (γ-Fe_2O_3); in some cases also ferrite particles[26] or chromium dioxide particles have been used.[27] Alternatively, the modification can be performed by binding paramagnetic cations on acid groups on the cell surface[28] or by the binding of ferritin[29] or magnetoferritin[30] on the cell surface. In most cases the attached magnetic particles or ions do not have a negative effect on the viability and phenotype alternation of modified cells. It should be taken into account that in specific cases the surface-bound particles can be internalized by the treated cells and will appear in protoplasm.

The individual modification procedures will be described in more detail below.

10.2.1 Interaction of Target Cells with Magnetic Nano- and Microparticles

Different types of magnetic microparticles as well as ionically and sterically stabilized magnetic nanoparticles have been used for magnetic cell modification. Magnetic modification of microbial cells can be performed using appropriate magnetic fluid. In the simplest way, perchloric acid stabilized magnetic fluid was mixed with baker's or brewer's yeast cells washed with and suspended in acetate buffer, pH 4.6 or in glycine–HCl buffer, pH 2.2; alternatively tetramethylammonium hydroxide stabilized magnetic fluid was used for baker's yeast cells modification in 0.1 M glycin–NaOH buffer,

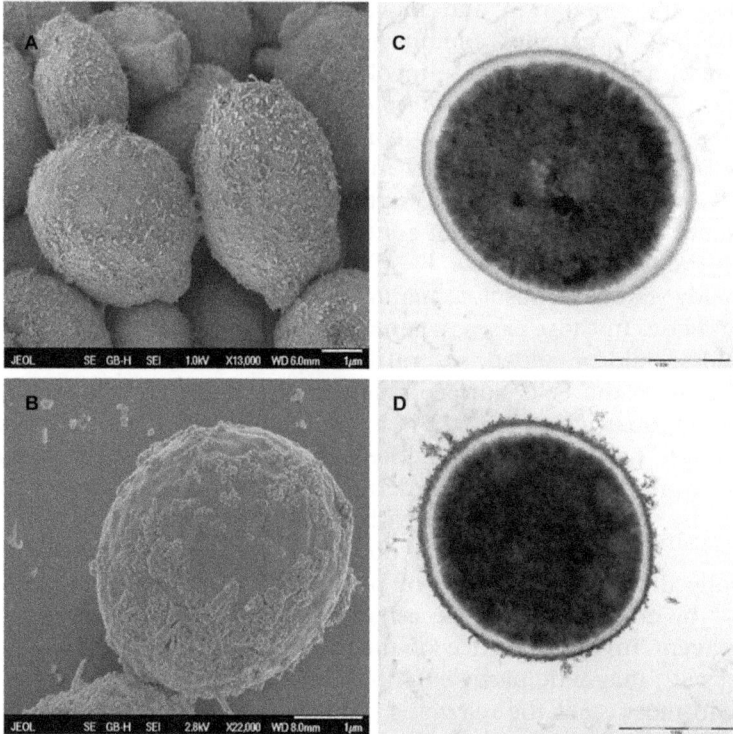

Figure 10.1   SEM micrographs of ferrofluid-modified *Saccharomyces cerevisiae* cells
                   showing attached magnetic nanoparticles and their aggregates on the
                   cell surface (A, B; bars: 1 µm). TEM micrographs of native *Saccharomyces
                   cerevisiae* cells (C; bar: 1 µm) and ferrofluid-modified cell with attached
                   magnetic iron oxide nanoparticles on the cell wall (D; bar: 1 µm).
                   Reproduced, with permission, from Ref. 31.

pH 10.6. After a short time period magnetic particles precipitated on the cell
surface (Figure 10.1). After washing the magnetically modified cells were
used as whole-cell biocatalysts for hydrogen-peroxide degradation or sucrose
hydrolysis. Alternatively the modified cells were heated in boiling water bath
to kill the cells, resulting in the formation of a stable adsorbent for the removal of selected organic and inorganic xenobiotics.[8,31,32]

A different procedure has to be used when working with dried *Kluyveromyces fragilis* (fodder yeast) and *Chlorella vulgaris* cells. The cells were
thoroughly washed several times with 0.1 M acetic acid to remove a substantial portion of soluble macromolecules that otherwise caused spontaneous precipitation of magnetic fluid. After washing and suspending the
cells in acetic acid solution the addition of perchloric acid stabilized magnetic fluid resulted in the formation of magnetically modified yeast and
algae cells (see Figure 10.2).[33,34]

Another procedure was based on the attachment of submicrometer,
acicular maghemite particles on the yeast cells; the binding occurred

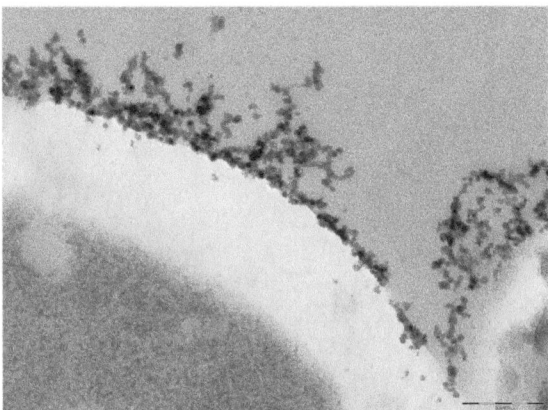

Figure 10.2 TEM picture of magnetically modified dried fodder yeast (*Kluyvero-myces fragilis*) cells (bar - 200 nm).
Reproduced, with permission, from Ref. 33.

irrespective of the solution pH and surface charge and was essentially ir-reversible.[9] Also, magnetite microparticles were used to capture bacterial cells; cell adsorption was best in the pH range 3–6 in the absence of calcium and magnesium, but the pH range was extended up to pH 10 in the presence of these two cations.[11,35]

Algae in water was separated by using a superconducting high-gradient magnet after magnetization of algae by means of attaching the colloidal particles of hydrous iron(III) oxide. It was found that the percentage of re-covery of algae depends on the added amount of iron oxide particles. The percentage of recovery was also found to reach almost 100% when a very small amount of particles was added to the algae suspension. These results demonstrated the feasibility of the magnetic separation of algae with no addition of polymer surfactant.[36]

An extremely simple procedure for the magnetic modification of yeast and algae cells has been developed recently that is based on the use of micro-wave-synthesized magnetic iron oxides nano- and microparticles. Two very cheap starting chemicals are used (ferrous sulfate heptahydrate and sodium or potassium hydroxide); after their mixing and formation of mixed iron hydroxides the suspension underwent the microwave treatment (a regular kitchen microwave oven can be used successfully) and microparticles of magnetic iron oxides formed.[14,37] Mixing of magnetic particles with algae cells (*Chlorella vulgaris*) and yeast cells (*Saccharomyces cerevisiae*) suspen-sions caused cell flocculation and magnetically responsive cells aggregates (usually *ca.* 100–300 μm in diameter) were formed[14,38] (Figure 10.3).

An interesting procedure for coating yeast cells by magnetite nano-particles *via* electrostatic interactions has been described recently. First, poly(allylamine hydrochloride) (PAH) was coated onto the hydrated yeast cells whose surfaces were negatively charged in water, then the cells were

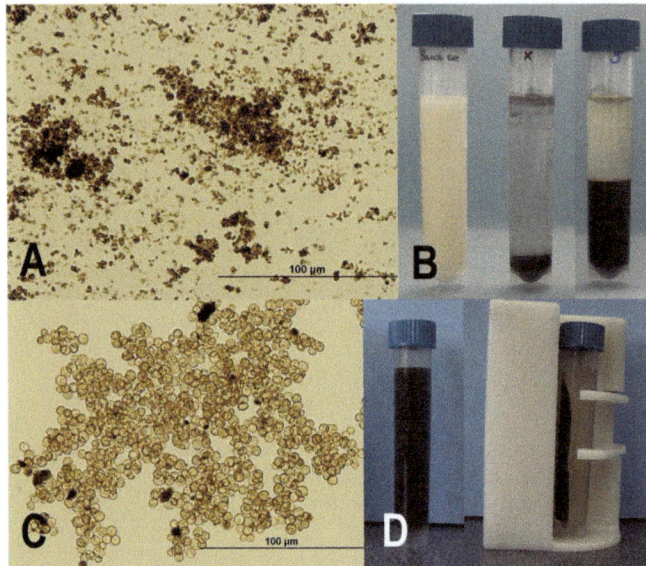

Figure 10.3 Optical microscopy of magnetic iron oxides microparticles prepared
by microwave assisted synthesis (A); process of magnetic modification
of yeast cells (left tube – *S. cerevisiae* cells suspension; middle tube –
sedimented iron oxides microparticles for magnetic modification;
right tube – sedimented magnetically modified yeast cells) (B); optical
microscopy of *S. cerevisiae* cells modified by iron oxides microparticles
(C); magnetic separation of magnetically modified yeast cells (D).
Reproduced, with permission, from Ref. 14.

coated with poly(sodium polystyrene sulfonate) (PSS). After repeating the
procedure to build PAH/PSS/PAH coatings on the cells, magnetite nanoparticles were deposited on the cells before the deposition of two additional
polyelectrolyte layers. The final products had the layer structures of PAH/
PSS/PAH/magnetic nanoparticles/PAH/PSS, which preserve the viability of
the yeast cells. Magnetic nanoparticles formed a multilayered coating on the
outer side of the yeast cell walls. Using yeast cells expressing GFP it was
shown that magnetic modification had little effect on the fluorescence
emission.[39] In another procedure (poly)allylamine hydrochloride stabilized
positively charged magnetic nanoparticles (average diameter around 15 nm)
were used for the magnetization of living *Chlorella pyrenoidosa* cells. The
single-step magnetization procedure is very simple and consisted of the
dropwise introduction of the aqueous suspension of algae cells into nanoparticles solution followed by intensive shaking for 10 min. TEM images
demonstrated the uniform layer of magnetic nanoparticles on the cell walls
with the thickness around 90 ± 20 nm,[40] (Figure 10.4).

Magnetic modification of mammalian cells (*e.g.*, stem cells) can be performed using superparamagnetic iron-oxide particles (SPIO). Different iron-
oxide nanoparticles coated by dextran were tested, especially contrast agents

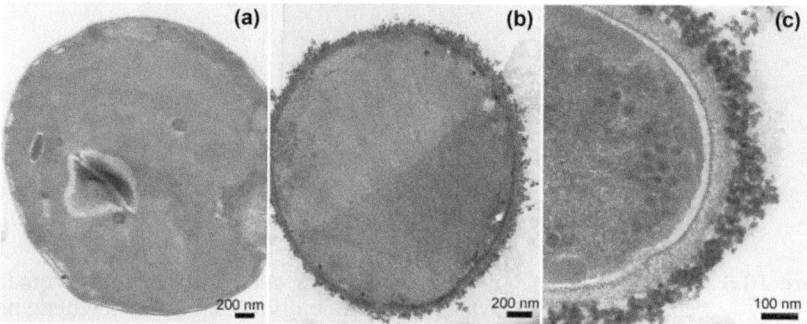

Figure 10.4   TEM images of the thin sections of (a) bare and (b) and (c) PAH-stabilized magnetic nanoparticles coated *C. pyrenoidosa* cells. Reproduced, with permission, from Ref. 40.

for magnetic resonance imaging based on dextran-coated monocrystalline iron oxide nanoparticles (MION). Different derivatives of iron oxides nanoparticles were used for cell labeling to track their migration *in vivo*. Superparamagnetic MION are widely used because of their small size (4–7 nm) and their known magnetic and biochemical properties, which enable the contrast agent to be shuttled into the cell; nanoparticles can be taken up by cells during cultivation by endocytosis. Dendrimer-encapsulated superparamagnetic iron oxides have also been used for magnetic labeling and *in vivo* tracking of stem cells.[41,42]

Another protocol for rapid magnetic modification of leukemia K562 cells *via* their decoration with cationic magnetic nanoparticles (CMNPs) has been developed. The CMNPs (3-aminopropyltriethoxysilane-treated Fe_3O_4) were synthesized; after the incubation of precultured leukemia K562 cells in the presence of CMNPs for a period of time, the CMNPs-modified living cells were prepared. In the next step, these cells were rapidly isolated from the medium and immobilized firmly on the electrode surface *via* a magnetic field.[43] In another procedure (poly)allylamine hydrochloride (PAH) stabilized positively charged magnetic nanoparticles (average diameter around 15 nm) were prepared using a simple synthesis. HeLa cells were magnetically labeled with the particles after 3 min of incubation. High-magnification SEM images demonstrate that the nanoparticles are concentrated as a monolayer on the surface of the cell.[44]

Magnetic cell labeling has also been employed during the magnetofection process. Magnetofection is a simple and highly efficient transfection method that uses magnetic fields to concentrate magnetic particles containing nucleic acid into the target cells. Magnetofection is based on three steps: formulating a magnetic vector, its addition to the medium covering cultured cells and applying a magnetic field in order to direct the vector towards the target cells. The simplest approach to form magnetic derivatives of nucleic acids employs magnetic nanocomposites covered with charged biocompatible polymers that enable formation of ionic complexes with nucleic acids.

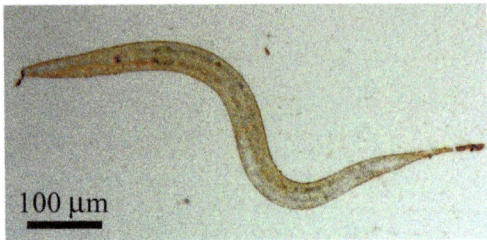

Figure 10.5   Optical microscopy image of a *Caenorhabditis elegans* nematode coated with poly(allylamine hydrochloride)-stabilized magnetic nanoparticles.
Reproduced, with permission, from Ref. 49.

Magnetofection can be performed with viral and synthetic nucleic acid vectors, and can be used to overexpress nucleic acids or to silence endogenous gene expression. It can improve the efficacy of nucleic acid delivery by concentrating and/or retaining an applied vector dose both in primary cells in culture as well as in explanted tissue specimens and in living animals. Transfected cells can be separated from nontransfected ones using an appropriate magnetic separation technique.[35,45,46]

The decoration of red blood cells (RBCs) with aminated and carboxylated core–shell magnetic nanoparticles was studied and elucidated. It was demonstrated that only aminated nanoparticles could decorate the RBCs and their adsorption interaction is mainly ruled by electrostatic attraction between the positively charged amino groups on the particles and the abundant sialic acid groups on the outer surface of RBCs.[47] Another process for RBCs modification employed superparamagnetic iron oxide nanoparticles (SPIONs); for loading, the RBC's membrane was opened by swelling under hypo-osmotic conditions and subsequently resealed. SPIONs could be loaded into RBCs in a concentration sufficient to obtain strong contrast enhancement in MRI.[48]

Not only single cells, but also multicellular organisms can be magnetically modified. Magnetic iron oxides nanoparticles stabilized with poly(allylamine hydrochloride) were used for magnetic modification of soil nematode *Caenorhabditis elegans* (Figure 10.5).[49]

10.2.2   Covalent Immobilization of Cells on Magnetic Carriers

Covalent binding is an extensively used technique for the immobilization of biopolymers, but this technique is used not so often for the immobilization of living cells. Covalent binding of microbial cells on a magnetic carrier is usually possible *via* reactive groups on the surface or through the aid of a reactive binding that links the cells to the carrier. Various coupling agents (*e.g.*, aminosilane, carbodiimide, glutaraldehyde) may be employed to introduce a specific group on the carrier surface, which subsequently can interact with reactive groups on the cell surface. In typical examples of cells covalent immobilization, magnetic chitosan particles activated by glutaraldehyde have been used for *Saccharomyces cerevisiae* immobilization.[15] Alternatively, magnetic cellulose microparticles, after their activation with periodic acid, were used for the yeast cell immobilization.[15] Polyacrolein microspheres with magnetic properties carrying reactive aldehyde groups on their surface were used for covalent binding of fresh human red blood cells.[50]

In addition to magnetic biopolymer and synthetic polymer particles, different types of magnetic iron oxides have been used for covalent cell immobilization or modification.[51–53] Carboxylate- and amino-modified magnetic nanoparticles were used for covalent modification of *Flavobacterium* ATCC 27551. Under optimal conditions, the magnetic cells displayed specific activity ratios of 93% and 89% compared with untreated cells, after the covalent coupling with carboxylate and amino-modified magnetic nanoparticles, respectively.[51] Silanized magnetite (20–40 nm, activated by (3)-aminopropyltriethoxysilane followed by glutaraldehyde treatment) was covalently bound to cells of the alkalotolerant producer of cyclodextrin glucanotransferase (CGTase) *Bacillus circulans* ATCC 21783 in order to increase the produced enzyme activity. The highest CGTase production was achieved after 96 h of semicontinuous process using this type of immobilized cells when the specific enzyme activity was 8.4-fold higher compared to that of free cells. Magnetic nanoparticles linked to the cell walls by the covalent bond between the activated magnetite and the cells were very stable.[52]

10.2.3   Entrapment of Cells into Biocompatible Polymers

Microbial cells can be entrapped in natural or biocompatible synthetic carriers (gels). The carriers can be grouped according to the mechanism leading to the gel formation. Gels can be formed by polymerization (*e.g.*, polyacrylamide, polymethacrylate), crosslinking (*e.g.*, proteins), polycondensation (polyurethane, epoxy resins), thermal gelation (*e.g.*, gelatin, agar, agarose), ionotropic gelation (*e.g.*, alginate, chitosan) and precipitation (cellulose, cellulose triacetate). The gel is formed in the presence of the cells and appropriate magnetic materials. There are various methods available to obtain particles (beads) containing entrapped cells and magnetic particles:[54]

- Block polymerization with subsequent mechanical disintegration into particles. This is a simple method but it results in irregular particles of a wide size distribution.
- Molding of particles (beads) in a template form. This method results in a uniform preparation of immobilized cells but it is less suitable for the preparation of large quantities of immobilized cells.
- Bead formation in a two-phase system. Spherical beads can be prepared in large quantities by suspending an aqueous mixture of cells, magnetic

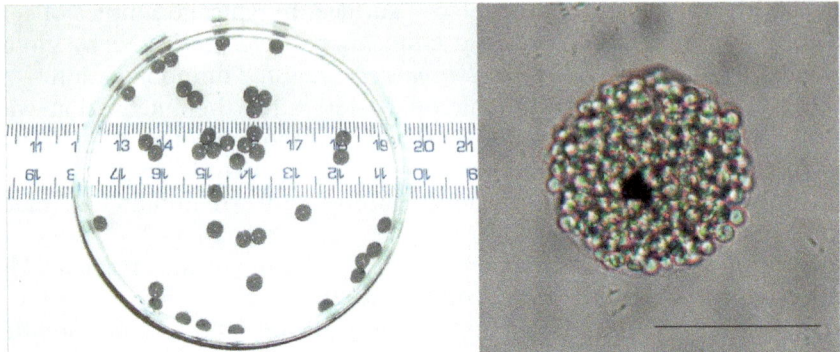

Figure 10.6   Magnetically responsive alginate beads containing entrapped *Sacchar-*
omyces cerevisiae cells and magnetite microparticles. Millimeter-sized
beads (left) and microbeads (right). The scale bar corresponds to 50 μm.
Reproduced, with permission, from Ref. 25.

particles and polymer in a hydrophobic phase under stirring and subsequently inducing gel formation.
- Bead formation of ionotropic polymers after dripping the mixture of
cells, magnetic particles and polymer into a medium containing an
appropriate hardening ion.

Such a procedure can be very mild, enabling magnetic modification of
living cells and subsequently employing their biological activities. One of the
major drawbacks of the entrapment technique is the possible diffusional
limitation as well as the steric hindrance, especially when the macromolecular compounds have to be treated with immobilized cells.[3] In a
typical example, magnetically responsive alginate beads containing entrapped *Saccharomyces cerevisiae* cells and magnetite microparticles were
prepared. Larger beads (2–3 mm in diameter) were prepared by dropping the
mixture into a calcium chloride solution, while microbeads (the diameter of
majority of particles ranged between 50 and 100 μm) were prepared using
the water in oil emulsification process. The immobilized cells were used as
whole-cell biocatalysts for hydrogen-peroxide degradation and sucrose hydrolysis (Figure 10.6).[25,55]

10.2.4   Crosslinking of Cells or Cell Walls

Microbial cell walls contain free amino and/or carboxyl groups, which can
easily be crosslinked by a bi- or multifunctional reagents such as glutaraldehyde or toluene diisocyanate. The cells are usually crosslinked in the
presence of an inert protein like gelatine, albumin, raw hen egg white and
collagen. Microbial cells can also be immobilized by ionic crosslinking
through a flocculation mechanism by addition of polyelectrolytes.

If magnetic particles are used throughout the crosslinking process, magnetic cells or cell walls derivatives can be prepared.[23]

10.2.5 Specific Interactions with Immunomagnetic Nano- and Microparticles

Immunomagnetic detection and modification of cells implies the use of magnetic microbeads or magnetic nanoparticles–antibody system causing the particles to be selectively attached to target cells when added to a cell suspension. After incubation, target cells with attached magnetic particles (and also excess particles) are isolated with the help of an appropriate magnetic separator. Both monoclonal and polyclonal antibodies (Abs) can be used in the course of magnetic modification. In the direct method, the appropriate antibodies are coupled to the magnetic particles, which are then added directly to the cells containing sample. Ideally, the antibody should be oriented with its F_c (fragment crystallizable) region towards the magnetic particle so that the F_{ab} (fragment antigen-binding) region is pointing outwards from the particle.[3,17]

The indirect method can also be used. In the first step, the cell suspension is incubated with primary antibodies that bind to the target cells. Prior sensitization of the target cells will ensure a proper orientation of the antibodies and an optimal number of interaction possibilities between magnetic particles and cells. Not only purified primary antibodies have to be used; crude antibody preparations or serum can be used, too. After incubation, the unbound antibodies are usually removed by washing. Thereafter, the magnetic particles with immobilized secondary antibodies are added, permitting the beads to bind rapidly and firmly to the primary antibodies on the target cells. Target cells–primary antibody complexes can be also captured by protein A or protein G immobilized on magnetic carriers. Alternatively, primary antibodies can be biotinylated or labeled with fluorescein and magnetic particles with immobilized streptavidin or antifluorescein antibodies are used for capturing the target cells.[3,17]

This binding mediated by a specific antigen–antibody reaction is preferably used in immunomagnetic separation of both eukaryotic and prokaryotic cells (*e.g.* microbial pathogens, stem and cancer cells, *etc.*). Both magnetic nanoparticles and microparticles can be used as antibodies carriers. After target-cell labeling the modified cells are usually separated from the sample. Target diamagnetic cells can be separated using two basic strategies, namely positive selection or depletion. The optimal separation strategy depends on the frequency of target cells in the cell sample, their phenotype compared with the other cells in the sample, the availability of reagents and a full consideration of how the target cells are to be used. Positive selection means that the desired target cells are magnetically labeled and isolated directly as the positive cell fraction (Figure 10.7). It is the most direct and specific way to isolate the target cells from a heterogeneous cell suspension.

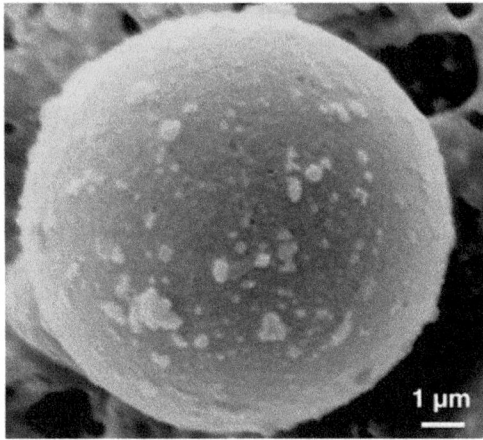

Figure 10.7 SEM photo of the separated CD34$^+$ cells labeled with immunomagnetic nanoparticles.
Reproduced, with permission, from Ref. 111.

Positive selection is particularly well suited for the isolation of rare cells. Both fractions – labeled and unlabeled – can be recovered and used. Depletion means that the unwanted cells are magnetically labeled and eliminated from the cell mixture, and the nonmagnetic, untouched fraction contains the cells of interest. Potential effects on the functional status of cells are minimized.[17,56]

10.2.6 Interaction with Magnetic Particles Bearing Immobilized Biologically Active Compounds

Magnetic selective labeling of target cells can employ different types of biological interactions. In addition to antigen–antibody interaction described in Section 2.5, the following interactions have been employed, namely:

- sugars and specific lectins or other sugar binding proteins;
- biotin and avidin or streptavidin;
- phosphatidylserine and annexin V;
- vancomycin and its receptors on the surface of target bacteria;
- folic acid and folate receptor overexpressed on the surface of human cancer cells.

Different types of lectins, such as those produced from *Triticum vulgaris* and *Agaricus bisporus*, or *concanavalin A*, immobilized on magnetic microspheres, were used to magnetically label specific bacterial pathogens, such as *Escherichia coli*. Recovered cell populations were free from environmental impurities and a high percentage of the culturable cells was extracted.

Specific cell recovery was found to be variable, but the use of lectins may offer some promise as an alternative cell discriminator.[57–59]

Pseudomonas aeruginosa has a specific galactophilic lectin on the outer cell membrane. Pigeon ovalbumin (POA), a phosphoprotein, contains high levels of terminal galactose units and can recognize the above-mentioned lectin. Magnetic nanoparticles with immobilized POA can be used to magnetically label *P. aeruginosa* specifically. Such a strategy enables rapid detection, separation and characterization of *P. aeruginosa* from clinical samples without the need to perform culturing steps.[60] It is also possible to immobilize appropriate sugars on magnetic particles directly and subsequently to use them to bind to specific cells; in a typical example, *Escherichia coli* strain ORN178 containing the mannose binding protein FimH in its fimbriae was magnetically labeled using magnetic particles bearing mannose. The labeled cells were magnetically separated.[61,62]

In another procedure, cell membrane proteins were first biotinylated and then bound to streptavidin magnetic particles. HeLa and TE671 cells were used as model cells; their viability was not changed and this magnetic labeling method was not toxic to cells. Partial internalization of magnetic particles was observed; uptake of these particles did not affect the cell viability.[63]

Magnetic (nano)particles with immobilized annexin V have been employed for simple and efficient separation of apoptotic cells from normal culture. This procedure is based on the fact that annexin V is a Ca^{2+}-dependent phospholipids-binding protein with high affinity for negatively charged phosphatidylserine (PS), which is redistributed from the inner to the outer plasma membrane leaflet in apoptotic or dead cells. Once on the cell surface, PS becomes available for binding to annexin V and any of its magnetic conjugates.[56,64,65]

In order to separate magnetically target bacteria, vancomycin (an antibiotic) was bound to the surface of FePt nanoparticles to capture Gram-positive bacteria *via* molecular recognition between vancomycin and the terminal peptide, D-Ala-D-Ala, on the surface of Gram-positive bacteria. Interestingly, this affinity adsorbent also exhibited selective binding to Gram-negative bacteria at a very low concentration.[66]

Folic acid immobilized on magnetic nanoparticles could be used to facilitate uptake to specific cancer cells overexpressing folate receptor for cancer therapy and diagnosis.[67]

10.2.7 Application of Magnetic Quantum Dots

Quantum dots are nanocrystals made of semiconductor material that are small enough to display quantum mechanical properties; in bioapplications they are particularly significant for optical applications due to their high extinction coefficient. Composite materials composed of magnetic particles and quantum dots can be successfully used also in magnetic cell labeling.[20,68–70]

Nanocomposite nanoparticles consisting of polymer-coated γ-Fe_2O_3 superparamagnetic cores and CdSe/ZnS quantum dots (QDs) shell with the average diameter of 30 nm were modified with carboxylic groups to increase their miscibility in aqueous solution. To demonstrate their utility anticycline E antibodies were immobilized on their surface and then bound to MCF-7 breast cancer cells containing cycline, a protein that is specifically expressed on the surface of breast cancer cells. The separated breast cancer cells were easily observed by fluorescence imaging microscopy due to the strong luminescence of the luminescent/magnetic nanocomposite particles.[68] Other possible applications of magnetic quantum dots can be found elsewhere.[71]

10.2.8 Application of Magnetoliposomes

Magnetoliposomes consist of vesicles composed of a phospholipid membrane encapsulating magnetic nanoparticles. These systems have several important applications, such as in MRI contrast agents, drug and gene carriers, and cancer-treatment device.[72] Different types of magnetoliposomes and other magnetic lipidic vesicles have been used as cell labels.[73]

Mesenchymal stem cells (MSCs), which can differentiate into multiple mesodermal tissues, have been magnetically labeled using cationic magnetoliposomes (leading to the concentration of 20 pg of magnetite per cell), in order to enrich them magnetically from bone marrow. The magnetoliposomes exhibited no toxicity against MSCs in proliferation and differentiation to osteoblasts and adipocytes. During subsequent culture, a substantially higher density of cells was obtained, compared to culture prepared without magnetoliposome treatment.[21]

Alternatively, another methodology for enriching and proliferating MSCs from bone marrow aspirates has been developed using antibody-conjugated magnetoliposomes (AMLs). The AMLs were liposomes conjugated to anti-CD105 antibody (immunoliposomes) and containing magnetite nanoparticles (10 nm diameter). AMLs successfully labeled MSCs, which could be separated by magnetic force. The MSCs proliferated and formed colonies.[74] Other immunomagnetic systems based on magnetoliposomes bearing specific antibodies have been used for magnetic labeling of target cells followed by their positive selection from a cell mixture. Anti-CD34 poly(ethylene glycol)-grafted (PEG) immunomagnetoliposomes were prepared and used for CD34 + cells separation.[75,76] In an *in vivo* application example, tumor-specific magnetoliposomes were conjugated with an antibody fragment to give specificity to target glioma cells. After injection of magnetoliposomes to mice and exposure to the alternating magnetic field, the temperature of tumor tissue increased to 43 °C and the growth of the tumor was found to be arrested over 2 weeks. Magnetoliposomes could target the glioma cells *in vitro* and *in vivo*, and could be efficiently applicable to the hyperthermia of tumors.[77]

10.2.9   Binding of Ferritin and Magnetoferritin on the Cell Surface

Ferritin is a globular protein complex consisting of 24 protein subunits and is the primary intracellular iron-storage protein in both prokaryotes and eukaryotes, keeping iron in a soluble and nontoxic form. Ferritin can be *in vitro* converted into magnetoferritin containing magnetic iron oxides within the protein cavity.[78]

The lymphocytes were incubated with cationized horse spleen ferritin (N,N-dimethyl-1,3-propanediamine derivative of the native horse spleen ferritin) exhibiting a net positive charge at pH 7.5. Under these conditions, the cationized ferritin readily formed ionic bonds with the anionic sites on the cell membrane; the labeled cells were used during the experiments with analytical magnetapheresis.[29]

A secondary antibody staining method was used to couple selectively magnetoferritin or native ferritin to human lymphocytes. (Magneto)ferritin surface was modified by biotin conjugation in preparation for avidin–biotin binding to the antibody complex. The biotinylated (magneto)ferritin was bound to specific biotinylated antibodies *via* an avidin bridge. The detailed study of the labeled cells movement in a ferrograph was performed; it was shown that the magnetic moment of magnetoferritin was sufficient for immunomagnetic isolation of lymphocytes from mononuclear cell preparations in the modified ferrograph.[30]

Recombinant magnetoferritin formed from human heavy-chain ferritin protein subunits can be used to target and visualize tumor tissues over-expressing transferrin receptor 1 (TfR1). The iron oxide core catalyzes the oxidation of peroxidase substrates in the presence of hydrogen peroxide to produce a color reaction that is used to visualize tumor tissues.[79]

10.2.10   Binding of Paramagnetic Cations on the Cell Surface

Lanthanides, especially erbium in the form of erbium chloride ($ErCl_3$), have been used for magnetic labeling of a variety of cells. Erbium ions have a high affinity for the external cell surface and preserve their exceptionally high atomic magnetic dipole moment (9.3 Bohr magnetons) in various chemical structures. The mechanism of Er binding to the cell surface is mostly ionic, with many different Er binding sites, such as carboxyl groups in glycoproteins, differing in affinity and binding capacity. The other well-recognized lanthanide binding sites are the Ca receptor sites on the cell wall. Both Gram-positive and -negative bacteria can be magnetically modified.[3,28,80]

10.2.11   Biologically Driven Precipitation of Paramagnetic Compounds on the Cell Surface

Specific micro-organisms such as *Desulfovibrio* (Gram negative sulfate-reducing bacteria) can precipitate heavy metals on their surfaces as a consequence of their metabolism and growth. This, therefore, gives them the ability to accumulate such metals in large quantities from external surroundings. Micro-organisms growing on glycerol-3-phosphate enzymatically produce phosphate anions in the vicinity of the cells. Heavy-metal phosphates are usually insoluble and that is why they precipitate on the cell surface. Other strains of micro-organisms reduce sulfate ions into sulfide under anaerobic conditions. As a consequence of this phenomenon, insoluble paramagnetic salts precipitate on the cells if paramagnetic cations are present in the medium. Microbial cells modified in this way can be magnetically separated using a high-gradient magnetic separation technique.[3,22]

10.3 Application of Magnetically Modified Cells

As mentioned in the previous section, different procedures are available to convert originally diamagnetic cells into their magnetic derivatives. In this section the selection of important applications of magnetically modified cells will be presented.

10.3.1 Magnetically Modified Cells in Cell Biology, Medicine and Related Areas

Labeling living cells with magnetic nano- or microparticles creates opportunities for numerous imaging and therapeutic applications such as cell manipulation, cell patterning for tissue engineering, magnetically assisted cell delivery or magnetic resonance imaging (MRI)-assisted cell tracking. The unique advantage of magnetic-based methods is to activate or monitor cell behavior by a remote stimulus, the magnetic field. Cell-labeling methods using magnetic particles have been widely developed, showing no adverse effect on cell proliferation and functionalities, while conferring magnetic properties to various cell types.[81]

Magnetic sorting of eukaryotic cells has become a standard method for cell separation in many different fields, both on the small and large scale. The isolation of almost any cell type is possible from complex cell mixtures, such as peripheral blood, haematopoietic tissue (spleen, lymph nodes, thymus, bone marrow, *etc.*), nonhaematopoietic tissue (solid tumors, epidermis, dermis, liver, thyroid gland, muscle, connective tissue, *etc.*) or cultured cells.[56,82]

Two basic variants for magnetic cell separation exist that differ in two main features, namely the composition and size of the magnetic particles used for cell labeling (nanoparticles or microparticles), and the mode of magnetic separation (high-gradient magnetic separation or batch magnetic separation).[56]

The most popular magnetic separation system employing magnetic nanoparticles is the MACS system (Miltenyi Biotec, Germany) which is characterized by the use of nanosized superparamagnetic particles made

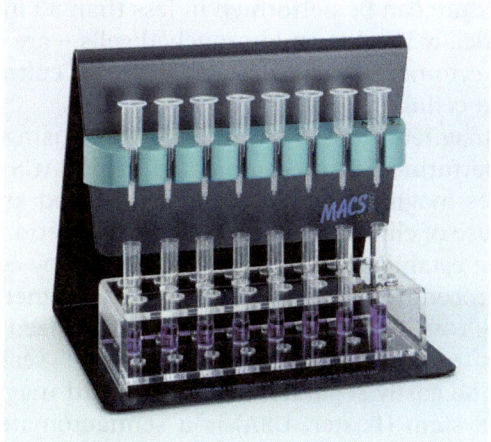

Figure 10.8 A typical example of laboratory-scale high-gradient magnetic separators. OctoMACS Separator (Miltenyi Biotec, Germany) can be used for simultaneous isolation of magnetically labeled cells or mRNA. Reproduced, with permission, from Ref. 112.

from an iron oxide core and a dextran coating, ranging from 20 to 150 nm in diameter, and forming stable colloidal solutions. Magnetic separation is performed in a separation column filled with a matrix of ferromagnetic steel wool or iron spheres that is placed inside the high-gradient magnetic separation system (Figure 10.8). The separator contains a strong permanent magnet creating a high-gradient magnetic field on the magnetizable column matrix; high magnetic gradients up to approx. 10^4 T m^{-1} are generated in the vicinity of the ferromagnetic matrix. The magnetic force is then sufficient to retain the target cells labeled with a very small number of magnetic nanoparticles. Columns of different size are commercially available.[56,82,83]

Magnetic cell separation using the MACS system is performed in three basic steps as follows:[56,82]

- Magnetic labeling of target cells in a cell suspension is performed by immunomagnetic nanoparticles (MicroBeads) that typically are directly covalently conjugated to a monoclonal antibody or other ligand specific for a certain cell type.
- Magnetic separation of magnetically labeled cells; the cell suspension is passed through the separation column that contains a ferromagnetic matrix and is placed in a MACS separator. Labeled target cells are retained in the column *via* magnetic forces, whereas unlabeled cells flow through. By simply rinsing the column with buffer, the entire untouched cell fraction is obtained.
- Elution of the labeled cell fraction; after removing the column from the magnetic field of the MACS separator, the retained labeled cells can easily be eluted with a buffer.

The entire procedure can be performed in less than 30 min, and both cell fractions – magnetically labeled and untouched cells – are ready for further use, such as flow cytometry, molecular biology, cell culture, transfer into animals, or clinical cellular therapy.[56,82]

The large-scale magnetic separation of target cells using magnetic nanoparticles can be performed in the automated CliniMACS device (Miltenyi Biotec) that enables magnetic cell selection in a closed and sterile system (Figure 10.9). The use of clinical-grade isolation or depletion of cells is now a standard technique established in many cellular therapy centers.

An alternative procedure is based on the use of magnetically responsive microparticles, such as Dynabeads, bearing immobilized primary or secondary antibodies or (strept)avidin. On the small scale, labeled cells (Figure 10.10) can be easily separated using standard magnetic separators. The Isolex@ 300 System (Baxter, USA) is a semiautomated magnetic cell separation system designed to select and isolate CD34 positive cells, *ex vivo,* from mobilized peripheral blood using anti-CD34 monoclonal antibody and superparamagnetic Dynabeads microspheres.[56]

The positively selected cells may, in many cases, not show any interference from the larger magnetic particles and may also be analyzed or used with the

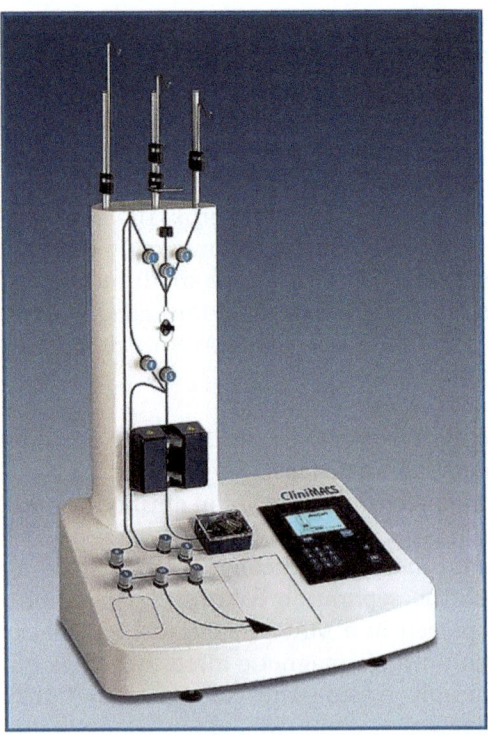

Figure 10.9   The automated system for clinical isolation of human cell subsets (CliniMACS, Miltenyi Biotec, Germany).
Reproduced, with permission, from Ref. 113.

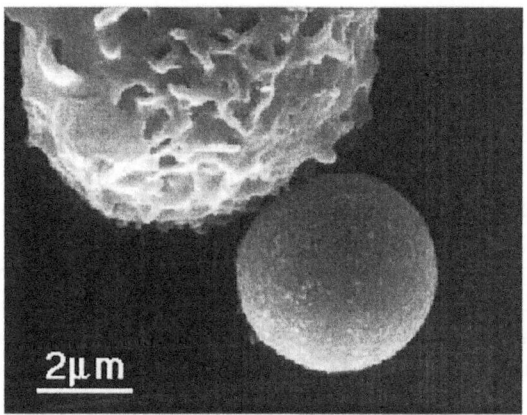

Figure 10.10   A scanning electron micrograph showing a human neutrophil adhering to an anti-CD45-coated Dynabeads M-450.
Reproduced, with permission, from Ref. 114.

particles attached on them. In some cases, however, it is necessary to remove larger immunomagnetic particles from the cells after their isolation. The detachment process can be performed in several ways, namely:[17,56]

- Incubating captured cells overnight in cell-culture medium and subsequent mechanical treatment (*e.g.*, firm pipetting flushing the suspension 5–10 times through a narrow-tipped pipette).
- Proteolytic enzymes can be used to release isolated cells from magnetic particles by selective cleavage of the protein epitope or antibody involved in the immunomagnetic binding.
- Application of an antibody that reacts with the Fab fragments of primary monoclonal Abs on magnetic beads and thus enables direct dissociation of the antigen–Ab binding thereby producing cells without Abs remaining on the surface and with unchanged antigen expression (*e.g.*, DEACHaBEAD from Invitrogen).
- Synthetic peptides that bind specifically to the antigen binding site of primary Abs compete with the target cell–magnetic particles complexes and enable to obtain target cells with unchanged antigen expression (used, *e.g.*, by Baxter Healthcare).
- Carbohydrate units on the Fc part of the Abs allow reversible attachment of the Abs to the magnetic particles with immobilized–B(OH)$_3$ groups. After selective isolation of the target cells sorbitol is added that replaces the Ab on the magnetic bead.
- A complex primary Ab–DNA linker can be immobilized on magnetic particles and after cell binding the DNA linker can be split enzymatically using DNase.
- In specific cases, decrease of pH can cause immunomagnetic particles release.

Immunomagnetic separation has found very important applications in food, medical, veterinary and environmental microbiology. One of the most important tasks in those disciplines is the detection of important pathogenic bacteria in various matrices (*e.g.*, food, clinical samples, soil, water, mud, *etc.*). Standard microbiology procedures for their detection usually require four stages and at least four different growth media; hence the total time from sampling the analyzed material to obtaining a result can be measured in days. One of the possibilities for shortening the isolation and detection period is to replace the selective enrichment stage (usually taking 24 h) with a nongrowth-related procedure. This can be achieved by specific immunomagnetic separation (IMS) of the target bacteria directly from the sample or the pre-enrichment medium (Figure 10.11). Isolated cells can then be identified by standard microbiology, molecular biology and microscopy techniques. IMS can be effectively combined with the polymerase chain reaction (PCR); the main purpose of IMS is to remove the PCR-inhibitory compounds from a sample without loss of sensitivity through dilution, and the concentration of target cells. The oligonucleotide primers should be specific either for the target genus (*e.g.*, detection of different strains of *Salmonella*) or for the individual strain of interest.[17,56]

Immunomagnetic separation is not only faster but also usually gives a higher number of positive samples. Also, sublethally injured microbial cells can be isolated using IMS. Several types of commercially available immunomagnetic particles for pathogen detection are available, such as Dynabeads anti-*Salmonella*, Dynabeads anti-*E. coli* O157, Dynabeads EPEC/VTEC O26, Dynabeads EPEC/VTEC O103, Dynabeads EPEC/VTEC O111, Dynabeads EPEC/VTEC O145, Dynabeads anti-*Legionella* and Dynabeads anti-*Listeria* (all from Invitrogen).[56] In addition, important protozoan parasites

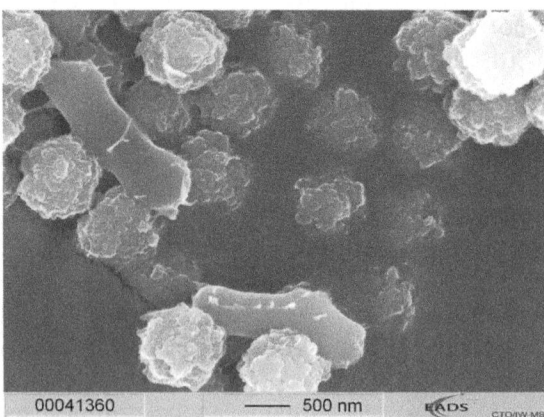

Figure 10.11   Electron microscopy of *Legionella pneumophila* bound to immunomagnetic beads (Dynabeads My One Streptavidin (Invitrogen) with bound biotinylated polyclonal anti-*Legionella* antibody).
Reproduced, with permission, from Ref. 115.

(*Cryptosporidium* and *Giardia*) have been detected in water samples using immunomagnetic procedure.[84]

An interesting combination of immunomagnetic separation of pathogenic bacteria with their photokilling has been developed. The immunomagnetic particles carried a titanium dioxide layer on their surface. After irradiation by low-power UV light, the growth of captured pathogenic bacteria (*e.g.* *Streptococcus pyogenes*, multiantibiotic-resistant *S. pyogenes*, and methicillin-resistant *Staphylococcus aureus* (MRSA)).[85] Bactericidal effect also exhibited magnetite nanoparticles with attached poly(4-vinyl-*N*-alkylpyridinium); the bactericidal efficiencies against *Staphylococcus aureus* were up to 93%.[86]

Magnetic particles with immobilized annexin V interact with cells having disrupted distribution of phosphatidylserine in the cells membranes. This situation appears in certain pathologies or in the course of their aging (*e.g.*, during blood conservation *in vitro*) and during apoptosis. The high sensitivity of this method allows its use to detect damaged red blood cells in various cases (*e.g.*, during storage *in vitro*, during parasitic diseases such as malaria or during neurodegenerative diseases) and for general detection of apoptic cells.[87]

Magnetically labeled cells have also found many other applications. During cells transplantation it is necessary to track and monitor the grafted cells in the transplant recipient. To screen cells both *in vitro* and *in vivo*, superparamagnetic iron-oxide nanoparticles, such as MRI contrast agent Endorem, dextran-based magnetic nanoparticles MicroBeads (Miltenyi Biotec) or biocompatible magnetic fluids have been used to label the stem cells; nanoparticles can often be taken up by cells during cultivation by endocytosis. The magnetically labeled cells enable either *in vitro* detection by staining for iron to produce ferric ferrocyanide (Prussian blue), or *in vivo* detection using MRI visualization, due to the selective shortening of the T_2-relaxation time, leading to a hypointense (dark) signal. MRI can be used to evaluate the cells engraftment, the time course of cell migration and their survival in the targeted tissue.[42,56,88–90] In order to simplify the preparation of magnetically labeled cells, a device for magnetosonoporation of target cells has been developed that employs ultrasound treatment.[91]

Recently poly(vinyl alcohol) coated magnetite nanoparticles were used for magnetic labeling of sperm cells. At pH 7.4 and in glucose-free modified Tyrode's solution the magnetic nanoparticles entered the sperm cells within 3 h. A relatively high concentration of the nanoparticles was internalized into bovine sperm cells and their internalization exhibited little negative effects on the motility and the ability to undergo the acrosome reaction of the sperm cells.[92]

Magnetically labeled cells have been successfully used for magnetic-force-based tissue engineering to develop functional substitutes for damaged tissues. Labelled cells can be manipulated by using a magnet that enables to seed labeled cells onto a low-adhesive culture surface through the use of magnetic force to form a tissue construct. By using this technique, complex cell patterns (curved, parallel, or crossing motifs) can be successfully fabricated from several cell types. Magnetically labeled keratinocytes were accumulated using a magnet, and stratification was promoted by a magnetic force to form a sheet-like 3D construct.[93] An excellent review describing various aspects of magnetic tissue engineering has been published recently.[94]

Cancer cells loaded with magnetic nanoparticles can be at least partially damaged during the magnetic-fluid hyperthermia because tumor cells are more susceptible to damage from heat. Healthy cells, but not cancer cells, can survive temperatures of up to 42 °C. Hyperthermia treatment kills cancerous cells by elevating their temperatures to the therapeutic temperature range of 42–45 °C. This approach can destroy tumors with minimal damage to healthy tissues and, therefore, limit negative side effects. During *in vivo* experimental procedures a suspension containing magnetic nanoparticles is usually directly injected into tumors. When placed in an alternating magnetic field, the nanoparticles dissipate heat and destroy the tumors. This minimally invasive procedure prevents unnecessary heating in healthy tissues because only the magnetic nanoparticles absorb the magnetic field energy. In many cases attaching specific ligands to the nanoparticle surface that recognize specific receptors in the cancerous cells improves the process.[95]

Magnetocytolysis of magnetically labeled cells has been tested after their exposure to an alternating magnetic field *in vitro*. This resulted in the destruction of the cells (magnetocytolysis). Cell-specific magnetocytolysis *in vivo* was achieved by injecting mice intravenously with hepatospecific magnetic nanoparticles and application of an alternating magnetic field (1 h at 200 A/m). Magnetocytolysis did not cause liver necrosis and neither was it accompanied by any increase in body or liver temperature, nor damage to any other tissue. The effects of magnetocytolysis were proportional to the amount of injected magnetic nanoparticles, field strength and its application time.[96]

Further possible applications of magnetically responsive cells have been summarized in several review papers.[81,94,97–99]

10.3.2 Magnetically Modified Cells in Biotechnology

Magnetic modification of living microbial cells can lead to the formation of magnetically responsive whole-cell biocatalysts. Both direct modification of cell walls with appropriate water-based magnetic fluids or magnetic microparticles, as well as target-cell entrapment into biocompatible (bio)polymer gels in the presence of magnetic particles can be performed in a simple way.[14,25,31,55] The intracellular enzyme activities have not decreased substantially after the modification, as shown by hydrogen peroxide degradation and sucrose hydrolysis by intracellular enzymes catalase and invertase present in magnetically modified *Saccharomyces cerevisiae* cells.[25,31] Detailed studies have confirmed that magnetic iron oxides nanoparticles have negligible toxicity on living bacterial cells and thus they can be used in different parts of biotechnology processes.[100]

Rhodococcus erythropolis IGST8 cells decorated with magnetic Fe_3O_4 nanoparticles (45–50 nm in diameter) were used for the biodesulfurization of dibenzothiophene (DBT) and for the postreaction separation of the bacteria from the reaction mixture. Using scanning electron microscopy it was found that the magnetic nanoparticles substantially coated the surfaces of the bacteria. Notably, it was also found that the decorated cells had 56% higher DBT desulfurization compared to the undecorated cells. Based on the fact that the nanoparticles enhance membrane permeability of black lipid membranes, the authors proposed that magnetic nanoparticles increased the permeability of the bacterial membrane, thus facilitating the mass transport of the reactant and product.[101]

Magnetically modified yeast cells can be used as a part of cost-effective biosensor system in microfluidics configurations. Such a screening method has used viable, genetically modified green fluorescent protein (GFP) reporter yeast cells that were magnetically functionalized by biocompatible positively charged magnetic nanoparticles with diameters around 15 nm and held within a microfluidic device. The GFP reporter yeast cells was used to detect genotoxicity by monitoring the exposure of the cells to a well-known genotoxic chemical (methyl methane sulfonate); effective fluorescence emitted from the produced GFP was measured. The magnetically enhanced retention of the yeast cells, with their facile subsequent removal and reloading, allowed for very convenient and rapid toxicity screening.[39]

An amperometric whole-cell herbicide biosensor based on magnetic retention of living microalgae cells *Chlorella pyrenoidosa* functionalized with magnetic nanoparticles on the surface of a permanent magnet-equipped screen-printed electrode was used as a sensing element for the fast detection of herbicides. The magnetic functionalization did not affect the viability and photosynthesis activity-mediated triazine herbicide recognition in microalgae. Atrazine (from 0.9 to 74 μM) and propazine (from 0.6 to 120 μM) could be detected.[102]

Three *Acinetobacter baylyi* ADP1 chromosomally based bioreporters, which were genetically engineered to express bioluminescence in response to salicylate, toluene/xylene and alkanes, were functionalized with 18 ± 3 nm iron oxide nanoparticles to acquire magnetic function. The modified cells were all viable and functional as good as the native cells in terms of sensitivity, specificity and quantitative response. The salicylate sensing modified cells were applied to sediments and garden soils, and used for semiquantitative detection of salicylate in those samples by discriminate recovering of modified cells with a permanent magnet.[103]

10.3.3 Magnetically Modified Cells as Xenobiotics Adsorbents

Different types of microbial cells (bacteria, yeasts, algae) were magnetically modified to prepare magnetic adsorbents. In most cases adsorption of magnetic particles and magnetic fluid treatment were used.[3,104]

Yeast biomass represents an important and promising material for xenobiotics biosorption. The yeast cells of the genus *Saccharomyces* are nonpathogenic, easily available and enable simple manipulation. *Saccharomyces cerevisiae* cells (both baker's and brewer's yeasts) were magnetically modified by contact with perchloric acid stabilized magnetic fluid.[8,105,106] In order to have stabilized product enabling work for a long period of time, dead yeast cells are preferred. The fodder yeast cells (*Kluyveromyces fragilis*) are usually prepared in the dried form, which enables their simple magnetic modification and preparation of an inexpensive adsorbent. Also in this case magnetic-fluid treatment was efficiently used.[33] The same modification process was used to prepare magnetic *Chlorella vulgaris* cells.[34]

Magnetically modified *Saccharomyces*, *Kluyveromyces* and *Chlorella* cells were used for the adsorption of water-soluble dyes from water solutions. The maximum adsorption capacities vary greatly, depending on the dyes structure. Great differences can also be observed even for dyes belonging to the same group (see Table 10.1). In most cases the adsorption process can be described by the Langmuir adsorption isotherm.[104]

Magnetically modified baker's and brewer's yeast cells were also tested as efficient adsorbents of Hg^{2+} and Cu^{2+} ions. The adsorption equilibrium data were well fitted to the Langmuir isotherm; the yeast biomass could be easily regenerated by nitric acid with high effectiveness.[32,107] Magnetically modified yeast cell walls were used for the adsorption of Cu^{2+}, Cd^{2+} and Ag^+ ions.[23]

Magnetically modified bacterial cells (especially of the genus *Pseudomonas*) were intensively studied as possible adsorbents for heavy metal ions.[104] *Pseudomonas putida* strain isolated from heavy-metal ions contaminated samples exhibited high affinity for Cu^{2+} ions; pretreatment of the cells with diluted hydrochloric acid led to the increase of the adsorption capacity.

Table 10.1 Comparison of maximum adsorption capacities Q_{max} (mg/g) of magnetically modified yeast and algae cells for tested dyes.[104]

Dyes	Color Index Number	Maximum adsorption capacities of magnetically responsive microbial cells (mg/g)			
		Saccharomyces cerevisiae[106]	*Saccharomyces cerevisiae* subsp. *uvarum*[8]	*Kluyveromyces fragilis*[33]	*Chlorella vulgaris*[34]
Acridine orange	46 005	82.8		62.2	
Amido black 10B	20 470		11.6	29.9	
Aniline blue	42 755	430.2	228.0		257.9
Bismarck brown	21 000			75.7	201.9
Congo red	22 120		93.1	49.7	156.7
Crystal violet	42 555	85.9	41.7	42.9	42.9
Malachite green	42 000	19.6			
Safranin O	50 240	90.3	46.6	138.2	115.7
Saturn blue LBRR	34 140			33.0	24.2

EDTA solution (0.1 M) efficiently removed the adsorbed copper ions from the adsorbent.[10,108] Another bacterial strain (*Rhodopseudomonas spheroides*) was used for the adsorption of selected chlorinated hydrocarbons, such as heptachlor, aldrin and *p,p'*-DDT from water samples;[109] the same bacterial strain was also used for the adsorption of pesticides such as lindane.[110]

Cultures of *Desulfovibrio* sp. grown anaerobically were suspended in a solution containing iron sulfate together with the compounds to be separated. A high concentration of paramagnetic sulfide precipitate was formed near the cell wall of the micro-organism that enabled magnetic separation of the product. The produced sulfides efficiently adsorbed various organic xenobiotics (*e.g.*, pesticides).[3,22]

10.4 Conclusion and Outlook

In this review chapter we have focused our attention on a very interesting interdisciplinary topic, namely functionalization of cells with magnetic labels. Different procedures for magnetic modification of originally diamagnetic prokaryotic and eukaryotic cells have been developed. Some of them have become widely used tools in various areas of biosciences. Immunomagnetic techniques are routinely used in food, clinical and environmental microbiology, cell biology, medicine, *etc.* A search in the Web of Science has shown 3858 papers containing the terms *immunomagnetic** or *immuno-magnetic** in the title, abstract or keywords; the same search in Scopus has found 5178 papers (search performed at August 11th, 2013). In most cases, immunomagnetic particles have been used for cell magnetic modification. Those high numbers of papers clearly document the importance of magnetic cell labeling.

The currently established protocols for magnetic modification of specific cells types are usually flexible and transferable to other cell types. Despite the existence of a large number of magnetization procedures, the research on optimizing magnetization of cells will continue, employing novel functionalized magnetic (nano)particles. Further progress can be especially expected in the application of magnetically modified cells as parts of selective biosensors, cell tissue constructs, and in large-scale biotechnology and environmental technology processes (*e.g.*, as magnetically responsive whole-cell biocatalysts or adsorbents).

We believe that the list of available magnetically modified cells will increase in the near future, which will lead to a broader application of these interesting materials. There is still plenty of room for the further work in this interesting area.

Acknowledgements

This research was supported by the Grant Agency of the Czech Republic (Projects No. 13-13709S and P503/11/2263) and by the Ministry of Education of the Czech Republic (project LD13023).

References

1. R. F. Fakhrullin and Y. M. Lvov, *ACS Nano*, 2012, **6**, 4557–4564.
2. I. Safarik, K. Pospiskova, K. Horska and M. Safarikova, *Soft Matter*, 2012, **8**, 5407–5413.
3. I. Safarik and M. Safarikova, *China Particuol.*, 2007, **5**, 19–25.
4. J. X. Xie, X. D. Zhang, H. Wang, H. Z. Zheng and Y. M. Huang, *Trends Anal. Chem.*, 2012, **39**, 114–129.
5. D. Schüler, ed., *Magnetoreception and Magnetosomes in Bacteria*, Springer, Berlin, Heidelberg, 2007.
6. F. F. T. De Araujo, M. A. Pires, R. B. Frankel and C. E. M. Bicudo, *Biophys. J.*, 1986, **50**, 375–378.
7. C. S. Owen, *Biophys. J.*, 1978, **22**, 171–178.
8. M. Safarikova, L. Ptackova, I. Kibrikova and I. Safarik, *Chemosphere*, 2005, **59**, 831–835.
9. R. R. Dauer and E. H. Dunlop, *Biotechnol. Bioeng.*, 1991, **37**, 1021–1028.
10. K. F. Sze, Y. J. Lu and P. K. Wong, *Resour. Conserv. Recycl.*, 1996, **18**, 175–193.
11. I. C. MacRae and S. K. Evans, *Water Res.*, 1983, **17**, 271–277.
12. M. Wainwright, I. Singleton and R. G. J. Edyvean, *Biorecovery*, 1990, **2**, 37–53.
13. P. K. Wong and K. Y. Fung, *Enzyme Microb. Technol.*, 1997, **20**, 116–121.
14. K. Pospiskova, G. Prochazkova and I. Safarik, *Lett. Appl. Microbiol.*, 2013, **56**, 456–461.
15. V. Ivanova, P. Petrova and J. Hristov, *Int. Rev. Chem. Eng.*, 2011, **3**, 289–299.
16. Z. Al-Hassan, V. Ivanova, E. Dobreva, I. Penchev, J. Hristov, R. Rachev and R. Petrov, *J. Ferment. Bioeng.*, 1991, **71**, 114–117.
17. I. Safarik and M. Safarikova, *J. Chromatogr. B*, 1999, **722**, 33–53.
18. I. Safarik, M. Safarikova and S. J. Forsythe, *J. Appl. Bacteriol.*, 1995, **78**, 575–585.
19. O. Olsvik, T. Popovic, E. Skjerve, K. S. Cudjoe, E. Hornes, J. Ugelstad and M. Uhlen, *Clin. Microbiol. Rev.*, 1994, 7, 43–54.
20. I. L. Medintz, H. Mattoussi and A. R. Clapp, *Int. J. Nanomed.*, 2008, **3**, 151–167.
21. A. Ito, E. Hibino, H. Honda, K. Hata, H. Kagami, M. Ueda and T. Kobayashi, *Biochem. Eng. J.*, 2004, **20**, 119–125.
22. A. S. Bahaj, D. C. Ellwood and J. H. P. Watson, *IEEE Trans. Magn.*, 1991, **27**, 5371–5374.
23. M. Patzak, P. Dostalek, R. V. Fogarty, I. Safarik and J. M. Tobin, *Biotechnol. Techniques*, 1997, **11**, 483–487.
24. D. Brady, P. Nigam, R. Marchant, L. McHale and A. P. McHale, *Biotechnol. Lett.*, 1996, **18**, 1213–1216.
25. I. Safarik, Z. Sabatkova and M. Safarikova, *J. Agric. Food Chem.*, 2008, **56**, 7925–7928.

26. D. Y. Lee, Y. I. Oh, D. H. Kim, K. M. Kim, K. N. Kim and Y. K. Lee, *IEEE Trans. Magn.*, 2004, **40**, 2961–2963.
27. M. N. Widjojoatmodjo, A. C. Fluit, R. Torensma and J. Verhoef, *J. Immunol. Methods*, 1993, **165**, 11–19.
28. M. Zborowski, P. S. Malchesky, T. F. Jan and G. S. Hall, *J. Gen. Microbiol.*, 1992, **138**, 63–68.
29. M. Zborowski, C. B. Fuh, R. Green, L. Sun and J. J. Chalmers, *Anal. Chem.*, 1995, **67**, 3702–3712.
30. M. Zborowski, C. B. Fuh, R. Green, N. J. Baldwin, S. Reddy, T. Douglas, S. Mann and J. J. Chalmers, *Cytometry*, 1996, **24**, 251–259.
31. M. Safarikova, Z. Maderova and I. Safarik, *Food Res. Int.*, 2009, **42**, 521–524.
32. H. Yavuz, A. Denizli, H. Gungunes, M. Safarikova and I. Safarik, *Sep. Purif. Technol.*, 2006, **52**, 253–260.
33. I. Safarik, L. F. T. Rego, M. Borovska, E. Mosiniewicz-Szablewska, F. Weyda and M. Safarikova, *Enzyme Microb. Technol.*, 2007, **40**, 1551–1556.
34. M. Safarikova, B. M. R. Pona, E. Mosiniewicz-Szablewska, F. Weyda and I. Safarik, *Fresenius Environ. Bull.*, 2008, **17**, 486–492.
35. Y. S. Antequera, O. Mykhaylyk, E. Hammerschmid and C. Plank, *Human Gene Therapy*, 2007, **18**, 1048–1048.
36. S. Takeda, T. Furuyoshi, I. Tari, A. Nakahira, Y. Kakehi, T. Kusaka, S. Ogawa, J. Katayama, Y. Inno, S. Nishijima, K. Fujino and K. Ohmatsu, *Nippon Kagaku Kaishi*, 2000, 661–663.
37. B. Z. Zheng, M. H. Zhang, D. Xiao, Y. Jin and M. M. F. Choi, *Inorg. Mater.*, 2010, **46**, 1106–1111.
38. G. Prochazkova, I. Safarik and T. Branyik, *Biores. Technol.*, 2013, **130**, 472–477.
39. R. F. Fakhrullin, J. Garcia-Alonso and V. N. Paunov, *Soft Matter*, 2010, **6**, 391–397.
40. R. F. Fakhrullin, L. V. Shlykova, A. I. Zamaleeva, D. K. Nurgaliev, Y. N. Osin, J. Garcia-Alonso and V. N. Paunov, *Macromolec. Biosci.*, 2010, **10**, 1257–1264.
41. P. Jendelova, V. Herynek, J. DeCroos, K. Glogarova, B. Andersson, M. Hajek and E. Sykova, *Magn. Reson. Med.*, 2003, **50**, 767–776.
42. E. Sykova and P. Jendelova, *Ann. NY Acad. Sci.*, 2005, **1049**, 146–160.
43. X. E. Jia, L. Tan, Y. P. Zhou, X. F. Jiang, Q. J. Xie, H. Tang and S. Z. Yao, *Electrochem. Commun.*, 2009, **11**, 141–144.
44. M. R. Dzamukova, A. I. Zamaleeva, D. G. Ishmuchametova, Y. N. Osin, A. P. Kiyasov, D. K. Nurgaliev, O. N. Ilinskaya and R. F. Fakhrullin, *Langmuir*, 2011, **27**, 14386–14393.
45. U. Schillinger, T. Brill, C. Rudolph, S. Huth, S. Gersting, F. Krotz, J. Hirschberger, C. Bergemann and C. Plank, *J. Magn. Magn. Mater.*, 2005, **293**, 501–508.
46. I. Safarik and M. Safarikova, *Chem. Papers*, 2009, **63**, 497–505.

47. T. D. Mai, F. d'Orlye, C. Menager, A. Varenne and J.-M. Siaugue, *Chem. Commun.*, 2013, **49**, 5393–5395.
48. M. Brahler, R. Georgieva, N. Buske, A. Muller, S. Muller, J. Pinkernelle, U. Teichgraber, A. Voigt and H. Baumler, *Nano Lett.*, 2006, **6**, 2505–2509.
49. R. T. Minullina, Y. N. Osin, D. G. Ishmuchametova and R. F. Fakhrullin, *Langmuir*, 2011, **27**, 7708–7713.
50. S. Margel, U. Beitler and M. Ofarim, *Immunol. Commun.*, 1981, **10**, 567–575.
51. S. M. Robatjazi, S. A. Shojaosadati, R. Khalilzadeh and E. V. Farahani, *Biocatal. Biotransform.*, 2010, **28**, 304–312.
52. M. Safarikova, N. Atanasova, V. Ivanova, F. Weyda and A. Tonkova, *Process Biochem.*, 2007, **42**, 1454–1459.
53. S. M. Robatjazi, S. A. Shojaosadati, R. Khalilzadeh, E. V. Farahani and M. Zeinoddini, *Biotechnol. Lett.*, 2013, **35**, 67–73.
54. P. Brodelius and E. J. Vandamme, in *Biotechnology (Enzyme Technology)*, eds. H.-J. Rehm and G. Reed, Verlag Chemie, 1987, vol. 7a, pp. 405–464.
55. I. Safarik, Z. Sabatkova and M. Safarikova, *J. Magn. Magn. Mater.*, 2009, **321**, 1478–1481.
56. I. Safarik and M. Safarikova, in *Magnetic Nanoparticles: From Fabrication to Biomedical and Clinical Applications*, ed. N. T. K. Thanh, CRC Press/Taylor and Francis, 2012, pp. 215–242.
57. M. J. Payne, S. Campbell and R. G. Kroll, *Food Microbiol.*, 1993, **10**, 75–83.
58. J. Porter and R. W. Pickup, *J. Microbiol. Methods*, 1998, **33**, 221–226.
59. J. Porter, J. Robinson, R. Pickup and C. Edwards, *J. Appl. Microbiol.*, 1998, **84**, 722–732.
60. J. C. Liu, W. J. Chen, C. W. Li, K. K. T. Mong, P. J. Tsai, T. L. Tsai, Y. C. Lee and Y. C. Chen, *Analyst*, 2009, **134**, 2087–2094.
61. K. El-Boubbou, D. C. Zhu, C. Vasileiou, B. Borhan, D. Prosperi, W. Li and X. F. Huang, *J. Am. Chem. Soc.*, 2010, **132**, 4490–4499.
62. M. Behra, N. Azzouz, S. Schmidt, D. V. Volodkin, S. Mosca, M. Chanana, P. H. Seeberger and L. Hartmann, *Biomacromolecules*, 2013, **14**, 1927–1935.
63. V. H. B. Ho, K. H. Mueller, N. J. Darton, D. C. Darling, F. Farzaneh and N. K. H. Slater, *Exp. Biol. Med.*, 2009, **234**, 332–341.
64. E. K. Dirican, O. D. Ozgun, S. Akarsu, K. O. Akin, O. Ercan, M. Ugurlu, C. Camsari, O. Kanyilmaz, A. Kaya and A. Unsal, *J. Assist. Reprod. Genetics*, 2008, **25**, 375–381.
65. K. Makker, A. Agarwal and R. K. Sharma, *Indian J. Exp. Biol.*, 2008, **46**, 491–497.
66. H. W. Gu, P. L. Ho, K. W. T. Tsang, C. W. Yu and B. Xu, *Chem. Commun.*, 2003, 1966–1967.
67. N. Saltan, H. M. Kutlu, D. Hur, A. Iscan and R. Say, *Int. J. Nanomed.*, 2011, **6**, 477–484.

68. D. S. Wang, J. B. He, N. Rosenzweig and Z. Rosenzweig, *Nano Lett.*, 2004, **4**, 409–413.

69. I. L. Medintz, H. T. Uyeda, E. R. Goldman and H. Mattoussi, *Nature Mater.*, 2005, **4**, 435–446.

70. A. Kale, S. Kale, P. Yadav, H. Gholap, R. Pasricha, J. P. Jog, B. Lefez, B. Hannoyer, P. Shastry and S. Ogale, *Nanotechnology*, 2011, **22**, Article No. 225101.

71. M. Q. Chu, X. Song, D. Cheng, S. P. Liu and J. Zhu, *Nanotechnology*, 2006, **17**, 3268–3273.

72. E. R. Cintra, F. S. Ferreira, J. L. Santos, J. C. Campello, L. M. Socolovsky, E. M. Lima and A. F. Bakuzis, *Nanotechnology*, 2009, **20**, Article No. 045103.

73. L. B. Margolis, V. A. Namiot and L. M. Kljukin, *Biochim. Biophys. Acta*, 1983, **735**, 193–195.

74. A. Ito, E. Hibino, K. Shimizu, T. Kobayashi, Y. Yamada, H. Hibi, M. Ueda and H. Honda, *J. Biomed. Mater. Res. Part B-Appl. Biomater.*, 2005, **75B**, 320–327.

75. J. C. Domingo, M. Mercadal, J. Petriz and M. A. De Madariaga, *J. Microencapsul.*, 2001, **18**, 41–54.

76. J. C. Domingo, M. Mercadal, J. Petriz, J. Garcia and M. A. de Madariaga, *Cell. Molec. Biol. Lett.*, 1999, **4**, 583–597.

77. B. Le, M. Shinkai, T. Kitade, H. Honda, J. Yoshida, T. Wakabayashi and T. Kobayashi, *J. Chem. Eng. Jpn.*, 2001, **34**, 66–72.

78. Z. Mitroova, L. Melnikova, J. Kovac, M. Timko and P. Kopcansky, *Acta Phys. Pol. A*, 2012, **121**, 1318–1320.

79. K. L. Fan, C. Q. Cao, Y. X. Pan, D. Lu, D. L. Yang, J. Feng, L. N. Song, M. M. Liang and X. Y. Yan, *Nature Nanotechnol.*, 2012, 7, 459–464.

80. M. Zborowski, Y. Tada, P. S. Malchesky and G. S. Hall, *Appl. Environ. Microbiol.*, 1993, **59**, 1187–1193.

81. F. Gazeau and C. Wilhelm, in *Magnetic Nanoparticles: From Fabrication to Biomedical and Clinical Applications*, ed. N. T. K. Thanh, CRC Press/Taylor and Francis, 2012, pp. 353–367.

82. M. Apel, U. A. O. Heinlein, S. Miltenyi, J. Schmitz and J. D. M. Campbell, in *Magnetism in Medicine*, eds. W. Andra and H. Nowak, Wiley-VCH Verlag GmbH & Co. KGaA, Weinheim, 2nd edn, 2007, pp. 571–595.

83. S. Miltenyi, W. Muller, W. Weichel and A. Radbruch, *Cytometry*, 1990, **11**, 231–238.

84. J. A. Castro-Hermida, I. Garcia-Presedo, A. Almeida, M. Gonzalez-Warleta, J. M. C. Da Costa and M. Mezo, *Water Res.*, 2008, **42**, 3528–3538.

85. W. J. Chen, P. J. Tsai and Y. C. Chen, *Small*, 2008, **4**, 485–491.

86. J. Lin, S. Y. Qiu, K. Lewis and A. M. Klibanov, *Biotechnol. Prog.*, 2002, **18**, 1082–1086.

87. J. Roger, J. N. Pons, R. Massart, A. Halbreich and J. C. Bacri, *Eur. Phys. J.-Appl. Phys.*, 1999, **5**, 321–325.

88. S. Kobukai, R. Baheza, J. G. Cobb, J. Virostko, J. P. Xie, A. Gillman, D. Koktysh, D. Kerns, M. Does, J. C. Gore and W. Pham, *Magn. Reson. Med.*, **63**, 1383–1390.

89. T. Schlorf, M. Meincke, E. Kossel, C. C. Gluer, O. Jansen and R. Mentlein, *Int. J. Molec. Sci.*, 2011, **12**, 12–23.

90. K. Andreas, R. Georgieva, M. Ladwig, S. Mueller, M. Notter, M. Sittinger and J. Ringe, *Biomaterials*, 2012, **33**, 4515–4525.

91. B. S. Qiu, D. H. Xie, P. Walczak, X. B. Li, J. Ruiz-Cabello, S. Minoshima, J. W. M. Bulte and X. A. M. Yang, *Magn. Reson. Med.*, 2010, **63**, 1437–1441.

92. S. B. D. Makhluf, R. Qasem, S. Rubinstein, A. Gedanken and H. Breitbart, *Langmuir*, 2006, **22**, 9480–9482.

93. K. Ino, A. Ito and H. Honda, *Biotechnol. Bioeng.*, 2007, **97**, 1309–1317.

94. A. Ito and M. Kamihira, *Prog. Molec. Biol. Transl. Sci.*, 2011, **104**, 355–395.

95. M. Latorre and C. Rinaldi, *Puerto Rico Health Sci. J.*, 2009, **28**, 227–238.

96. A. Halbreich, E. V. Groman, D. Raison, C. Bouchaud and S. Paturance, *J. Magn. Magn. Mater.*, 2002, **248**, 276–285.

97. H. Markides, M. Rotherham and A. J. El Haj, *J. Nanomater.*, 2012, **Volume 2012**, Article ID 614094.

98. Y. Pan, X. W. Du, F. Zhao and B. Xu, *Chem. Soc. Rev.*, 2012, **41**, 2912–2942.

99. V. Mailander and K. Landfester, *Biomacromolecules*, 2009, **10**, 2379–2400.

100. M. E. Kafayati, J. Raheb, M. T. Angazi, S. Alizadeh and H. Bardania, *Iran. J. Biotechnol.*, 2013, **11**, 41–46.

101. F. Ansari, P. Grigoriev, S. Libor, L. F. Tothill and J. J. Ramsden, *Biotechnol. Bioeng.*, 2009, **102**, 1505–1512.

102. A. I. Zamaleeva, I. R. Sharipova, R. V. Shamagsumova, A. N. Ivanov, G. A. Evtugyn, D. G. Ishmuchametova and R. F. Fakhrullin, *Anal. Methods-UK*, 2011, **3**, 509–513.

103. D. Y. Zhang, R. F. Fakhrullin, M. Ozmen, H. Wang, J. Wang, V. N. Paunov, G. H. Li and W. E. Huang, *Microb. Biotechnol.*, 2011, **4**, 89–97.

104. I. Safarik, K. Horska and M. Safarikova, in *Microbial Biosorption of Metals*, eds. P. Kotrba, M. Mackova and T. Macek, Springer 2011, pp. 301–320.

105. R. B. Azevedo, L. P. Silva, A. P. C. Lemos, S. N. Bao, Z. G. M. Lacava, I. Safarik, M. Safarikova and P. C. Morais, *IEEE Trans. Magn.*, 2003, **39**, 2660–2662.

106. I. Safarik, L. Ptackova and M. Safarikova, *Eur. Cells Mater.*, 2002, **3**(Suppl. 2), 52–55.

107. L. Uzun, N. Saglam, M. Safarikova, I. Safarik and A. Denizli, *Separat. Sci. Technol.*, 2011, **46**, 1045–1051.

108. H. Chua, P. K. Wong, P. H. F. Yu and X. Z. Li, *Water Sci. Technol.*, 1998, **38**, 315–322.

109. I. C. MacRae, *Water Res.*, 1986, **20**, 1149–1152.

110. I. C. MacRae, *Water Res.*, 1985, **19**, 825–830.
111. W. Chen, H. B. Shen, X. Y. Li, N. Q. Jia and J. M. Xu, *Appl. Surf. Sci.*, 2006, **253**, 1762–1769.
112. I. Safarik and M. Safarikova, in *Encyclopedia of Separation Science*, eds. I. D. Wilson, R. R. Adlard, C. F. Poole and M. R. Cook, Academic Press, London, 2000, pp. 2163–2170.
113. I. Safarik and M. Safarikova, *Mon. Chem.*, 2002, **133**, 737–759.
114. J. Y. Shao, H. P. Ting-Beall and R. M. Hochmuth, *Proc. Natl. Acad. Sci. USA*, 1998, **95**, 6797–6802.
115. U. Reidt, B. Geisberger, C. Heller and A. Friedberger, *J. Lab. Automat.*, 2011, **16**, 157–164.

CHAPTER 11

Cell Surface Engineering Using a Layer-by-Layer Nanofilm for Biomedical Applications

MICHIYA MATSUSAKI AND MITSURU AKASHI*

Department of Applied Chemistry, Graduate School of Engineering, Osaka University, 2-1 Yamadaoka, Suita, Osaka 565-0871, Japan
*Email: akashi@chem.eng.osaka-u.ac.jp

11.1 Introduction

Nearly all tissue cells in the body reside in the micrometer-sized fibrous meshwork of the extracellular matrix (ECM). The ECM is typically composed of fibronectin (FN), collagen and laminin, and provides complex biochemical and physical signals.[1-3] The ECM not only acts as storage for growth factors and cytokines but also induces cell–cell contacts and cell–matrix interactions.[4] Accordingly, control of the cellular microenvironment using artificial ECM and growth factors will be important *in vitro* technique to control cell growth, cytokine expression, stem-cell differentiation, and cellular assembly. To control cellular microenvironment, the patterning of substrate surfaces[5] and the chemical modification of cell surfaces[6,7] have been generally employed. However, there were limitations due to cytotoxicity, complicated methodology or less application.

We focused on the layer-by-layer (LbL) assembly technique of polymer or protein films to modify cell surfaces. The polyelectrolyte LbL assemblies were applied for the surface modification of cells in 2001 for the first time and the red blood cells and bacteria were used for fabrication of hollow

RSC Smart Materials No. 9
Cell Surface Engineering: Fabrication of Functional Nanoshells
Edited by Rawil F Fakhrullin, Insung S Choi and Yuri Lvov
© The Royal Society of Chemistry 2014
Published by the Royal Society of Chemistry, www.rsc.org

microcapsules using this technique.[8] The LbL assemblies have generally been used for the surface modification of micro-organisms and yeasts[9] due to the toxicity for fragile mammalian cells[10] because they do not have a strong cell wall structure. Thus, there were few reports of surface modification of mammalian cells with LbL films and further application for biomedical field.

We achieved fabrication of LbL nanofilms onto the surface of various mammalian cells by employing different driving force to LbL films. The surface modification with LbL nanofilms was valuable for various biomedical applications. In this chapter, we describe LbL assemblies using biological affinity interaction and their application for tissue engineering, especially development of three-dimensional (3D) tissue constructs.

11.2   LbL Assembly

Decher *et al.* reported in 1991 that the stepwise immersion of substrates such as mica and glass into aqueous solutions of positively and negatively charged polymers produced multilayered ultrathin polymer films with a controllable nanometer thickness.[11–13] This method is called layer-by-layer (LbL) assembly, and both basic research and applications have been widely developed in the field of polymer Science, (Figure 11.1). Since this technique involves the simple immersion of a substrate into an oppositely charged polymer solution, researchers have extended LbL assembly to include the

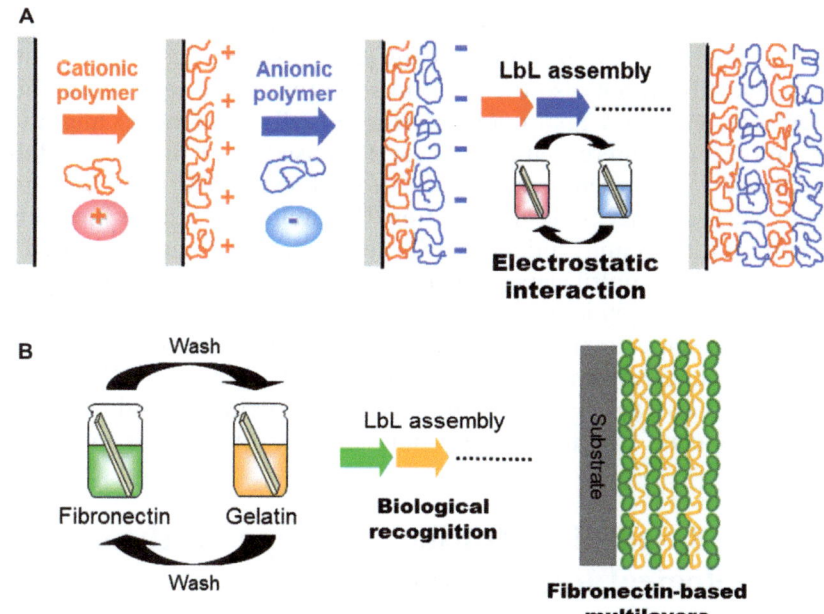

Figure 11.1   Schematic illustration of LbL assemblies through electrostatic interaction (A) and biological recognition (B).

deposition of not only water-soluble linear charged polymers, but also viruses,[14] proteins,[15–18] silica colloids,[19,20] metal nanoparticles,[21–24] dyes,[25–27] metal oxides,[28–30] amphiphiles,[31,32] clays,[33–35] and polystyrene nanospheres.[36–38] At the beginning, only electrostatic interactions were utilized for LbL assembly. Subsequently, other interactions such as covalent bond,[39–45] hydrogen bonding,[46–52] charge transfer,[53,54] hydrophobic,[55–57] host-gest,[58–61] and coordination bond[62–68] interactions have been investigated to facilitate polymer association for ultrathin film deposition. The most important concept for LbL assembly is how to use the interactions between the polymeric materials. It follows that other weak interactions between Macromolecules, must be crucial for the further development of LbL research.

Biologically specific recognition such as protein–ligand, antibody–antigen, and protein–polysaccharide interactions are composed of multiple weak interactions such as electrostatic, hydrogen bonding, and hydrophobic interactions, and provide high specificity to the target Macromolecules. These interactions would extend the scope of the LbL technique in constructing functional thin-film assemblies, because nonionic polymers and even polymers with the same polarity can be built into the same assemblies simultaneously *via* the biological interactions. A few reports have appeared on LbL assemblies based on such biological interactions, avidin–biotin,[69–72] antibody–antigen,[73,74] lectin–polysacccharides,[75,76] and ECM Macromolecules.[77]

We have also reported for the first time ECM multilayers focused on FN using biologically specific recognition.[78] FN is a highly flexible multifunctional glycoprotein, and is well known to interact with not only a variety of ECM proteins and glycosaminoglycans (GAGs) such as collagen, gelatin, heparin, and fibrin, but also the integrin receptor on the cell membrane.[79] Furthermore, FN plays an important role in cell attachment, migration, proliferation, and differentiation.[2] These FN-based ECM multilayers fabricated through biological interactions showed extremely high biological activities such as enzymatic degradation,[80] stability in culture medium,[81] cytocompatibility,[82] and inducing cell–cell adhesion[83] as compared to polyelectrolyte FN multilayers. We have developed a simple and unique bottom-up approach, "hierarchical cell manipulation", using FN-based ECM multilayers to fabricate 3D-cellular multilayers[78] and their applications as a tissue model for tissue engineering and pharmaceutical assays.[84–86]

LbL assemblies constructed through weak interactions would extend not only the scope of LbL technology, but also the breadth of biomedical applications.

11.3 FN-Based LbL Multilayers Prepared through Biological Recognition

Various FN-based LbL multilayers using gelatin (G) and α-elastin (E) as an ECM protein, heparin (Hep) and dextran sulfate (Dex) as a GAG or its derivative, and α-poly(L-lysine) (PLL) as a cationic polymer were fabricated by both electrostatic interaction and FN biological recognition.[80] Although G, E, Hep, and Dex have a negative charge under physiological conditions, they interacted with negatively charged FN (pI 5.4) because FN has a binding domain for each component (biological recognition).[79,87,88] On the other hand, FN/PLL and PLL/Dex polyelectrolyte multilayers were also obtained through the conventional electrostatic interaction. A comparative evaluation between FN-specific and electrostatic assembly was performed since we are especially interested in the difference of the driving forces for LbL assembly. The biodegradability and stability of ultrathin polymer films is one of the most important requirements for biomedical applications. Thus, we compared the enzymatic degradation and stability as biological properties of these two kinds of FN-based nanofilms. Figure 11.2 shows the results of the LbL assembly and enzymatic degradation behaviors. Here, we chose elastase because it is well known as a serine protease possessing the characteristic ability to degrade FN within a short time at 37 °C.[89] Interestingly, the FN/Dex films with larger thickness showed more quickly degradation behavior as compared to that of FN/PLL films. It seems that the inhibition of the FN active site from additional Dex was minimized because the excess adsorption of Dex onto the FN layer was inhibited by electric repulsion. van Tassel and coworkers also reported that the active sites of FN on poly(sodium styrenesulfonate) (PSS)-terminated films were significantly more accessible as compared with the poly(allylamine hydrochloride) (PAH)-terminated films.[90]

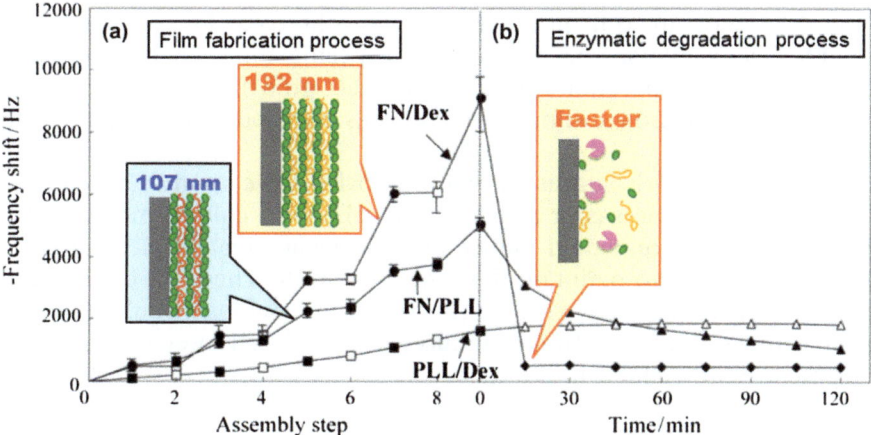

Figure 11.2   Frequency shift of the quartz crystal microbalance (QCM) stepwise assembly of FN (closed circle)/Dex (open square), FN (closed circle)/PLL (closed square) and PLL (closed square)/Dex (open square) in 50 mM PB(pH 7.4) at 37 °C ($n=3$) (a). Enzymatic degradation of FN/Dex (closed diamond), FN/PLL (closed triangle) and PLL/Dex (open triangle) films in aqueous solution with elastase at 37 °C analyzed by QCM (b). Abbreviation: FN is fibronectin, Dex is dextran sulfate, PLL is α-poly(L-lysine), PB is phosphate buffer, QCM is a quartz crystal microbalance.

They stated that the end-segments of FN carry strong negative charges, whereas the middle segment is almost neutral; therefore, they suppose that, on a negatively charged surface, FN would adsorb in a side-on orientation. The excess adsorbed cationic PLL on the FN layer may have inhibited the elastase access and activity.

The weight remaining percentage of FN/G, FN/Dex, FN/PLL, and PLL/Dex films in Eagle's MEM with 10% fetal bovine serum (FBS) was evaluated at 37 °C. The FN-specific assembly films, FN/G and FN/Dex (biological recognition), were surprisingly stable in the cell-culture medium for over 150 min, although the polyelectrolyte nanofilms (FN/PLL and PLL/Dex) showed a weight increase in the first 15 min; thereafter the weight gradually decreased. These results clearly show that FN-specific assembly films are more stable and useful for cell-culture media. We hypothesized that some serum proteins in the culture medium adsorbed onto the polyelectrolyte film surfaces by electrostatic interactions in the first 15 min, and protein exchange or desorption would occur subsequently.

We also evaluated the cytocompatibility of FN-specific and polyelectrolyte nanofilms by preparing these nanofilms onto cell surfaces.[81] The morphologies of L929 fibroblasts after preparing polyelectrolyte multilayers clearly changed from a spread to a round-shaped morphology, and cell viability decreased with increasing thickness of these films, suggesting low cytocompatibility. On the other hand, FN-specific multilayers showed extended cell morphologies similar to control cells (without films), and cytotoxicity was not observed, independent of their thicknesses. Moreover, a clear difference in cell proliferation was observed for polyelectrolyte and FN-specific films. The cells with FN-specific films on their surfaces showed good proliferation profiles, but cell growth was not observed using the polyelectrolyte films even the cells survived during this culture period.

These results clearly suggested that the driving force strongly affected the thickness, enzyme access or activity, and stability in culture medium. The biological recognition would induce biologically adequate molecular structure of these films to maintain the functions of FN. These FN-based nanofilms constructed through FN-specific domain interactions would be useful as an artificial ECM nanofilm to induce stable cell–cell interactions for development of 3D-cellular multilayers.

11.4 Top-Down Approach for Development of 3D Tissue Constructs

There are basically two kinds of approaches, top-down and bottom-up approaches, for construction of 3D-tissue constructs (Figure 11.3). A top-down approach has been reported historically, especially biodegradable scaffold method. Biodegradable scaffolds and hydrogels consisting of biodegradable polymers, such as poly(lactic acid), poly(glycolic acid), alginate and collagens

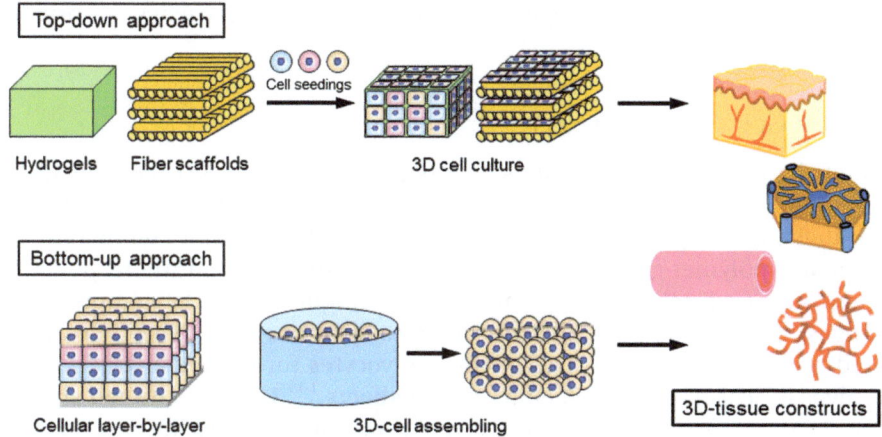

Figure 11.3 Schematic images of top-down and bottom-up approaches to construct 3D-tissue structures.

have been used for the construction of 3D constructs containing living cells.[91,92] The topological control of biodegradable porous scaffolds,[93] especially nanofiber scaffolds by electrospinning[94] or self-assembling amphiphilic peptides[95] has attracted much attention due to their high porosity and the controlled alignment of the fibers to control cellular function and the development of 3D-engineered tissues. The cells encapsulated in the scaffolds can grow actively and finally they formed random cell aggregations inside it. Although their growth rate can be controlled using growth factors in culture media, 3D-engineered tissues possessing precisely controlled cell type, cell alignment, and cell–cell interaction have not been developed yet. The nanofiber scaffolds can contribute to tentative cell alignment or adhesion due to the morphology, but it is difficult to maintain these effects because the nanofibers are covered completely with the cultured cells and expressed ECMs from the cells. Accordingly, a conventional approach using biodegradable matrices such as hydrogels or fiber scaffolds seems to have several limitations in developing 3D-tissue constructs that satisfy the above requirements. A bottom-up approach using multiple cell types as pieces of tissue recently attracted much attention to solve these problems. Accordingly, in this chapter, we focused on a bottom-up approach as follows.

11.5 Bottom-Up Approach for Development of 3D-Tissue Constructs

Bottom-up approaches such as cell sheets,[96] magnetic liposomes,[97] hierarchical cell manipulation,[78] polymeric aqueous two-phase systems,[98] and printing of cells and polymers[99] have been reported in constructing a complex tissue structure. These bottom-up approaches are generally categorized by two groups, a cell-based method and cell and a polymer-based method.

Here, characteristics, advantages, and issues of both approaches are briefly summarized.

11.5.1   Cell LbL Approach

One of the most general methods as a bottom-up approach is cellular LbL. Okano and coworkers have reported fabrication of monolayer of cell sheets and their stacking to make multilayered structures using temperature-responsive polymer-grafted culture dishes.[96] Using temperature-responsive dishes, cultured cells can be harvested as fragile sheets by temperature changes, thereby avoiding the use of proteolytic enzymes. They have prepared multiple-sheet structures of cardiomyocytes for the reconstruction of 3D-myocardical tissues by stacking of cell sheets.[100] Because the cell sheets have ECM at the bottom of the sheet, they can stack easily due to the interaction with cell-membrane receptors. It is one of the leading methods to fabricate multilayered cellular structures.

Although the cell sheet is an intriguing method, the use of the fragile sheets is not easy for the layman. The other method to bind cells to form 3D structures, the magnetic force, has been reported from Ito and coworkers.[97] They applied a magnetic force to construct a heterotypic, layered coculture system of hepatocytes and endothelial cells that was not limited by cell type. Magnetic cationic liposomes with positively charged surfaces were adsorbed onto cell surfaces, and then magnetic forces were applied to make multi-layered structures. This method is able to fabricate multilayered structures of cells independent of cell types because the driving force is the magnetic force. However, although the remaining amount of magnetic liposomes is very small, they might induce warning of toxicity.

Takayama and coworkers have reported new approach, biphasic system for noncontact cell printing on cells.[98] Aqueous solutions of polyethylene glycol (PEG) and Dextran (Dex) above certain concentrations segregate and form a two-phase system with Dex always forming the bottom phase. They controlled the interfacial tension of each phase to achieve cell printing on cells, and reported cellular niches to support neuronal differentiation of mouse embryonic stem cells (mESC) and show that the density of printed mESC is an important factor for guiding mESC differentiation to neurons. This is a unique technique to prepare bilayered cellular constructs, and thus it is expected to apply for over three layers.

11.5.2   Hierarchical Cell Manipulation

For development of the 3D-cellular multilayers, direct fabrication of nano-meter-sized cell-adhesive materials like ECM fibrous scaffolds onto the surface of a cell membrane is crucial, because the insufficient ECM is se-creted onto cell surfaces in the early stage of cell culture. We focused on the LbL technique that is an appropriate method to prepare nanometer-sized films on a substrate by alternate immersion into interactive polymer solutions.[11,12] The preparation of nanometer-sized multilayer films composed of ECM components on the surface of the first layer of cells will provide a cell-adhesive surface for the second layer of cells. Rajagopalan *et al.* demonstrated a bilayer structure composed of hepatocytes and other cells by preparing a polyelectrolyte (PE) multilayer consisting of chitosan and DNA onto the hepatocyte surface.[101] However, chitosan cannot dissolve in neutral buffer, and the use of PE films as a cell-adhesive material is limited due to the cytotoxicity of polycations.[102,103] The appropriate choice of natural ECM components for preparation of the nanofilms is important to avoid cytotoxicity, and the typical ECM presents cell-adhesive moieties such as RGD (arginine–glycine–aspartic acid) and other functional moieties.[104] We selected FN and G to prepare nano-ECM films on the cell surface. FN is a flexible multifunctional glycoprotein, and plays an important role in cell attachment, migration, differentiation, and so on.[2] FN is well known to interact not only with a variety of ECM proteins such as collagens (gelatins) and glycosaminoglycans but also with the $\alpha_5\beta_1$ integrin receptor on the cell surface.[105] Although FN and G have a negative charge under physiological conditions, they interacted with each other because FN has a collagen binding domain,[2] indicating different driving force of PE films using polycations. Thus, the FN-G nanofilms are expected to provide a suitable cell-adhesive surface similar to the natural ECM for the second layer of cells, without any cytotoxicity.

The fabrication of 3D-cellular multilayers composed of cells and FN-G nano-ECM films was performed according to the process shown in Figure 11.4. The LbL assembly of FN and G onto the cell surface was analyzed quantitatively using a QCM as the assembly substrate, and with a phospholipid bilayer membrane as a model cell membrane.[78] A phospholipid bilayer composed of 1,2-dipalmitoyl-*sn*-glycero-3-phosphatidylcholine (DPPC) and 1,2-dipalmitoyl-*sn*-glycero-3-phosphate (DPPA) was prepared onto the base layer, a 4-step assembly of poly(diallyldimethylammonium chloride) (PDDA) and PSS, according to Krishna *et al.*'s report.[106] The mean thickness of the LbL assembly after 1, 7, and 23-steps was calculated to be 2.3, 6.2, and 21.1 nm, respectively. The top and cross sections of the confocal laser scanning microscopy (CLSM) 3D-merged images indicated a homogeneous assembly of fluorescently labeled FN-G nanofilms on the surface of mouse L929 fibroblasts. For the quantitative studies on the thickness of the multilayers on the cell surface, the fluorescence intensity of rhodamine-labeled FN (Rh-FN) was estimated by a line scan. The fluorescence intensity of the Rh-FN-G nanofilms increased linearly on increasing the LbL assembly step number, similar to the frequency shift of the QCM analysis, indicating a clear increase in the film thickness on the cell surface. These results demonstrated the fabrication of FN-G nanofilms on the cell surface.

We fabricated a bilayer of mouse L929 fibroblast cells with or without FN-G nanofilms by using a cover glass as a substrate. When the 7-step assembled FN-G nanofilms were prepared on the surface of the first L929 cell layer, the second layer cells were then observed on the first cell layer. However, when the nanofilm was not prepared or the 1-step-assembled

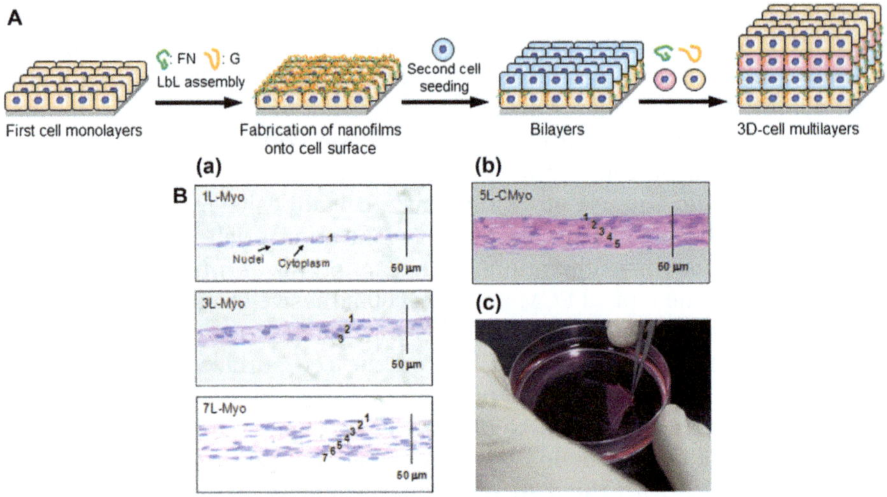

A

First cell monolayers   Fabrication of nanofilms   Bilayers   3D-cell multilayers
                                     onto cell surface

(a)   (b)

B

Figure 11.4   Schematic illustration of hierarchical cell manipulation (A). HE staining images of one to seven-layered Myo (a), five-layered CMyo (b) and
photograph of the peeling process of four-layered mouse L929 fibroblasts from a cover glass (B).
Reproduced with permission from Ref. 107.

nanofilm (only FN) was assembled on the first cell layer, then the bilayer
architecture was not observed. These results suggested 2.3 nm of FN film
was inadequate and at least approximately 6 nm of FN-G nanofilm was required as a stable adhesive surface for the second cell layer.

The four-layered (4L) cellular multilayers were clearly observed after four
repetitions of these steps by confocal laser scanning microscopy (CLSM)
and hematoxylin and eosin (HE) staining images. This hierarchical cell
manipulation technique can be applied to various types of cells, *e.g.* myo-
blast cells (Myo), cardiac myocytes (CMyo), smooth muscle cells (SMC),
hepatocytes (Hep), and endothelial cells (EC). The FN-G nanofilms did not
show any cytotoxicity, and the obtained 3D-cell architectures presented high
intercellular adhesion to easily peel off from the substrate (Figure 11.4Bc).

11.5.3   Cell Functions in 2D *vs.* 3D Structures

The cellular functions in the 3D-tissue constructs are expected to be higher
than those of monolayered cells in general cultures. Some researchers have
reported the functions of layered cellular architectures *in vitro*.[101] However,
the basic properties induced by 3D-cellular structures, such as the layer
number or the cell types, have not yet been clarified.

We evaluated the structural stability of layered constructs consisting of
normal human dermal fibroblasts (NHDFs) and human umbilical vein
endothelial cells (HUVECs) in relation to their layer number.[83] To evaluate
the general effects of 3D-layered structures on cellular stress or

inflammation, we purposely fabricated biologically meaningless layered structures consisting of endothelial cells and fibroblasts. General culture media, Dulbecco's modified Eagle's medium (DMEM) containing 10% FBS, was employed in this study to avoid influence of specific growth factors. Interestingly, the HUVECs adhered homogeneously on the surface of 4L-NHDFs, and tight junction formation was widely observed at the centimeter scale, while heterogeneous HUVEC domain structures were observed on the monolayered (1L) NHDFs. The production of heat shock protein 70 (Hsp70) and interleukin-6 (IL-6) from the cellular structures were investigated to elucidate any 3D-structural effect on cellular function (Figure 11.5).

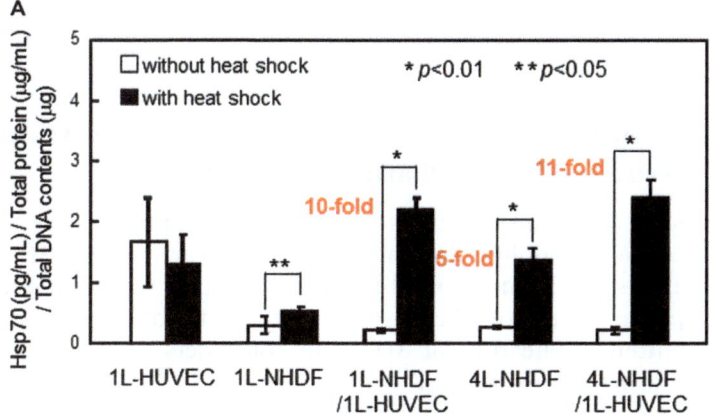

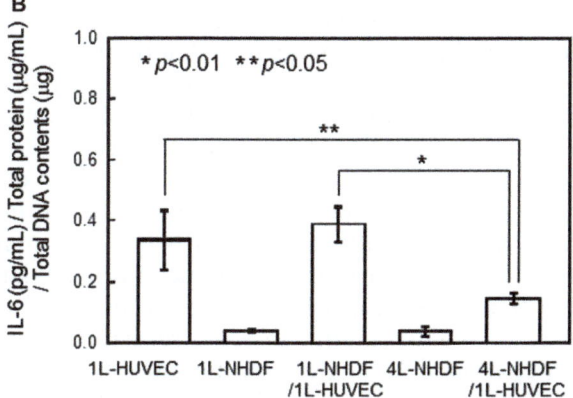

Figure 11.5 Production of heat-shock protein 70 (Hsp70) *versus* the total protein from nonheat-shocked and heat shocked varied layered constructs composed of NHDFs and HUVECs ($n = 4$) (A). The heat shock conditions were 20 min of incubation at 45 °C, followed by a 2-h recovery period. IL-6 secretion *versus* the total protein from the layered structures ($n = 3$) (B). Asterisk denotes statistically significant difference using a two sample *t*-test (*$p < 0.01$, **$p < 0.05$) for each comparison. Reproduced with permission from Ref. 83.

The Hsp70 expression of the HUVECs decreased after adhesion onto the 4L-NHDF structure as compared with the HUVEC monolayer. Surprisingly, the Hsp70 production response to heat shock increased drastically by approximately 10-fold as compared with a nonheat shock by 3D-structure formation, whereas the monolayer structures showed no change. Moreover, the production of the inflammatory cytokine IL-6 decreased significantly depending on the layer number of NHDFs. These results suggested that 4L-NHDF provided a more favorable environment for HUVECs than a cell culture plastic disk to induce high thermosensitivity and to suppress the inflammatory response from the substrate. Since the FN-G nanofilms prepared on the cell surface did not affect cellular functions,[81] we concluded that the layered construct would be an analogous environment to natural tissues. These findings could be important not only for tissue engineering, but also for basic cell biology.

11.5.4 Potential Application of 3D Constructs as a Human Tissue Model for Pharmaceutical Assays

The constructed 3D-tissue structures are desired to be applicable as a human tissue model for pathophysiological assays. However, since a 3D human tissue model has not been reported much, application reports are also few (here, cell spheroids are not taken as an object). We tried to demonstrate the *in vitro* potential application of 3D-tissue constructs as a human-tissue model for pharmaceutical assays.

The 3D blood-vessel models consisting of human SMCs and HUVECs were fabricated by hierarchical cell manipulation.[82] The 5L blood-vessel models consisting of 4L-umbilical artery SMC (UASMC) and outermost 1L-HUVEC showed histological staining images similar to that of human blood vessels, and the surface of 4L-UASMC/1L-HUVEC revealed high biocompatibility avoiding adsorption and activation of platelets in human platelet-rich plasma (PRP). We tried to use the 5L-artery models consisting of human aortic SMCs and ECs for analyses of production and diffusion of nitric oxide (NO) from ECs in response to the peptide hormone, bradykinin.[84]

The NO produced from ECs diffuses into SMCs through their cell membranes, and activates guanylate cyclase to produce intracellular cyclic guanosine monophosphate (cGMP), which induces a signaling pathway mediated by kinase proteins leading to SMC relaxation.[108] Accordingly, quantitative, kinetic, and spatial analyses of the extracellular delivery of NO molecules from the EC layer to the SMC layers upon drug stimulation are important for pharmaceutical and biomedical evaluations of hypertension and diabetes. Recently, a significant correlation between NO production and diabetes mellitus was clarified, *e.g.* reduced NO production in both type 1 and 2 diabetes[109] and also an increase of NO production related to improvement of type 2 diabetes were reported.[110] So far, pharmaceutical assays of NO production have been performed by *in vivo* animal experiments, but low reproducibility and the difference of NO production depending on animal types are unsolved issues. Thus, development of a convenient and versatile method for the *in vitro* quantitative and spatial analyses of NO diffusion inside the artery wall instead of animal experiments is important for biological and pharmaceutical applications.

Briefly, we demonstrate for the first time spatial and quantitative analyses of NO diffusion from the EC layer to the SMC layers using a 3D artery model with sensor particles[111] allocated into each cellular layer. The 3D structural effect of ECs and SMCs on NO production from the ECs was clarified in relation to the direction of interaction between these cells. Furthermore, a graded concentration change of NO from the uppermost EC layer to the underlying SMC layers was elucidated by 3D analysis using CLSM. This method does not require special instruments or techniques and is convenient and effective for drug screening, and thus it has the possibility to be a solution of general animal experiments.

11.5.5 3D-Cell Assembling Approaches

The cellular LbL methods are useful to construct multilayered structures one by one, but it takes a long time to fabricate thick tissue constructs because almost half a day incubation is necessary for stable cell adhesion on cell surfaces. Thus, 3D cell-assembling approaches have recently attracted attentions. Takeuchi *et al.* have reported the molding technique of cell beads, collagen gel beads covered with adhered cells.[112] The cell beads are molded into a designed silicone chamber to form the macroscopic 3D-tissue constructs. In the mold, the cell beads can adhere to each other *via* the cells coated on the collagen gel beads. Finally, since the collagen gel beads were degraded enzymatically by secreted enzymes from the cells, the molded tissues can be released from the mold. They reported improvement of secretion amount of albumin from human hepatocellular carcinoma (HepG2) cells by coculture with fibroblast cells in the constructed 3D-tissues. The construction of complex macroscopic 3D tissues using multiple cell types is expected using this method.

11.5.6 Cell Accumulation Technique

We are currently developing a simple and rapid bottom-up approach, called the cell-accumulation technique, by a single-cell coating using FN-G nanofilms.[85] Since the FN-G nanofilms prepared on individual cell surfaces provide an interactive property with the $\alpha_5\beta_1$ integrin receptor of the cell membrane, the cell–cell adhesion of all seeded cells in three dimensions can be induced at the same time (Figure 11.6A).

The fabrication of approximately 6 nm of FN-G films onto a dispersed single-cell surface was performed based on our previous reports for the preparation of polyelectrolyte LbL nanofilms onto silica particles.[111] Fluorescent images of the L929 mouse fibroblast cells clearly demonstrated a ring-shaped morphology, suggesting the successful preparation of homogeneous

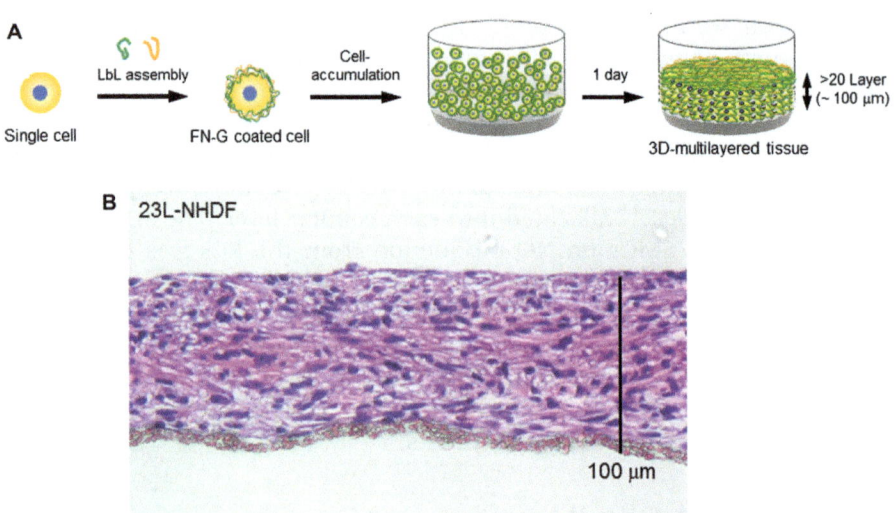

Figure 11.6 Cartoon of cell-accumulation technique by FN-G nanofilm fabrication (A). HE staining image of 23L-NHDF tissues after 24 h of incubation (B).
Reproduced with permission from Ref. 85.

fluorescently labeled FN-G nanofilms onto the surface of dispersed cells. Furthermore, we also confirmed over 91% cell viability for NHDFs, HUVECs, and HepG2 cells after preparing the FN-G films. After coating with the FN-G nanofilms, 2×10^6 NHDFs were incubated in a cell culture insert for one day. The 8L-dense constructs of 35 ± 4 µm thicknesses were successfully obtained, whereas a tattered and porous structure was barely obtained using noncoated cells. The thickness of the obtained tissue constructs increased on increasing the volume of culture media, and thus the current maximum thickness was over 100 µm, more than 20L structures (Figure 11.6B). More importantly, the construction of approximately 20L structures per one day after only a nanofilm coating within 1 h has never been reported previously, and this approach has tremendous versatility for various cells.

We demonstrated fabrication of vascularized 3D-tissue arrays were constructed in 12 and 24 microwell plates using a cell-accumulation technique.[85] The construction of the thick multilayered tissues with endothelial tube networks by embedding HUVECs in 3D-tissues composed of NHDFs has been performed by sandwich culture. After 2–7 days of incubation, highly developed capillary networks and a tubular morphology of the HUVECs were clearly observed by CLSM analyses (Figure 11.7). A dense and homogeneous vascularized network in the multilayered tissues of 1 cm width and 50 µm height was confirmed in 12 and 24 microwell cell-culture inserts. The occupied area percentage and distance of this capillary network of HUVECs was calculated to be approximately $63 \pm 12\%$ and 50–150 µm by

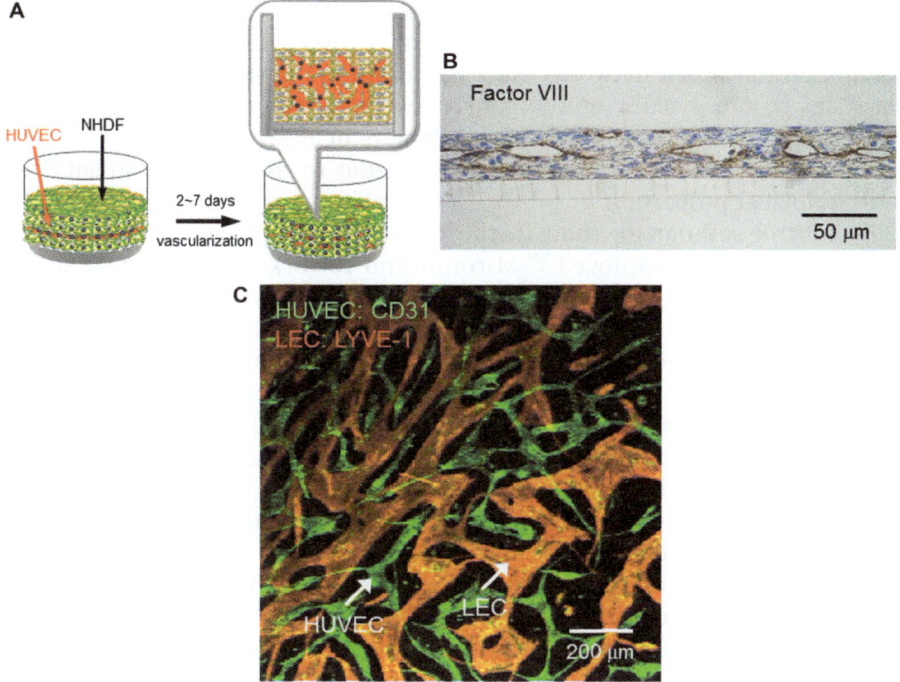

Figure 11.7 Schematic illustration of HUVEC tube formation by sandwich culture with NHDF tissues (A). Immunological staining image with antifactor VIII antibody for 4L-NHDF-1L-HUVEC-4L-NHDF tissues after 7 days of incubations (B). CLSM image of HUVEC and LEC networks immunologically staining with anti-CD31 and anti-LYVE-1 antibodies, respectively (C).

image analysis. This would provide a benefit for the surrounding cells, because cells in living tissues require endothelial tube networks within a perimeter of 100–150 μm to supply oxygen. Moreover, lymph capillary network structures could also be obtained by employing lymph endothelial cells (LECs) instead of HUVECs. When we use both blood and LECs for the vascularization process, individual networks in 3D-NHDF tissue constructs were successfully obtained. The obtained vascularized constructs were stable during long-term culture for at least 3 weeks. These microvascular and -lymph networks in multiwell plates will be a powerful tool for pharmaceutical and pathophysiological applications.

11.5.7 Inkjet-Printing Approaches

The cell-based approaches allow us to build a 3D-tissue construct with high cell density that shows cell–cell interactions. However, the obtained constructs are still far from the real *in vivo* tissues. In particular, the mechanical property and precise control of the cell location are significantly lower and poorer than those of real tissues because it is caused by ECM components. Complex ECM fiber networks have an important role for stiffness and elasticity of our tissues and organs. Development of composite tissues consisting of not only cells but also polymers or ECM proteins will be important for construction of 3D tissues analogous to similar mechanical and morphological properties.

To control cellular location in the matrix, inkjet printing of cells and polymers has been employed.[99] Mironov and coworkers demonstrated 3D construction of complex structures by printing of *in situ* crosslinkable polymers and ECM solutions containing cells.[113,114] Nakamura and coworkers have reported to write lines about 50 μm wide by printing cells suspended in sodium alginate onto a thin film of calcium chloride due to formation of alginate gels through a coordinate bond with a calcium ion.[115] Both methods reported fabrication of free-standing tubular constructs containing living cells, and their mechanical and morphological properties were clearly higher than those of the products from cell-based approaches. However, the encapsulated cells were isolated inside the matrix and no functions as a tissue were observed until their death after a couple of day incubations. Furthermore, the obtained 3D constructs were not stable for long-term cultures (more than 1 week) due to the enzymatic degradation or hydrolysis. Variety of *in situ* crosslinkable polymers is also limited. The challenge here is in the field of developing new biomaterials to allow high cell adhesion, cell–cell interaction, and long-term stability.

We also tried to fabricate a chip of 3D-cellular multilayered constructs using a technique combining our hierarchical cell manipulation and automatic inkjet printing for drug screening (Figure 11.8). The 440 microarrays of simplified human 3D-tissue structures, micrometer-sized multilayers with different layer numbers and cell types were constructed automatically.[116] For high-throughput drug evaluations, 3D liver tissue chips, which have high liver functions, would be a powerful and important microarray because the liver plays a central role in drug metabolism and toxicity. We integrated simplified 3D liver structures, such as monolayered (1L) to three-layered (3L) structures composed of HepG2 and HUVEC, onto a chip. We chose these cell types because the most common cell types in the liver are hepatocytes and liver endothelial cells; they account for more than 80% of the liver's mass. Comprehensive high-throughput assays of liver functions (albumin secretion, cytochrome-P450 (CYP450) enzyme secretion and CYP450 metabolism activities) were then performed. We found for the first time that these simplified 3D liver structures, a sandwich of endothelial cells and hepatocytes, revealed the highest functions as compared to other 3D structures. Therefore, 3D human tissue chips prepared by this rapid and automatic system will be an innovative technology for *in vitro* evaluations of drugs, cosmetics, and chemicals, instead of animal testing.

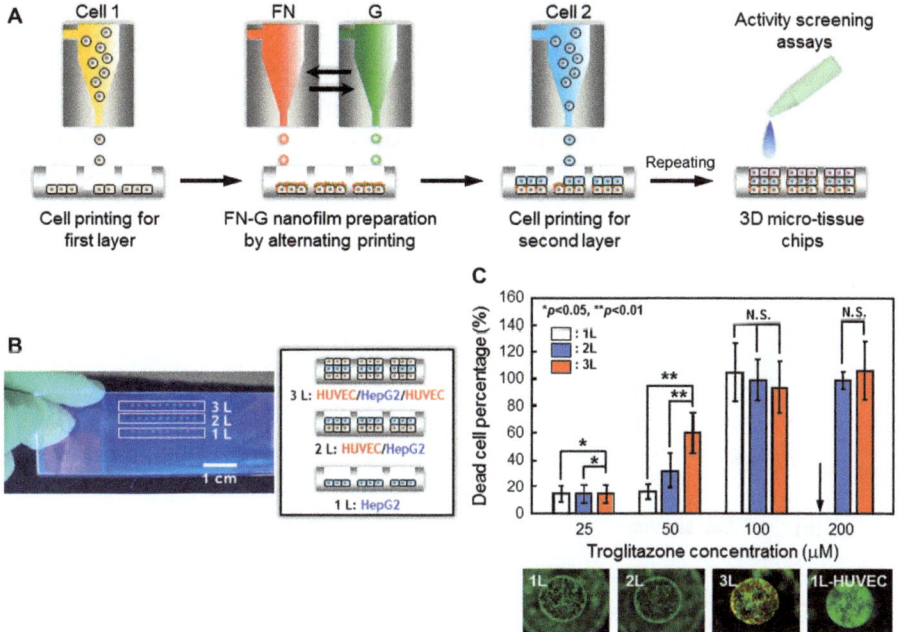

Figure 11.8 Schematic illustration of the development of 3D microtissue arrays by the LbL printing of single cells and proteins (A). Photograph of 1L (HepG2 monolayer), 2L (HUVEC/HepG2) and 3L (HUVEC/HepG2/HUVEC) hepatic layered coculture arrays prepared in 440 microwell plates under UV light. These tissues were immunofluorescently labeled with an antihuman albumin antibody after 48 h of incubation. The inset shows an illustration of these hepatic tissue structures (B). The layer number and dose-dependent cytotoxicity of troglitazone in coculture after 2 days of incubation ($n = 3$, over 10 wells per image) (C). The data were normalized to untreated cultures (100% activity). The bottom fluorescence images are live/dead images of 1L to 3L constructs and HUVEC monolayer at 50 μM concentration. The arrow indicates undetectable samples due to the detachment of the dead cells. *, $P < 0.05$; **$P < 0.01$. N.S. means no significant difference.
Reproduced with permission from Ref. 116.

11.6 Protection Effects of LbL-Assembled Films Prepared on Cell Surfaces from Physical Stress

Various tissue engineering approaches have been discovered such as three-dimensional culture inside biodegradable hydrogels.[117,118] LbL assembly technique,[78,85,86,101] microfluidic system,[119] cell sheet technique,[120] and inkjet printing of cells and polymers.[113–115] However, these advances in tissue engineering have created various stresses for cells from physical, chemical, and environmental stimuli. For example, Boland and coworkers reported the formation of nanometer-sized pores on cell membranes due to shear stress during the inkjet-printing process.[121] Accordingly, we have to carefully consider cell viability and function during tissue engineering.

As we described before, we recently discovered a simple and unique bottom-up approaches "hierarchical cell manipulation" and "cell accumulation technique" to develop 3D-tissue constructs by the fabrication of nano-ECM films onto the cell surfaces. Since the coating of LbL-assembled

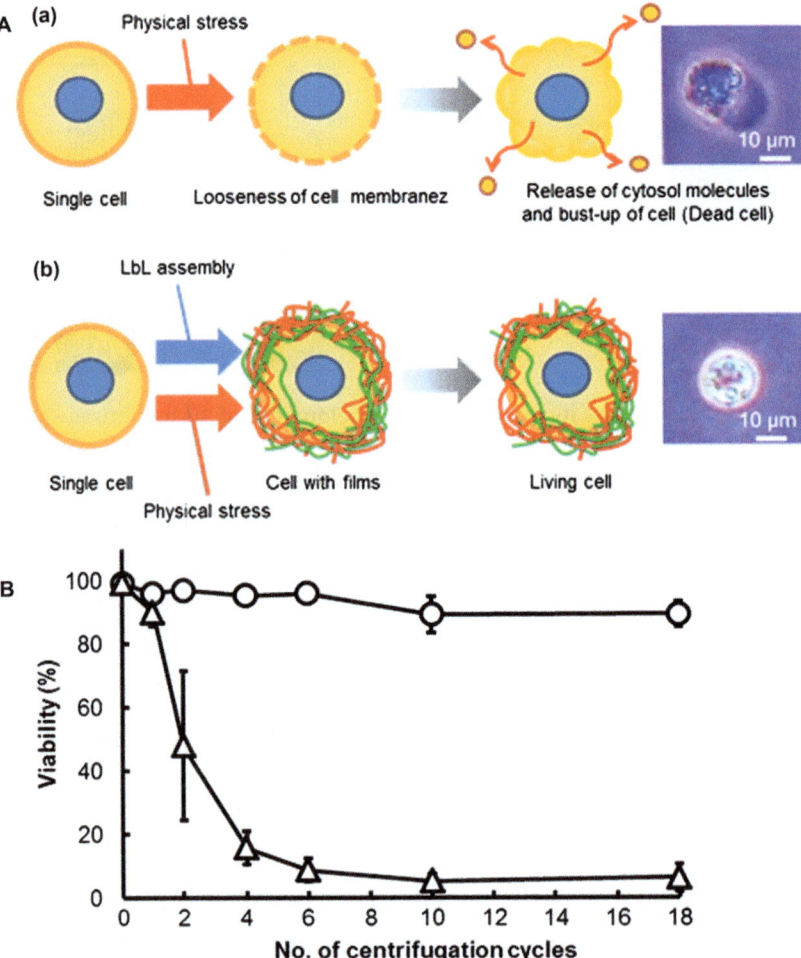

Figure 11.9   Schematic illustration of the effect of physical stress on (a) uncoated cells or (b) cells with films by LbL assembly (A). Cell viability of HepG2 with or without FN-G films for 9 step LbL assembly (B). Cell viability of HepG2 with (○) or without (△) FN-G films at each centrifugation was plotted ($n = 3$). Cell viability was measured by trypan blue staining and was calculated by the ratio of the number of living cells to the total number of cells. Error bars denote standard deviations. Reproduced with permission from Ref. 122.

nanofilms onto the cell membranes requires many cycles of centrifugation (physical stress), the physical stress may damage the cell membranes, causing the leakage of cytosol molecules or ultimately cell death. However, there are no reports on physical stress during LbL assembly processes.

We investigated for the first time the effect of LbL assembly centrifugation processes (gravity stress) on cell viability and leakage of cytosol enzymes.[122] After 2 steps of LbL assembly using Tris-HCl buffer solution without polymers or proteins (4 centrifugation cycles including washing), hepatocyte carcinoma (HepG2) cells showed extremely high cell death and viability was *ca.* 15% (Figure 11.9). Their viability ultimately decreased to 6% after 9-step LbL assembly (18 cycles of centrifugation), which is the typical number of steps involved in preparing LbL nanofilms. On the other hand, significantly higher viability (>85%) of HepG2 cells was obtained after 9-step LbL assembly employing FN-G or type IV collagen (Col IV)-laminin (LN) solution combinations, which are typical components of ECM, to fabricate 10-nm thick LbL films. When LbL films of synthetic polymers created *via* electrostatic interactions were employed instead of ECM films described above, the viability of the HepG2 cells after the same 9 steps slightly decreased to 61%. The protective effects of LbL films were strongly dependent on their thickness and the critical thickness was over 5 nm. Surprisingly, high viability of over 85% was achieved even under extreme physical stress conditions (10 000 rpm). We evaluated the leakage of lactate dehydrogenase (LDH) during the LbL-assembly processes to clarify the protective effect, and a reduction in LDH leakage was clearly observed when using FN-G nanofilms. Moreover, the LbL films do not inhibit cell growth during cell culture, suggesting that these coated cells can be useful for other experiments.

LbL nanofilm coatings, especially ECM nanofilm coatings, will be important techniques for protecting cell membranes from physical stress during tissue engineering.

11.7 Conclusions

Here, we described an overview of a novel technique to modify cell surface characteristics safely by LbL nano-ECM films for tissue engineering application, especially development of 3D-tissue constructs. The nano-ECM LbL films constructed through biological recognition, which is not a usual driving force like electrostatic interaction, allowed us to coat cell surface without any damage for the first time. Since the nano-ECM showed fiber formation on cell membranes the same as natural ECM in the body,[81] the artificial nano-ECM protected cell surfaces from physical stresses probably the same as real ECM.[122] We discovered the nano-ECM LbL films acted as glue between cell surfaces, due to biological interaction between integrin molecules and FN, for fabrication of 3D-cell multilayers. Moreover, sandwich culture of HUVEC between 4L-NHDF provided blood capillary formation in the 3D-tissue constructs and we found that the capillary networks contributed to higher cell viability than that of 3D tissues without any capillary networks (unpublished data). Combination with inkjet printing technology gave automatic fabrication of a chip that integrated over 400 micrometer-sized 3D-tissue constructs, for high-throughput drug-screening assay.

We believe that such a simple cell coating technique has opened up a new window as a key technique for tissue engineering, pharmaceutical and pathophysiological applications.

Abbreviations

ECM	Extracellular matrix
FN	Fibronectin
LbL	Layer-by-layer
3D	Three-dimensional
GAG	Glycosaminoglycan
Hep	Heparin
Dex	Dextran sulfate
PLL	α-poly(L-lysine)
PSS	Poly(sodium styrenesulfonate)
QCM	Quartz crystal microbalance
PAH	Poly(allylamine hydrochloride)
FBS	Fetal bovine serum
mESC	mouse embryonic stem cell
PE	Polyelectrolyte
RGD	arginine-glysine-aspartic acid
DPPC	1,2-dipalmitoyl-*sn*-glycero-3-phosphatidylcholine
DPPA	1,2-dipalmitoyl-*sn*-glycero-3-phosphate
PDDA	Poly(diallyldimethylammonium chloride)
CLSM	Confocal laser scanning microscopy
Rh-FN	Rhodamine-labeled FN
CMyo	Cardiac myocyte
SMC	smooth muscle cell
Hep	Hepatocyte
EC	Endothelial cell
NHDF	Normal human dermal fibroblast
HUVEC	Human umbilical vein endothelial cell
HSP70	Heat shock protein 70
IL-6	Interleukin-6
UASMC	Umbilical artery smooth muscle cell
PRP	Platelet-rich plasma
NO	Nitric oxide
cGMP	cyclic guanosine monophosphate
HepG2	human hepatocellular carcinoma cell line
LEC	Lymph endothelial cell
CYP450	Cytochrome-P450

Col IV   Type IV collagen
LN   Laminin
LDH   Lactate dehydrogenase

References

1. N. J. Boudreauand and P. L. Jones, *Biochem. J.*, 1999, **339**, 481.
2. R. O. Hynes, *Fibronectins*, Springer-Verlag Inc., New York, 1990.
3. E. W. Raines, *Int. J. Exp. Pathol.*, 2000, **81**, 173.
4. M. P. Lutolf and J. A. Hubbell, *Nature Biotech.*, 2005, **23**, 47.
5. C. S. Chen, M. Mrksich, S. Huang, G. M. Whitesides and D. E. Ingber, *Science*, 1997, **276**, 1425.
6. R. A. Chandra, E. S. Douglas, R. A. Mathies, C. R. Bertozzi and M. B. Francis, *Angew. Chem. Int. Ed.*, 2006, **45**, 896.
7. P. Gong, W. Zheng, Z. Huang, W. Zhang, D. Xiao and X. Jiang, *Adv. Funct. Mater.*, 2013, **23**, 42.
8. B. Neu, A. Voigt, R. Mitröhner, S. Leporatti, C. Y. Gao, E. Donath, H. Kiesewetter, H. Möhwald, H. J. Meiselman and H. Bäumler, *J. Microencapsulat.*, 2001, **18**, 385.
9. A. Diaspro, D. Silvano, S. Krol, O. Cavalleri and A. Gliozzi, *Langmuir*, 2002, **18**, 5047.
10. N. G. Veerabadran, P. L. Goli, S. S. Stewart-Clark, Y. M. Lvov and D. K. Mills, *Macromol. Biosci.*, 2007, 7, 877.
11. G. Decher and J. D. Hong, *Makromol. Chem. Macromol. Symp.*, 1991, **46**, 321.
12. G. Decher, *Science*, 1997, **277**, 1232.
13. *Multilayer Thin Films*, (Eds: G. Decher and J. B. Schlenoff), Wiley-VCH, Weinheim, 2003.
14. Y. Lvov, H. Haas, G. Decher, H. Möhwald, A. Mikhailov, B. Mtchedlishvily, E. Morgunova and B. Vainshtein, *Langmuir*, 1994, **10**, 4232.
15. Y. Lvov, K. Ariga and T. Kunitake, *Chem. Lett.*, 1994, 2323.
16. Y. Lvov, I. Ichinose, K. Ariga and T. Kunitake, *J. Am. Chem. Soc.*, 1995, **117**, 6117.
17. K. Ariga, M. McShane, Y. Lvov, Q. Ji and J. P. Hill, *Expert Opin. Drug Deliv.*, 2011, **8**, 633.
18. T. Komatsu, X. Qu, H. Ihara, M. Fujihara, H. Azuma and H. Ikeda, *J. Am. Chem. Soc.*, 2011, **133**, 3246.
19. K. Ariga, Y. Lvov, M. Onda, I. Ichinose and T. Kunitake, *Chem. Lett.*, 1997, 125.
20. Y. Lvov, K. Ariga, M. Onda, I. Ichinose and T. Kunitake, *Langmuir*, 1997, **13**, 6195.
21. H. C. Yang, K. Aoki, H.-G. Hong, D. D. Sackett, M. F. Arendt, S.-L. Yau, C. M. Bell and T. E. Mallouk, *J. Am. Chem. Soc.*, 1993, **115**, 11855.
22. T. Yonezawa, S. Onoue and T. Kunitake, *Adv. Mater.*, 1998, **10**, 414.

23. D. S. Shchukin, E. A. Ustinovich, G. B. Sukhorukov, H. Möhwald and D. V. Sviridov, *Adv. Mater.*, 2005, **17**, 468.

24. C. M. Anders and N. A. Kotov, *J. Am. Chem. Soc.*, 2010, **132**, 14496.

25. Y. Sun, X. Zhang, C. Sun, Z. Wang, J. Shen, D. Wang and T. Li, *J. Chem. Soc. Chem. Commun.*, 1996, 2379.

26. K. Ariga, Y. Lvov and T. Kunitake, *J. Am. Chem. Soc.*, 1997, **119**, 2224.

27. M. R. Linford, M. Auch and H. Möhwald, *J. Am. Chem. Soc.*, 1998, **120**, 178.

28. F. Caruso, R. A. Caruso and H. Möhwald, *Science*, 1998, **282**, 1111.

29. B. Zebi, A. S. Susha, G. B. Sukhorukov, A. L. Rogach and W. J. Parak, *Langmuir*, 2005, **21**, 4262.

30. S. H. Hu, C. H. Tsai, C. F. Liao, D. M. Liu and S. Y. Chen, *Langmuir*, 2008, **24**, 11811.

31. I. Ichinose, K. Fujiyoshi, S. Mizuki, Y. Lvov and T. Kunitake, *Chem. Lett.*, 1996, 257.

32. U. Sohling and A. J. Schouten, *Langmuir*, 1996, **12**, 3912.

33. S. W. Keller, H.-N. Kim and T. E. Mallouk, *J. Am. Chem. Soc.*, 1994, **116**, 8817.

34. N. A. Kotov, T. Haraszti, L. Turi, G. Zavala, R. E. Geer, I. Dékány and J. H. Fendler, *J. Am. Chem. Soc.*, 1997, **119**, 12184.

35. F. Hua, T. Cui and Y. Lvov, *Nano Lett.*, 2004, **4**, 823.

36. T. Serizawa and M. Akashi, *Chem. Lett.*, 1997, 809.

37. T. Serizawa, S. Kamimura and M. Akashi, *Colloids Surf.*, 2000, **164**, 237.

38. M. A. Correa-Duarte, A. Kosiorek, W. Kandulski, M. Giersig and L. M. Liz-Marzán, *Chem. Mater.*, 2005, **17**, 3268.

39. Z. Feng, Z. Wang, C. Gao and J. Shen, *Adv. Mater.*, 2007, **19**, 3687.

40. C. J. Ochs, G. K. Such, Y. Yan, M. P. van Koeverden and F. Caruso, *ACS Nano*, 2010, **4**, 1653.

41. M. Li, S. Ishihara, M. Akada, M. Liao, L. Sang, J. P. Hill, V. Krishnan, Y. Ma and K. Ariga, *J. Am. Chem. Soc.*, 2011, **133**, 7348.

42. U. Manna, J. Dhar, R. Nayak and S. Patil, *Chem. Commun.*, 2010, **46**, 2250.

43. J. Seo, P. Schattling, T. Lang, F. Jochum, K. Nilles, P. Theato and K. Char, *Langmuir*, 2010, **26**, 1830.

44. D. Usov and G. B. Sukhorukov, *Langmuir*, 2010, **26**, 12575.

45. M.-H. Park, S. S. Agasti, B. Creran, C. Kim and V. M. Rotello, *Adv. Mater.*, 2011, **23**, 22839.

46. W. B. Stockton and M. F. Rubner, *Macromolecules*, 1997, **30**, 2717.

47. L. Wang, Z. Q. Wang, X. Zhang, J. C. Shen, L. F. Chi and H. Fuchs, *Macromol. Rapid Commun.*, 1997, **18**, 509.

48. S. L. Clark and P. T. Hammond, *Langmuir*, 2000, **16**, 10206.

49. P. He, N. Hu and J. F. Rusling, *Langmuir*, 2004, **20**, 722.

50. J. F. Quinn and F. Caruso, *Langmuir*, 2004, **20**, 20.

51. F. Huo, H. Xu, L. Zhang, Y. Fu, Z. Wang and X. Zhang, *Chem. Commun.*, 2003, 874.

52. Y. Lu, Y. J. Choi, H. S. Lim, D. Kwak, C. Shim, S. G. Leee and K. Cho, *Langmuir*, 2010, **26**, 17749.
53. Y. Shimazaki, M. Mitsuishi, S. Ito and M. Yamamoto, *Langmuir*, 1997, **13**, 1385.
54. F. Wang, N. Ma, Q. Chen, W. Wang and L. Wang, *Langmuir*, 2007, **23**, 9540.
55. L. Xu, Z. Zhu and S. A. Sukhishvili, *Langmuir*, 2011, **27**, 409.
56. M. Zhang, L. Su and L. Mao, *Carbon*, 2006, **44**, 276.
57. L. A. Al-Hariri, A. Reisch and J. B. Schlenoff, *Langmuir*, 2011, **27**, 3914.
58. O. Crespo-Biel, B. Dordi, D. N. Reinhoudt and J. Huskens, *J. Am. Chem. Soc.*, 2005, **127**, 7594.
59. A. Ikeda, T. Hatano, S. Shinkai, T. Akiyama and S. Yamada, *J. Am. Chem. Soc.*, 2001, **123**, 4855.
60. T. Kida, T. Minabe, S. Nakano and M. Akashi, *Langmuir*, 2008, **24**, 9227.
61. S. Gao, D. Yuan, J. Lü and R. Cao, *J. Colloid Inter. Sci.*, 2010, **341**, 320.
62. D. G. Kurth and R. Osterhout, *Langmuir*, 1999, **15**, 4842.
63. M. Schütte, D. G. Kurth, M. R. Linford, H. Cölfen and H. Möhwald, *Angew. Chem. Int. Ed.*, 1998, **37**, 2891.
64. M. Ginzburg, J. Galloro, F. Jäkle, K. N. Power-Billard, S. Yang, I. Sokolov, C. N. C. Lam, A. W. Neumann, I. Manners and G. A. Ozin, *Langmuir*, 2000, **16**, 9609.
65. D.-J. Qian, C. Nakamura, T. Ishida, S.-O. Wenk, T. Wakayama, S. Takeda and J. Miyake, *Langmuir*, 2002, **18**, 10237.
66. M. Altman, A. D. Shukla, T. Zubkov, G. Evmenenko, P. Dutta and M. E. van der Boom, *J. Am. Chem. Soc.*, 2006, **128**, 7374.
67. W. Zhao, B. Tong, J. Shi, Y. Pan, J. Shen, J. Zhi, W. K. Chan and Y. Dong, *Langmuir*, 2010, **26**, 16084.
68. I. Welterlich and B. Tieke, *Macromolecules*, 2011, **44**, 4194.
69. G. Decher, B. Lehr, K. Lowack, Y. Lvov and J. Schmitt, *Biosens. Bioelectron.*, 1994, **9**, 677.
70. P. He, T. Takahashi, T. Hoshi, J. Anzai, Y. Suzuki and T. Osa, *Mater. Sci. Eng.*, 1994, **C2**, 103.
71. H. Ebato, J. N. Herron, W. Muller, Y. Okahata, H. Ringsdorf and P. A. Suci, *Angew. Chem. Int. Ed.*, 1992, **31**, 1087.
72. J. Anzai, Y. Kobayashi, N. Nakamura, M. Nishimura and T. Hoshi, *Langmuir*, 1999, **15**, 221.
73. C. Bourdillon, C. Demaille, J. Miroux and J. M. Savéant, *J. Am. Chem. Soc.*, 1994, **116**, 10328.
74. C. Bourdillon, C. Demaille, J. Miroux and J. M. Savéant, *J. Am. Chem. Soc.*, 1995, **117**, 11499.
75. J. Anzai, Y. Kobayashi and N. Nakamura, *J. Chem. Soc., Perkin Trans.*, 1998, **2**, 461.
76. Y. Zhu, W. Tong and C. Gao, *Soft Matter*, 2011, **7**, 5805.

77. R. F. Mhanna, J. Vörös and M. Zenobi-Wong, *BioMacromolecules*, 2011, **12**, 609.
78. M. Matsusaki, K. Kadowaki, Y. Nakahara and M. Akashi, *Angew. Chem. Int. Ed.*, 2007, **46**, 4689.
79. E. Ruoslahti and M. D. Pierschbacher, *Science*, 1987, **238**, 491.
80. Y. Nakahara, M. Matsusaki and M. Akashi, *J. Biomater. Sci. Polymer Edn.*, 2007, **18**, 1565.
81. K. Kadowaki, M. Matsusaki and M. Akashi, *Langmuir*, 2010, **26**, 5670.
82. M. Matsusaki, K. Kadowaki, E. Adachi, T. Sakura, U. Yokoyama, Y. Ishikawa and M. Akashi, *J. Biomater. Sci. Polymer Edn.*, 2012, **23**, 63.
83. K. Kadowaki, M. Matsusaki and M. Akashi, *Biochem. Biophys. Res. Commun.*, 2010, **402**, 153.
84. M. Matsusaki, S. Amemori, K. Kadowaki and M. Akashi, *Angew. Chem. Int. Ed.*, 2011, **50**, 7557.
85. A. Nishiguchi, H. Yoshida, M. Matsusaki and M. Akashi, *Adv. Mater.*, 2011, **23**, 3506.
86. M. Matsusaki, *Bull. Chem. Soc. Jpn.*, 2012, **85**, 401.
87. P. G. Natali, D. Galloway, M. R. Nicotra and C. de Martino, *Connect. Tissue Res.*, 1981, **8**, 199.
88. L. Stanislawski, H. Serne, M. Stanislawski and M. Jozefowicz, *J. Biomed. Mater. Res.*, 1993, **27**, 619.
89. A. Bonnefoy and C. Legrand, *Thrombosis Res.*, 2000, **98**, 323.
90. A. P. Ngankam, G. Mao and P. R. van Tassel, *Langmuir*, 2004, **20**, 3362.
91. R. Langer and J. P. Vacanti, *Science*, 1993, **260**, 920.
92. K. Y. Lee and D. J. Mooney, *Chem. Rev.*, 2001, **101**, 1869.
93. J. Lee, M. J. Cuddihy and N. A. Kotov, *Tissue Eng.: Part B*, 2008, **14**, 61.
94. Y. Dzenis, *Science*, 2004, **304**, 1917.
95. S. Zhang, *Nature Biotechnol.*, 2003, **21**, 1171.
96. J. Yang, M. Yamato, C. Kohno, A. Nishimoto, H. Sekine, F. Fukai and T. Okano, *Biomaterials*, 2005, **26**, 6415.
97. H. Akiyama, A. Ito, Y. Kawabe and M. Kamihira, *Biomaterials*, 2010, **31**, 1251.
98. H. Tavana, B. Mosadegh and S. Takayama, *Adv. Mater.*, 2010, **22**, 2628.
99. B. Derby, *Science*, 2012, **338**, 921.
100. T. Shimizu, M. Yamato, Y. Isoi, T. Akutsu, T. Setomaru, K. Abe, A. Kikuchi, M. Umezu and T. Okano, *Circ. Res.*, 2002, **90**, e40.
101. P. Rajagopalan, C. J. Shen, F. Berthiaume, A. W. Tilles, M. Toner and M. L. Yarmush, *Tissue Eng.*, 2006, **12**, 1553.
102. D. Fischer, Y. Li, B. Ahlemeyer, J. Krieglstein and T. Kissel, *Biomaterials*, 2003, **24**, 1121.
103. M. Chanana, A. Gliozzi, A. Diaspro, I. Chodnevskaja, S. Huewel, V. Moskalenko, K. Ulrichs, H.-J. Galla and S. Krol, *Nano Lett.*, 2005, **5**, 2605.
104. H. K. Kleinman, D. Phlip and M. P. Hoffman, *Curr. Opin. Biotechnol.*, 2003, **14**, 526.
105. K. M. Yamada, *Ann. Rev. Biochem.*, 1983, **52**, 761.

106. G. Krishna, T. Shutava and Y. Lvov, *Chem. Commun.*, 2005, 2796.
107. M. Matsusaki, H. Ajiro, T. Kida, T. Serizawa and M. Akashi, *Adv. Mater.*, 2012, **24**, 454.
108. W. K. Alderton, C. E. Cooper and R. G. Knowles, *Biochem. J.*, 2001, **357**, 593.
109. S. Kurioka, K. Koshimura, Y. Murakami, M. Nishiki and Y. Kato, *Endocr. J.*, 2000, **47**, 77.
110. Y. Nakaya, A. Minami, N. Harada, S. Sakamoto, Y. Niwa and M. Ohnaka, *Am. J. Clin. Nutr.*, 2000, **71**, 54.
111. S. Amemori, M. Matsusaki and M. Akashi, *Chem. Lett.*, 2010, **39**, 42.
112. Y. T. Matsunaga, Y. Morimoto and S. Takeuchi, *Adv. Mater.*, 2010, **23**, H90.
113. V. Mironov, T. Boland, T. Trusk, G. Forgacs and R. R. Markwald, *Trends Biotechnol.*, 2003, **21**, 157.
114. V. Mironov, V. Kasyanov, X. Z. Shu, C. Eisenberg, L. Eisenberg, S. Gonda, T. Trusk, R. R. Markwald and G. D. Prestwich, *Biomaterials*, 2005, **26**, 7628.
115. Y. Nishiyama, C. Henmi, S. Iwanaga, H. Nakagawa, K. Yamaguchi, K. Akita, S. Mochizuki, K. Takiura and M. Nakamura, *Digital Fabrication 2006* (Society for Imaging Science and Technology, Springfield, VA, 2006), pp. 89–92.
116. M. Matsusaki, K. Sakaue, K. Kadowaki and M. Akashi, *Adv. Healthcare Mater.*, 2013, **2**, 534.
117. S. Levenberg, J. Rouwkema, M. Macdonald, E. S. Garfein, D. S. Kohane, D. C. Darland, R. Marini, C. A. van Blitterswijk, R. C. Mulligan, P. A. D'Amore and R. Langer, *Nature Biotechnol.*, 2005, **23**, 879.
118. D. R. Albrecht, G. H. Underhill, T. B. Wassermann, R. L. Sah and S. N. Bhatia, *Nature Methods*, 2006, **3**, 369.
119. D. Huh, B. D. Matthews, A. Mammoto, M. Montoya-Zavala, H. Y. Hsin and D. E. Ingber, *Science*, 2010, **328**, 1662.
120. T. Sasagawa, T. Shimizu, S. Sekiya, Y. Haraguchi, M. Yamato, Y. Sawa and T. Okano, *Biomaterials*, 2010, **31**, 1646.
121. X. Cui, D. Dean, Z. M. Ruggeri and T. Boland, *Biotechnol. Bioeng.*, 2010, **106**, 963.
122. A. Matsuzawa, M. Matsusaki and M. Akashi, *Langmuir*, 2013, **29**, 7362.

CHAPTER 12

Future of Cell Surface Engineering

RAWIL F. FAKHRULLIN,*[a] YURI M. LVOV[b] AND INSUNG S. CHOI[c]

[a] Bionanotechnology Group, Institute of Fundamental Medicine and Biology, Kazan (Idel buye/Volga region) Federal University, Kreml uramı 18, Kazan, Republic of Tatarstan, Russian Federation; [b] Institute for Micromanufacturing, Louisiana Tech University, 911 Hergot Ave., Ruston, LA 71272, USA; [c] Center for Cell-Encapsulation Research, Department of Chemistry, KAIST, Daejeon 305-701, Korea
*Email: kazanbio@gmail.com

This book is the first attempt to summarize the current progress in cell surface engineering. Most of the topics currently associated with this area were covered in the preceding chapters. Cell surface engineering is currently in its infancy, with the most interesting and promising ideas yet to be realized. Today, we are apparently witnessing the quiet revolution in our understanding of manipulations that can be performed under the living cell. The paradigm of cell surface engineering implies that the surfaces of single isolated cells can be deliberately modified by attaching the foreign particles or films onto the cellular membranes. Unlike in traditional approaches, where the cells are manipulated after being immobilized in larger microgels or united in uncontrollably grown flakes, here scientists benefit from the controllable and sequential modification of the cells. Cell surface engineering as an auxiliary area of research opens up new avenues in biomedicine, chemistry and materials science. Fabrication of surface-engineered cells,

RSC Smart Materials No. 9
Cell Surface Engineering: Fabrication of Functional Nanoshells
Edited by Rawil F Fakhrullin, Insung S Choi and Yuri Lvov
© The Royal Society of Chemistry 2014
Published by the Royal Society of Chemistry, www.rsc.org

also known as *"cyborg" cells* is an extremely interesting task *per se*, since the methods required for deposition of tiny particles onto cell surfaces are very elaborate. Within this concluding chapter we try to envisage the direction of the future work within this area, outlining the most promising and challenging topics that, in our opinion, will significantly attract the attention of the scientific community.

Speaking about the future of cell surface engineering we need to mention the rapidly developing area of functional nanomaterials and "smart" polymers as the engines of the development of novels techniques for cell surface modification and applications of surface-functionalized cells. Today, the typical procedure of cell surface engineering with nanomaterials relies on utilization of model nanomaterials, such as layer-by-layer (LbL) polymer multilayered films or rather simple nanomaterials, such as gold or magnetic nanoparticles. These contemporary techniques can be thus regarded as simple prototype approaches that can serve as a model to elaborate new, much more sophisticated techniques. Basically, cell surface engineering will benefit from progress in polymer science[1] and fabrication of novel multifunctional nanomaterials, such as Janus nanoparticles[2] exhibiting double functionalities or hybrid materials based on functional nanoparticles[3] or nanotubes.[4] In addition, the fabrication techniques also benefit from our understanding of biological processes found in Nature. The bioinspired approaches include bioinspired mineralization, such as silicification, and dopamine polymerization.[5]

Although representatives of all major taxonomical groups have been subjected to cell surface engineering, totally just a limited number of species was used. Therefore, the future work will be definitely based on using new species as templates for deposition of functional nanocoatings. The selection of cells will be outlined by the functional properties and perspective applications of the materials obtained through engineering of cell surfaces. The most intriguing results are expected from various types of human cells and other mammalian cells. Obviously, the clinical applications of surface-engineered human cells are extremely diverse. Here, we offer a typical scenario of the future evolution of cell surface engineering, mentioning the positive outcomes while bearing in mind the likely obstacles. Such nanofunctionalized human cells can be utilized in tissue engineering (see Chapter 11), however, other applications may include the directed therapeutic cells delivery directly to the target tissue/organ within the human body. For several human pathologies the replacement or addition of therapeutic cells into a certain organ is believed to be the effective way of treatment.[6] Magnetic surface functionalization appears to be a perfect tool to controllably guide the modified cells, however, there might be other ways of functionalization, such as fabrication of self-propelling coatings that would mimic the tiny motors navigating cells inside the blood vessels. The idea to fabricate self-navigating "cyborg" cells might be very fruitful, however, to do this the researchers will have to overcome several challenging issues, such as cell aggregation during surface functionalization and preserving the viability of the cells. Importantly, cells must be functionalized using such nanomaterials that would allow for avoiding the aggregation with the native cells inside the blood vessels and organs. It is likely that "smart" polymers might be employed to prepare such coatings over magnetic nanoparticles and cells that will change their functional properties according to the local environments if injected into the blood stream. Ideally, such surface-engineered cells will repel the native cells on the way to the target organ, emulating the "stealth" coatings, and after reaching it will settle down in a desired position. Crucially, the coatings are required to be nonantigenic, in other words, the injected nanocoated cells should not be targeted by the immune system. This can be achieved using low-immunogenic polymers as outer coatings. So far, such "smart" nanocoated cells have not been obtained due to the complexity of the coating that would provide the desired functionality. In future, we anticipate that these approaches will be elaborated successfully.

Just recently "cyborg" cells were successfully used to deliver nanoparticles into microscopic worms *Caenorhabditis elegans*.[7] Magnetic or silver nanoparticles were immobilized on the surfaces of bacteria or microalgae and administered into microworms, which fed on "cyborg" cells as a sole food source. This approach allowed for the directed delivery of nanoparticles into worms' intestines, as shown in Figure 12.1.

The current approach delivers the majority of nanoparticles into the digestive tract, however the minor aggregations were observed in other parts of the worms' bodies. Obviously, the future research will be focused onto the more precise targeting of nanoparticles delivery, *i.e.* when the "cyborg" cells are prepared in such a way that allows for the sequential release of nanoparticles at different positions inside the *C. elegans* worms. Furthermore, we anticipate the recruitment of other microscopic species both as templates

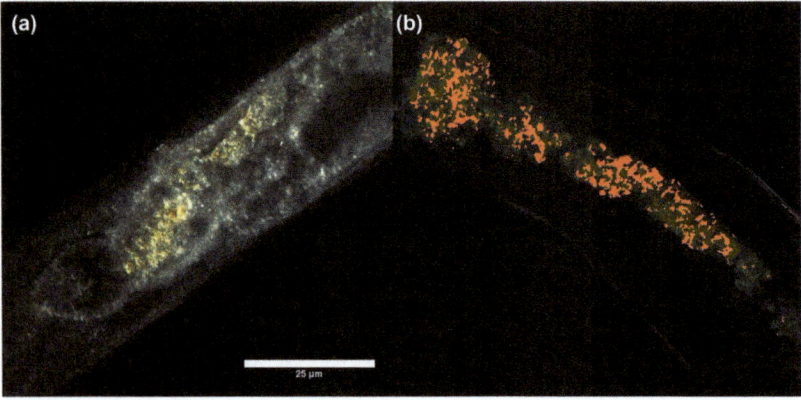

Figure 12.1   (a) Dark-field optical microscopy image demonstrating the delivery of magnetic nanoparticles into *C. elegans* using "cyborg" cells; (b) Hyperspectral mapping of magnetic nanoparticles (red areas) inside the worms.

for nanoparticle deposition and for applications in toxicity screening. Accordingly, new approaches for monitoring of nanoparticle distribution within the bodies of the model organisms are to be envisaged.

Apart from directed cell delivery, the methods of cell surface engineering will be appreciated by neurobiologists studying the mechanisms of neurotransmission. The selective and tuneable deposition of polymer shells onto excitable cells (such as neurons or muscle cells) may allow control of the ions flow and modulation of the action potential. It is likely that semipermeable or electrically conductive artificial coatings may help neuroscientists to better understand the molecular principles of the propagation of nerve impulses. Here, the cell surface engineers can study from Nature mimicking the myelin coatings of neuronal axons.

Another fascinating object for future cell surface engineering will be the motile microbial cells that are able to propel themselves using rapidly moving flagella. So far, the motility of microbial cells was not taken into consideration during the nanoshell deposition, moreover, to the best of our knowledge there are no reports demonstrating the preserved locomotion in nanocoated bacteria or microalgae. Current nanomaterial deposition techniques indiscriminately coat the whole cell with the nanoshell, thus forcing the flagella or cilia to stick to the cell surface and inhibiting their movements. The deposition of partial patch-like coatings[8] onto motile cells will help to prevent their aggregation. Such nanocoated motile cells are the ideal candidates for fabrication of remotely controlled biological microrobots that will be propelled using cellular machinery and controlled using electric or magnetic fields and cell-surface attached nanoparticles as antennae. Alternatively, the deposition of polymer nanocoatings or inorganic shells onto flagella may be utilized in investigation of the mechanical properties of cellular engines. These techniques can further be extended for modification of small motile invertebrates.

Cell surface-deposited nanomaterials can be regarded as a promising tool in fabrication of "smart" coats providing cells with protection from harsh environments. Micro-organisms are known for their ability to survive in extreme conditions, however, if one imagines the introduction of terrestrial micro-organisms into the soils of other planets as the first step of human colonization, that will require more than ordinary measures to protect the cells. Recently, nanoshells have been used to protect cells from various external stressors. Apart from this, the "space" micro-organisms will require mechanical protection and the surplus of nutrients for the very initial stages of colonization. In principle, this can be achieved using inorganic nanocontainers attached to cell walls. Recently, we used polyelectrolyte-mediated deposition of halloysite clay nanotubes on yeast cells.[9] The typical images of halloysite-coated "cyborg" cells are shown in Figure 12.2.

Halloysite nanotubes (50 nm diameter, up to 1 micrometer length) were shown to provide the encapsulated cells with nutrients (glucose) being steadily releasing into the media. Here the nanotubes were filled with glucose, but the other nutrients, vitamins and enzymes can be loaded into

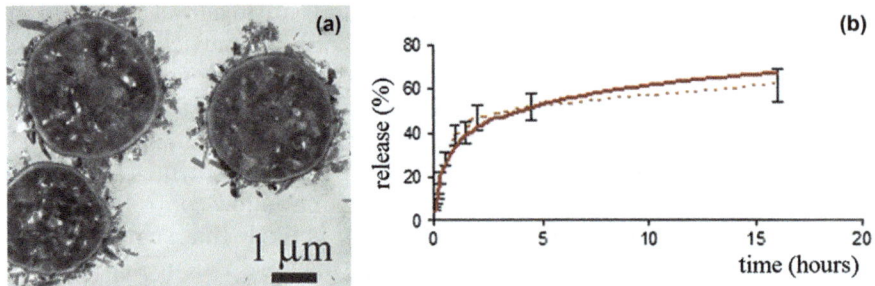

Figure 12.2   (a) Halloysite-modified yeast cells (TEM); (b) Release profile of nutri-
ent (glucose) loaded to halloysite from the HNTs/PEs-coated cells.
Reprinted with permission from Ref. 9. Copyright 2013 the Royal
Society of Chemistry.

halloysite as well. The payload was able to diffuse freely from the lumen,
more complex structures will utilize semipermeable functional stoppers,
triggering the release under certain conditions. This simple example demonstrates how functional nanomaterials may help the encapsulated cells to
survive in unusual environments. We assume that a similar approach might
be utilized in modifying the target cells with far more sophisticated nanocontainers incorporated into the hybrid polymer coating providing the encapsulated cells with a durable mechanical protection, pH-protection,
increased adhesion, thermostability at extreme temperatures and a storage
of water, nutrients, oxygen or any other gas/liquid required for the normal
cellular activity. In addition to the cytoprotective ability of the nanoshells, the
fabrication of nanoshells onto individual living cells will give a tool for
chemically controlling and tuning cellular activities. Advanced technologies
for cell surface coating make it possible to manipulate single cells and achieve
the cryptobiotic processes. In this sense, the so-called "artificial spores",
armed with functional cytoprotective nanoshells, will be realized.[10] Apart from
using such cells in rather extraordinary space applications, the same methodology may be appreciated in more ordinary biotechnological processes.

It can be envisioned that the future technology of cell surface engineering
will advance from simple deposition of nanomaterials onto cell surfaces to
cell surface initiated reactions that respond to the environment changes.
These environment-adaptable, configuration-changeable coating will open
up many unforeseen applications.

Acknowledgement

RFF acknowledges the support by RFBR 12-03-93939-G8, RFBR 12-04-33290
and RFBR 14-04-01474-a grants. ISC acknowledges the financial support by
the National Research Foundation of Korea (NRF) grant funded by the Korea
government (MSIP) (2012R1A3A2026403).

References

1. B. Le Droumaguet and J. Nicolas, *Polym. Chem.*, 2010, **1**, 563.
2. J. Hu, S. Zhou, Y. Sun, X. Fang and L. Wu, *Chem. Soc. Rev.*, 2012, **41**, 4356.
3. T. Govindaraju and M. B. Avinash, *Nanoscale*, 2012, **4**, 6102.
4. T. Chatterjee and R. Krishnamoorti, *Soft Matter*, 2013, **9**, 9515.
5. J. H. Park, S. H. Yang, J. Lee, E. H. Ko, D. Hong and I. S. Choi, *Adv. Mater.*, 2014, **26**, 2001.
6. F. R. Appelbaum, *Nature*, 2001, **411**, 385.
7. G. I. Däwlätşina, R. T. Minullina and R. F. Fakhrullin, *Nanoscale*, 2013, **5**, 11761.
8. A. J. Swiston, C. Cheng, S. H. Um, D. J. Irvine, R. E. Cohen and M. F. Rubner, *Nano Lett.*, 2008, **8**, 4446.
9. S. A. Konnova, I. R. Sharipova, T. A. Demina, Y. N. Osin, D. R. Yarullina, O. N. Ilinskaya, Y. M. Lvov and R. F. Fakhrullin, *Chem. Commun.*, 2013, **49**, 4208.
10. D. Hong, M. Park, S. H. Yang, J. Lee, Y.-G. Kim and I. S. Choi, *Trends Biotechnol.*, 2013, **31**, 442.

Subject Index

References in figures are given in *italic* type. References to tables are given in **bold** type.